GOLD NANOPARTICLES
FOR
PHYSICS, CHEMISTRY AND BIOLOGY

GOLD NANOPARTICLES
—————— FOR ——————
PHYSICS, CHEMISTRY AND BIOLOGY

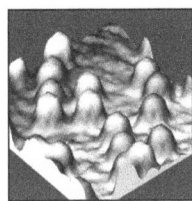  

CATHERINE LOUIS
OLIVIER PLUCHERY
Université Pierre et Marie Curie, France

 Imperial College Press

Published by

Imperial College Press
57 Shelton Street
Covent Garden
London WC2H 9HE

Distributed by

World Scientific Publishing Co. Pte. Ltd.
5 Toh Tuck Link, Singapore 596224
USA office: 27 Warren Street, Suite 401-402, Hackensack, NJ 07601
UK office: 57 Shelton Street, Covent Garden, London WC2H 9HE

British Library Cataloguing-in-Publication Data
A catalogue record for this book is available from the British Library.

GOLD NANOPARTICLES FOR PHYSICS, CHEMISTRY AND BIOLOGY

ISBN 978-1-84816-806-0

Typeset by Stallion Press
Email: enquiries@stallionpress.com

Printed in Singapore.

Contents

———————

Preface — Gold Nanoparticles for Physics, Chemistry and Biology

The fascination with gold is a story which spans millennia and this metal has played a role in almost every area of human existence. It has been a way of expressing wealth, it has been the cause of battles and wars, it has often been related to religious devotion, and has been linked with our most intimate feelings as a way of expressing love. These meanings are still important today. However, in recent years, a new type of fascination with gold has emerged in the scientific community that is not linked to any greed or emotion but to more rational concerns. Scientists have found a new interest in gold when it is divided into miniscule grains, such as gold nanoparticles. This scientific enthusiasm started in various fields of science over the last three decades. For instance, gold was thought to be chemically inactive, but it was discovered in 1987 that gold nanoparticles with sizes smaller than 5 nm are excellent catalysts. Bulk gold was thought to exhibit its 'eternal' yellow shining colour; it turns out, however, that gold nanoparticles are red or blue due to the so-called plasmon resonance, a property that has excited the interest of physicists since 1990 with biologists having now also joined the move.

This statement that various scientific communities are working on the same object with low awareness of each other is actually at the origin of the publication of this book. It is also backed up by the success of the French Network *Or-nano* (Gold-Nano) that we founded in 2006 (www.or-nano.org). This French network is sponsored by the CNRS (Centre National pour

la Recherche Scientifique) and gathers researchers and PhD students working on gold nanoparticles with various motives: from very fundamental
studies of the properties of gold nanoparticles to more applied topics, such
as catalysis, biosensors or medical imaging. *Or-nano* has organized annual
meetings, a summer school in 2008 and specialized discussions, all of which
keep attracting a great audience proving the need for the scientific survey
of gold nanoparticles proposed by the present book.

Gold Nanoparticles for Physics, Chemistry and Biology provides a
broad introduction to the fascinating and intriguing world of gold nanoparticles. **Chapter 1** relates the history of gold nanoparticles, which begins in
remote times with red ruby glass and reached a peak at the end of the seventeenth century. This section is an original work that has never been treated in
other scientific books. Basic properties of gold as an element are surveyed
in **Chapter 2** with a special emphasis on the relativistic effect that is responsible of many unusual properties of this metal. **Chapters 3 and 4** lead the
reader into the optical and thermal properties of gold nanoparticles by detail-
ing the plasmon resonance and giving the basis necessary to understand the
applications of these nano-objects as ultra small light emitters: nano-heaters
or nano-antennas. The preparation of gold nanoparticles with sometimes fascinating shapes is reviewed in **Chapter 5**. Very often applications demand,
however, that nanoparticles are deposited on a substrate and the preparation
methods have to be adapted or completely revisited. This crucial aspect is
treated in **Chapter 6**. The preparation of such supported gold nanoparticles
is the key to the catalytic properties of gold and **Chapter 7** reviews the
present knowledge on these aspects with the emblematic reaction of carbon
monoxide oxidation and many other hot topics. Fundamental studies of the
formation and reactivity of gold nanoparticles in a highly controlled environment such as ultra-high vacuum are treated in **Chapter 8**. **Chapter 9**
goes into more fundamental questions and presents state of the art *ab initio*
calculations to reveal the geometry of gold clusters made with ten or so gold
atoms and their non-metallic behaviour with the onset of a semiconductor-
like gap. Applications in the fields of biology and medicine are treated in
two chapters with two complementary approaches: **Chapter 10** reviews
the approach of physicists who engineer the plasmonic properties to design
smart biosensors and **Chapter 11** presents the approach of biologists who
seek in gold nanoparticles a new method for drug delivery and therapeutic

treatments. However nanoparticles also inspire fear because they can be considered as invasive and uncontrollable nano-objects and may lead to unpredictable consequences on health and the environment. That is the reason why the potential toxicity of gold nanoparticles is reviewed in **Chapter 12**. The book concludes in **Chapter 13** with a survey of the promises of gold nanoparticles and the technological applications that could become a part of everyday life in the future.

The book may be used as an advanced textbook by graduate students and young scientists who need an introduction to gold nanoparticles. It is also suitable for experts in the related areas of chemistry, biology, material science, optics and physics, who are interested in broadening their knowledge and gaining an overview of the subject. Each chapter gradually leads the reader from the basis of a topic to some selected scientific challenges in the area. It provides the necessary up-to-date background material and scientific literature to go further.

Finally, we are grateful to all who contributed to this work: first to the ten authors who have always been very responsive and enthusiastic about the idea of the book. We thank Imperial College Press for its strong support of the proposition of publishing such an interdisciplinary book based on gold nanoparticles. A special thanks goes to Catharina Weijman and Sarah Haynes who made the task of assembling the book easier. Thanks to Richard Holliday of the World Gold Council for his suggestions and advice. We also want to acknowledge particularly the help from Rachel Doherty and Philip Campbell for their contribution in improving the quality of the English of some parts of the text.

Catherine Louis
Olivier Pluchery
Paris, 20 June, 2012

Chapter 1
Gold Nanoparticles in the Past: Before the Nanotechnology Era

Catherine Louis

Laboratoire de Réactivité de Surface, UPMC-CNRS, 4 Place Jussieu, 75005 Paris, France. Email: catherine.louis@upmc.fr

1.1 The First Usage of Gold

The role played by gold in history relies on its outstanding qualities among metals, making it exceptionally valuable from the earliest civilisations until the present day. As quoted by Auric Goldfinger in a James Bond movie, gold is attractive due to 'its brilliance, its colour, its divine heaviness', and also due to its incorruptibility and scarcity. Its great malleability makes gold one the easiest of the metals to work with. Moreover it often occurs naturally in a fairly pure state.

The first uses of gold were linked to deities and royalty in early civilisations. The word 'gold' exists in all old languages, often connected with the image of the Sun, with light and life giving warmth, growth and hence power. In cultures like ancient Egypt, which deified the Sun, gold represented its earthly form. In fact, nothing has changed through history, and the same thinking about gold keeps going (golden crown of the kings, gold medals, wedding rings, cult objects, gold ingots, etc.).

1.1.1 *Quest for gold and gold production*

The earliest signs of crude metallurgy occurred 9000–7000 BCE (before the Common Era). For instance, in Alikosh in Iran and Cayönü Tepesi close to Ergani in Anatoly, humans first began using native copper and gold, meteoric iron, silver and tin to create tools and possibly jewellery ornamentation. Gold was most probably discovered as shining, yellow nuggets. Although it can be easily worked because of its ductility, it is not clear whether it was worked before copper.[a]

It is known that the Egyptians mined gold before 2000 BCE in Nubia. The *Turin Papyrus* drawn during the reign of Ramesses IV (1151–1145 BCE) is the earliest known topographic and geological map.[1] Along with specifics of the geology and topography, it shows an ancient gold-working settlement, gold-bearing quartz veins in Wadi Hammamat, a dry river bed in Egypt's Eastern desert. Large mines were also present across the Red Sea in what is now Saudi Arabia. By 325 BCE, the Greeks had mined in areas from Gibraltar to Asia Minor and Egypt. The Romans mined gold extensively throughout the empire, developing the technology of mining to new levels of sophistication. For example, they would divert streams of water in order to mine hydraulically, and even pioneered 'roasting', the technique of separating gold from rock.

Occasional passages on mining and metallurgy of metals can be found in the works of Theophrastus (Greek, 372–288 BCE), Vitruvius (Roman, 90–20 BCE), Strabo (Greek, 63/64 BCE–c. 24 CE), Pliny the Elder (Roman, 23–79 CE) and Discorides (Greek, 40–90 CE). One important surviving document is the *Leyden Papyrus X* of the Museum of Antiquities in the Netherlands: it is the working notebook of a goldsmith and jeweller, probably written in the early years of the fourth century. It gathers 111 recipes of refining, alloying and working of gold; some of them are reported in Hunt's paper[2] (accessible online, free of charge).

Another important date for the history of gold is 1492, with the discovery of America and the beginning of massive expeditions and exploration with the quest for the *El Dorado*, and the encounter with Native American people, in Central and South America, with their extensive displays of gold ornaments. The Aztecs regarded gold as literally the product of the gods, calling it 'the sweat of the sun'.

---

[a]One can read on some websites that the earliest traces of gold dated back to the Paleolithic period 40,000–10,000 BCE and were found in Spanish caves of Maltravieso; this is wrong according to Dr Antoni Canals y Salomó (Universidad de Tarragona), a paleontolongist, specialist of this cave.

Two hundred years later, in 1700, gold was discovered in Minas Gerais in Brazil, which became the largest producer by 1720, responsible for nearly two-thirds of the world's gold output, but the production was in rapid decline by 1760. 1799 is the year of the first discovery of gold in the United States, when a 17-pound nugget was found in North Carolina. For the next 25 years, North Carolina supplied all the domestic gold coined for currency by the US. In 1848, John Marshall found flakes of gold near Sacramento in California, triggering the California Gold Rush. In 1850, E.H. Hargraves, returning to Australia from California, found gold in his home country within a week. 1868 saw the next major discovery, in South Africa, where G. Harrison uncovered gold while digging up stones to build a house, and in 1898, South Africa became the world's top gold producer with a quarter of the world production.

Up to now, a total of 161,000 tonnes of gold have been mined in human history; this corresponds to the volume of a single cube 20 m on a side (equivalent to $8000\,m^3$). 75% of all gold ever produced has been extracted since 1910. The typical annual production in recent years has been around 2,500 tonnes per year. In 2009, the largest producers were China (12.8%), then Australia, South Africa and the United States (9.1% each). India is the world's largest consumer of gold (800 tonnes of gold every year), and the largest importer; in 2008 India imported around 400 tonnes of gold.

1.1.2 *Gold as jewels and artefacts*

The most ancient gold artefacts were found in necropolis, but not in Mesopotamia or Egypt as is often believed. The history of gold starts long before the invention of writing and the establishment of the first cities of Mesopotamia and Egypt (circa 2800 BCE). It starts around 4500 BCE with 'Old Europe' civilisation in south-eastern Europe that was at that time among the most sophisticated and technologically advanced regions in the world. A necropolis with 294 graves dating to 4600–4200 BCE was discovered in 1972 in Varna on the Black Sea coast, which is located in modern-day Bulgaria. The graves contained some 300 objects made of pure gold: sceptres, axes, bracelets, other decorative pieces and bull-shaped plates. These objects attest to the high-level skill of goldsmithing. They can be seen at the Varna Archaeological Museum and at the National Historical Museum in Sofia.

Three important discoveries of gold artefacts were found in tombs dated to circa 2500 BCE in three different geographical areas:

- The tomb of Djer at Abydos in Egypt. He was probably the third king of the First Dynasty (c.2800 BCE). Although the tomb had been robbed, a human arm was discovered near the entrance, still wearing four golden bracelets (shown in the Cairo Museum).
- The tomb of Queen Pu-Abi in southern Iraq. She was an important figure who lived about 2600–2500 BCE, during the First Dynasty of Ur of the Sumer civilisation. Among other excavations of the Royal Cemetery of Ur, discovered between 1922 and 1934 by Sir Leonard Woolley, her tomb had been untouched by looters. It revealed several gold ornaments and a profusion of gold tablewares, golden beads for necklaces and belts and golden rings and bracelets. The treasure was split between the British Museum in London, the Penn State Museum in Philadelphia, and the National Museum in Baghdad.
- The so-called Gold of Troy treasure hoard, also called the Treasure of Priam by Heinrich Schliemann who excavated it in 1873, on the ancient site of Troy in the area of the city of Çanakkale in Turkey. Dated to 2600-2450 BCE (i.e. 1,000 years before the Trojan war!), it showed a range of gold-work from jewellery to a gold 'gravy boat' weighing 600 g. Most of the treasure, which was first in Berlin, is now in the Pushkin Museum in Moscow.

A millennium later (1200 BCE), probably the much better known hoard of gold was found in the tomb of Tutankhamun in Egypt (1333–1324 BCE). It contained the largest discovered collection of gold and jewellery, including a gold coffin. At the same period, pre-Columbian goldsmiths started producing gold items in South America. Their art reached its zenith during the Chimu civilisation between the twelfth and fifteenth centuries, but was stopped by the mass looting of the 'conquistadors'.

1.1.3 *Gold for monetary exchanges and the gold standard*

Gold has been also widely used throughout the world, as a vehicle for monetary exchange, even before the establishment of a gold standard, a monetary system in which the standard economic unit of account is a fixed weight of gold.

Egyptian Pharaohs began to commission gold tokens around 2700 BCE, but these tokens of variable purity were used as gifts, not for commerce. Much later, circa 600 BCE, the first gold coins known were minted by King Alyattes in Lydia (present-day Turkey). As a matter of fact, they were made of electrum, a natural alloy of gold and silver arising from alluvial deposits of the river running through Sardis, the Lydian capital. At the same period, 600–500 BCE, another gold coin, the Ying Yuan, was used in the kingdom of Chu in China.

Gold coins were used in some of the great empires of earlier times, such as the Byzantine Empire. But after the ending of this empire, the 'civilised world' tended to use silver coins. Paper money was first introduced in China between the seventh and fifteenth centuries, and then in Europe in the seventeenth century. It was a promissory note, i.e. a receipt redeemable for gold and/or silver coins. In 1816, England ended its policy of bimetallic standard (gold and silver) and adopted a single gold standard while the rest of Europe remained on a silver or bimetallic standard. Between 1872 and 1900, most major countries abandoned silver or bimetallic systems and achieved gold convertibility. At the beginning of the First World War, the gold standard was at its pinnacle, with 59 countries having adopted this standard.

However, during the First World War, governments had to face the huge war effort and boosted banknote printing, while international trade dropped dramatically. At the end of the war, all the countries had left the gold standard. However, England returned to the gold standard between 1925 and 1931, and France was the last country to abandon the convertibility in 1936. After the Second World War, the Bretton Woods Agreements (22 July 1944) created a system of fixed exchange rates, and gold was replaced by the US dollar. Nevertheless, nowadays, gold remains a safe investment.

1.1.4 *Gold for human well-being: food, drinks and medicine*

Pure metallic gold is non-toxic and non-irritating when it is ingested. Metallic gold has been approved as a food additive in the EU (E175 in the *Codex Alimentarius*). As gold leaf, it is sometimes used as food decoration in

China, Japan, India and also in Europe (for instance in France on *'palet d'or'* chocolate). Gold leaves are also used as a component of alcoholic drinks, such as *'Goldschläger'*, *'Gold Strike'* and *'Goldwasser'*.

Since the discovery of gold, people have thought of it as having an immortal nature and have associated it with longevity, probably because of its resistance to chemical corrosion. Many ancient cultures, such as those in India and Egypt, used gold in medicine but mainly for its magico-religious power. However, gold played almost no role in rational therapeutics. An exception is China, with the earliest application of gold as a therapeutic agent back in 2500 BCE. Pliny the elder, in the first century, reported gold for healing fistulas and haemorrhoids. The uses of gold were limited because at that time people did not know how to dissolve it and make it soluble. It was with the medieval period and the European (al)chemists that gold became a prominent medicinal element, with the idea that the elixir of life, *Aurum potabile*, can restore youth. *Aurum potabile* was closely related with the discovery of *aqua regia* (a mixture of hydrochloric and nitric acids), the 'royal' solvent of gold. A gold cordial was advocated in the seventeenth century for the treatment of ailments caused by a decrease in the vital spirits, such as melancholy, fainting, fevers and falling sickness. Later, in the nineteenth century, a mixture of gold chloride and sodium chloride was used to treat syphilis.

The use of gold compounds in modern medicine began with the discovery in 1890 by the German bacteriologist Robert Koch that gold cyanide $K[Au(CN)_2]$ was bacteriostatic towards the tubercle bacillus. Gold therapy for tuberculosis was subsequently introduced in the 1920s, but soon proved to be ineffective. In contrast, gold therapy proved to be effective against rheumatoid arthritis. Since that time gold drugs have also been used to treat a variety of other rheumatic diseases such as juvenile arthritis, palindromic rheumatism and various inflammatory skin disorders such as pemphigus, urticaria and psoriasis.

Today, in allopathic medicine, only salts and radioisotopes of gold are of pharmacological value, as elemental metallic gold is inert. However, some forms of alternative or traditional medicine assign metallic gold a healing power. The ayurvedic medicine in India, dated back thousands of years and related to the medical use of metals and minerals, involves gold in such medicines. For instance, *Swarna Bhasma* comprises globular gold nanoparticles with an average size of about 60 nm. Gold is considered to be a rejuvenator and, as such, is taken by millions of Indians each year. A typical daily dose corresponds to one or two milligrams of gold incorporated into a mixture of herbs.

Metallic gold may also have a renewed potential in 'modern' medicine as colloidal gold nanoparticles, which could be used for imaging, diagnostics, drug delivery or radiotherapy (see Chapters 10 and 11).

The malleability and resistance to corrosion make gold perfect for dental use, although its softness requires that it is alloyed, most commonly with platinum, silver or copper. So gold in alloys is used in tooth restorations, such as crowns and permanent bridges. There are examples of its use by the Phoenicians, the Etruscans and the Romans for restoration and also for aesthetics reasons.

For more information on gold in medicine, the reader can refer to Refs. 3–8 (free access) from which most of the information above has been drawn.

1.1.5 *Gilding gold and gold-like lustre*

The use of gilded films of gold on oxide substrates to decorate glass, ceramic and mosaics may be dated from the Roman period circa the first century, as reported by Pliny the Elder, but wider use dates from the twelfth century. Gold foil coating is the most ancient technique used, and *tesserae* of mosaics (small block of material used in the construction of a mosaic) were the first supports used. In this process, a few micrometers of thick gold foil is pasted onto substrates of glass or ceramic with an adhesive agent, such as linseed oil or egg white, covered with glass powder and heated. The most ancient articles are probably the golden mosaics of the cupola of the mausoleum of Galla Placida built in Ravenna in 425–443, but the peak of gold gilded glass production is in the thirteenth and fourteenth centuries with the Mamelouk production in Egypt and Syria, and also in the nineteenth century.

Gilded films must be distinguished from lustre, which is a surface layer with a metallic appearance applied on glazed ceramics, i.e. on a surface of terracotta covered by a glassy layer. Lustre exhibits various colours, from gold to brown or red. However, in spite of the appearance, it does not contain any gold, but only silver and copper metal particles in various sizes and compositions, dispersed in a glassy matrix with a gradient of size and concentration.[9-11]

1.2 The First Uses of Gold Nanoparticles

The first use of gold nanoparticles is intimately related to the history of red-coloured glass. The production of red glass (opaque) starts with the very beginning of glassmaking in Egypt and Mesopotamia back in 1400–1300 BCE.[12] The colour of this red glass was given by the addition of copper. The origin of the red colour is debated, with some scientists stating that it is due to metal copper nanoparticles, while others state that it is due to cuprous oxide (cuprite) nanoparticles or to both. The origin of the coloration also depends on the sites and dates of production, the method of preparation and components of glass.[13] The production of copper red glass is a real challenge from a technological point of view because it requires a reducing atmosphere; for this reason, red glasses are less frequent than other colours.

Another way of making red glass involves the use of gold nanoparticles. According to most of the textbooks and technical encyclopedias on gold, glass and ceramics, the production of the so-called 'gold ruby glass' did not take place until the end of the seventeenth century. The discovery is attributed to Johann Kunckel (c.1637–1703, Brandenburg) and that of the gold preparation that is added to melted glass to give it the ruby red colour is attributed to Andreas Cassius of Leyden in 1685.[14] This is the so-called *Purple of Cassius*, which is a precipitate obtained from the dissolution of gold metal in *aqua regia* followed by the precipitation of metallic gold by a mixture of stannous and stannic chloride.

As a matter of fact, the story of gold ruby glass begins long before, and there is no break until the peak of its production at the end of the seventeenth century.

1.2.1 *The Lycurgus cup*

Hence, the first milestone in the history of gold ruby glass is a Roman opaque glass cup dated to the fourth century, the Lycurgus cup, which is exhibited at the British Museum in London[15] (Fig. 1.1). The carved decoration depicts a mythological scene that is the triumph of Dionysus over Lycurgus, a king

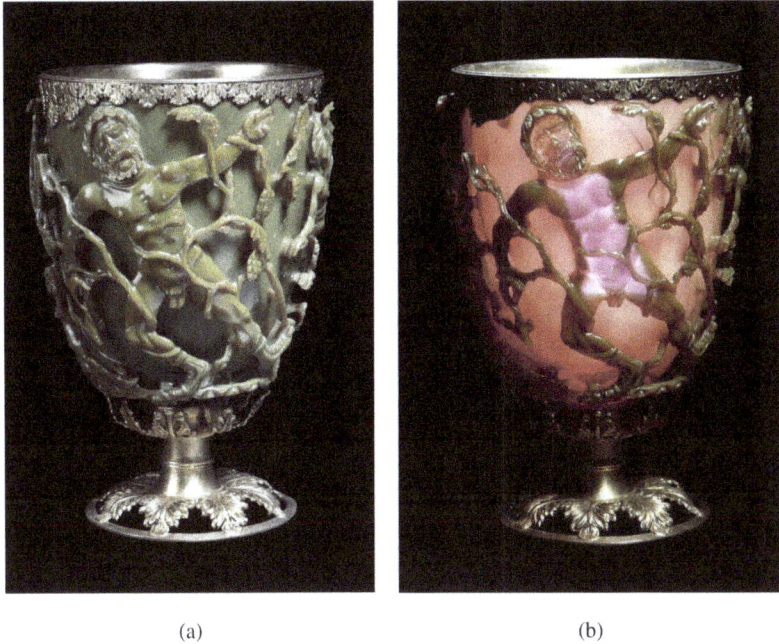

(a)   (b)

Fig. 1.1.   The Lycurgus cup, late Roman, fourth century CE, probably made in Rome (from the British Museum free image service). (a): illuminated from outside. (b): illuminated from inside.

of the Thracians (circa 800 BCE): one of Dionysus' maenads, Ambrosia, transformed into a vine by Mother Earth, holds Lycurgus captive while Dionysus instructs his followers to kill him.

This cup shows a green jade colour due to the diffusion of light when it is illuminated from outside (Fig. 1.1.a) and a deep ruby red one in transmission when it is illuminated from inside (Fig. 1.1.b) (See also Section 1.3.1). A detailed analysis of the Lycurgus cup, published in 1965 by Brill,[16] revealed the presence of minute amounts of gold (about 40 ppm) and silver (about 300 ppm) in glass. In 1980, a further analysis by Barber and Freestone[17] attested the presence of nanoparticles of 50–100 nm in diameter by electron microscopy, composed silver-gold alloy, with a ratio of silver to gold of about 70:30. Later on, Hornyak *et al.*[18] confirmed through a theoretical study that the deep red colour of the Lycurgus cup due to light absorption around 515 nm is consistent with the presence of silver-gold alloy with Ag:Au of 70:30. (See Chapter 3 for optical properties of gold nanoparticles.)

9

The British Museum experts believe that the colouring of glass using gold and silver was far from routine during the Roman period since only a limited number of other glasses appeared to have been coloured by gold.[19] Moreover, no other glass of this period replicates the dichroic optical effect of the Lycurgus cup. They conclude that the technology seems to have been very restricted and did not outlast the fourth century.

However, a very recent study by Verità and Santopadre[20] reports the chemical analyses of nine flesh-tone glass *tesserae* of mosaics, arising from nine important churches in Rome of the fourth to twelfth centuries. All of them reveal that the flesh colour originates from the presence of 10–30 ppm of gold or gold-silver alloy particles. Since a considerable number of flesh-colored glass *tesserae* were employed in mosaics of these churches, the authors conclude that the colour was obtained routinely rather than by chance, and that the Roman glassmakers mastered this complex coloration process. Since there is no evidence that the Romans were able to produce *aqua regia* to prepare gold chloride at that period, the authors propose that the Roman glassmakers may have used silver slags without knowing that they also contained gold, thus without knowing that gold was the actual colorant of glass; they also propose that the colour arises from the local dissolution of gold leaves and the formation of 'droplets' of gold ruby glass since these droplets are commonly found in the gold-foil *tesserae* of Roman mosaics.

1.2.2 *Medieval period*

There is written evidence that the (al)chemists[b] of the Middle Ages knew how to produce red-coloured glass with gold, although samples of such glass have yet to be found.[19,21] It should be noted that some textbooks and websites state that the red colour of stained glasses of medieval church windows is given by gold. However, in all cases analysed so far, the colorant found is copper.[19]

Al Razi (865–925), a Persian scholar, philosopher and alchemist, reports the earliest known written account of a gold ruby glass in his treatise *Secrets*

---

[b]Note that it is during the nineteenth century that a distinction is made between alchemists and chemists.

of Secrets. The instruction was to heat a very finely powdered batch of different elements including gold powder for three days in a closed furnace fuelled with very hard wood. In his paper, Sheybany[22] concludes that this may allow temperatures of 800–1000 °C to be reached in a reducing atmosphere. Al Razi believed he had fulfilled the objective of the transmutation of metals; in his treatise, he stated that this glass attracted gold and silver like a magnet and that it could convert 1,000 times its weight into gold.[22]

It is important to stress that the main goal of the medieval alchemists was the making of the philosopher's stone. In alchemical writings, the philosopher's stone is often described as a red substance, which is supposed to be the key to transmutation of 'impure' base metals into gold, the unique pure metal.

1.2.3 *Fifteenth and sixteenth centuries*

In the Bologna manuscript, *Segreti per colori*, written in the first middle of the fifteenth century, three recipes of gold ruby glass are described. However, according to Zecchin's paper[23] they are inconsistent. Later on, between 1458 and 1464, Antonio Averlino, also called Filarete, provided some technical information on glass coloration in his *Trattato di Architettura*, and wrotes 'It is also said that gold makes colour.'[23]

Georgius Agricola (1494–1555, Saxony), who is considered the founder of geology, is supposed to have described the preparation of gold ruby glass in *De natura fossilium* published in 1546[14,24]: 'A famous variety of dyeing glass is made from gold and this is used to tint the glass clear ruby red.' As a matter of fact, according to Zecchin[23] and von Kerssenbrock-Krosigk,[25] this sentence is wrong and results from a mistake in the first translation from Latin to English. However, there are several other writings that refer to gold ruby glass during the sixteenth century. Benvenuto Cellini (1500–1571), a famous sculptor and goldsmith in Florence, refers to a transparent red enamel discovered by an alchemist who was also a goldsmith.[26] Later, Andreas Libavius (c.1540–1616), a German chemist and physician, mentioned the red colour of gold dissolved in liquid to make red crystal in *Alchemia* published in 1597. According to Polak,[27] Andreas Libavius based himself in this on two earlier 'distillers', the Neopolitan Giambattista Porta (1535–1615), author of *Magiae Naturalis* (1588) and Gerhard

Dorn (c.1530–1584), the German author of *Clavis Totius philosophiae chymistica* (1567).

1.2.4 *Seventeenth century*

L'Arte Vetraria is the first print book exclusively devoted to glassmaking. It was published in 1612 by Antonio Neri (1576–1614), a Florentine priest, son of a physician. In Book 7, Chapter 129, one recipe mentions the use of gold to produce red glass. In short, the recipe, which is entirely reported in Franck's paper,[24] involves the calcination of gold with *aqua regia* in a furnace, which forms a red powder that is then added to glass. The recipe attests that the potential of using gold as a red colorant was fully understood in early seventeenth century.[28] The only known gold ruby vessels of Italian origin of that period are a series of ribbed bowls, ewers and bottles that King Frederick IV of Denmark brought back from a trip to Venice in 1708–1709. These artefacts are visible in Rosenborg castle in Copenhagen.

Antonio Neri's book was then translated into English in 1662 by Christopher Merrett (1614/5–1695); he added 147 pages of his own, from other authors and his own observations. In 1679, the first German edition of the Neri–Merrett book appeared, translated with further extensive addition by the famous Johann Kunckel (cited at the beginning of Section 1.2) under the title *Ars Vitraria Experimentalis*.

Other written sources were recently found by Zecchin in Murano archives.[23] A manuscript written by Giovanni Darduin (1585–1654), a glass-maker of Murano, provides a recipe of gold ruby glass among other glass recipes of his and of his father who died in 1599. Two other recipes of gold ruby glass were provided by Giusto Darduin (1661–1700) and one by Antonio dalla Rivetta (1628–1695). Zecchin could not establish the existence of a relationship between the Italian branch and the German one and Kunckel.[23] However, he suggests that a relationship may have existed with Bernard Perrot in France (see Section 1.2.4.3).

1.2.4.1 *Purple of Cassius*

As mentioned at the beginning of Section 1.2, the paternity of the purple gold precipitate used for colouring glass, the so-called *Purple of Cassius*, has been attributed to Cassius. As described earlier, the preparation involves gold being dissolved in *aqua regia*, then its precipitation as metallic gold by a mixture of stannic and stannous chlorides.

As a matter of fact, there were two Andreas Cassiuses, father (born circa 1605 in Schleswig and died in 1673 in Hamburg) and son (born in 1645 in Hamburg and died circa 1700 in Lübeck), both of whom were physicians. The son wrote *De Auro*, published in 1685, in which he gave his father's recipe of the *Purple of Cassius*, obtained by reducing a gold chloride aqueous solution with stannous chloride; the entire translation of the recipe can be found in Hunt's paper.[14] In a short book published in 1684, *Sole Sine Vest* ('Gold unclothed'), Johann Christian Orschall, who was a metallurgist and also interested in gold ruby glass, reported the anecdote that Cassius, the son, succeeded in making a very fine ruby flux and sold the secret in various places.[14] On the other hand, Cassius, the son, was aware that the formula of the preparation had been used before his father and that he may have been influenced by the work of Johann Rudolf Glauber. Johann Kunckel also mentioned that Cassius was not the true inventor of the *Purple of Cassius*, and that perhaps Glauber may have given him the idea.

Johann Rudolf Glauber (1604–1670), a native of Bavaria who settled in Amsterdam, was a pharmacist, living off the sales of his medicinal preparations (which was exceptional at this time). His writing in Part IV of *Prosperitatis Germaniae* published in 1659, i.e. a quarter of a century before the publication of *Cassius*, is considered as the first report that mentions that gold can be precipitated with a solution of tin compound. However, there is no evidence that Glauber made use of the purple precipitate for colouring glass.[14]

It is important to stress that the seventeenth century is still a period in which (al)chemists were obsessed not only with attempts to unlock the secrets of nature by simulating natural processes in laboratory conditions, but also with attempts to manufacture metals for mystical purposes. They believed that the colour of metals indicated their 'souls' or essence, and that if the colour could be extracted, it would possess the spirit of the metal and could perform alchemical transmutation. Great scientists such as Robert Boyle (1627–1691) and Isaac Newton (1642–1727) firmly believed in this principle. (Al)chemists also invested considerable efforts in making glass imitations of gemstones, and new methods of colouring glass and mixing batches were invented.[29,30] Coming back to Glauber, although he can be regarded as one of the founders of the chemical industry, he also related the production of gold ruby glass to alchemy. He claimed that the soul of gold is captured in the red colour of gold ruby glass, and he regarded the making of gold ruby glass as akin to the process of alchemical transmutation, in that the substance turned red before it was transformed into gold. He also believed that this was a demonstration of the multiplication of gold, because only a small amount of gold was required to colour a large amount of glass.[29,30]

1.2.4.2 *Kunckel glass*

As already mentioned, it is widely reported in textbooks that Johann Kunckel is the first important maker of gold ruby glass. If Neri and his predecessors had managed to produce gold ruby glass in small quantities, maybe for the purpose of imitating natural stone, Kunckel is recognised as the first glassmaker to be successful in producing gold ruby glass on a rather large scale. He was the son of an (al)chemist glassmaker, and himself was first an (al)chemist and apothecary; he taught at the University of Wittenberg in Saxony for about ten years, then he moved to Postdam in Brandenburg in 1678, where the great Elector, Friedrich Wilhelm, commissioned him to take charge of a glass factory. He started developing the production of gold ruby glass vessels by 1684. How Kunckel managed to produce gold ruby glass on such a large scale remains a mystery. Moreover, although some vessels can be dated to a period where Kunckel might have been the glassmaker (Fig. 1.2), none of them can be unambiguously attributed to him.[28]

From *Ars Vitraria Experimentalis* published in 1679, it is clear that Kunckel was unwilling to describe his recipe of gold ruby glass. Moreover, his factory was located at an isolated site, the Pfaueninsel, or Peacock island, between Berlin and Potsdam. His secret of fabrication was revealed later in *Laboratorium Chymicum* published posthumously in 1716. It is known that Daniel Crafft (1624–1697), who had worked as Glauber's assistant for about ten years and became a glassmaker, worked with Johann Kunckel in Dresden after 1673,[30] and that Kunckel had known Antonio Neri's book *L'Arte Vetraria* (1612), since he translated it into German and published it in 1679.

According to von Kerssenbrock-Krosigk,[28] between 1685 and 1705 enthusiasm for gold ruby was at its peak in Europe, and almost every central European sovereign owned gold ruby glass vessels. At that time, gold ruby

Fig. 1.2. Goblet, engraved in manner of Gottfried Spiller, about 1700, Postdam, Germany, gold ruby glass; blown, cut, engraved; h. 24.1 cm from the Collection of The Corning Museum of Glass, Corning, NY; gift of The Ruth Bryan Strauss Memorial Foundation (79.3.258) with permission of Corning Museum of Glass, Corning, NY, USA.

glass was considered as a genuinely new material and a decorative folly; a way of imitating semiprecious gemstone, like crystal whose fabrication had been discovered at the same period.

1.2.4.3 *Perrot glass*

What is not reported in textbooks is that 16 years before Kunckel started producing ruby red glass, Bernard Perrot was producing glass artefacts containing gold ruby glass in France (Fig. 1.3).[31] Bernardo Perrotto (1640–1709), an Italian glassmaker from Altare (Ligury), opened a glass shop in Orleans (France), and became Bernard Perrot. In 1668 he obtained the royal privilege from Louis XIV to colour glass in red. An exhibition dedicated to his glass work was held in Orleans in 2010. Chemical analysis of various samples of glass revealed his gold ruby glass was not produced from the *Purple of Cassius*; no tin but arsenic (0.6 to 2.9 wt%) was identified in the presence of 23 to 280 ppm gold.[32] It is not certain whether

Bernard Perrot produced gold ruby glass himself; it is possible that the glass or the recipe arose from Marc-Antoine Galaup de Chasteuil, an alchemist who later became involved in the famous 'affair of the poisons'![33] It is also possible that gold ruby glass came from the Italian branch, because the recipes provided by Giusto Darduin and Antonio dalla Rivetta (cited at the beginning of Section 1.2.4) also involve arsenic, and that according to Zecchin,[23] a relationship between the two may have been established.

1.2.5 *Gold ruby glass in the eighteenth century*

It is also widely reported in textbooks that the art of making gold ruby glass was lost on Kunckel's death and rediscovered during the nineteenth century. This is illustrated by the Werner Herzog movie from 1976, *Heart of Glass* (in German: *Herz aus Glas*). The film is set in eighteenth-century Bavaria. A factory produces ruby red glass blowings until the death of the master glass-blower. Knowledge of the glass-blowing method is lost causing depression to afflict the inhabitants of the town.

According to von Kerssenbrock-Krosigk,[28] during the eighteenth century, even though the peak interest for gold ruby glass was over in Germany, it persisted in some regions and occasionally arose in others. In Brandenburg, gold ruby glass continued to be produced, but the kings of Prussia and other statesmen in Germany favoured the hard-paste porcelain. The alchemist Johann Friedrich Böttger (1682–1719), who first became famous for producing gold 'transmutation', played a crucial role in the discovery of the hard porcelain in Europe and in the development of the first porcelain manufacture in 1707–1709 in Meissen in Saxony. Around 1713 he also experimented with gold ruby glass.

Already during Kunckel's lifetime, knowledge on gold ruby glass spread especially to Bavaria and Bohemia.[24,34] A connection between Kunckel and Hans Christoph Fidler (1677–1702), the crystal-maker of the electorate of Bavaria, who experimented with ruby glass in 1686–1688 at the Zákupy glasswork in Northern Bohemia, is well documented. The production of ruby red glass started in 1683 in Southern Bohemia and in Silesia about 1700.[34] Several rare pieces dated from about 1700 are preserved at the Museum of Applied Art in Prague. The use of gold ruby for windows is also reported in the records of William Peckitt, a leading glassmaker and producer of stained glass during the eighteenth century in Yorkshire.[24] The interest in gold ruby glass throughout the eighteenth century is also attested by the incorporation of part of Neri, Merrett and Kunckel's books in major publications of different languages, such as the *Encyclopedia Britannica*.[24]

1.2.6 *Gold ruby glass and cranberry glass in the nineteenth century*

Hence, the seventeenth and eighteenth centuries laid the foundations for the great practical and theoretical interest in coloured glass, which took place during the nineteenth century.[24]

In the first half of the nineteenth century during the Biedermeier period (1820–1850), gold ruby glass was mostly produced in Bohemia. The fashion for these glass artefacts swept across Europe and to the United States, first with exports from Bohemia then with development of local production.

For instance, in France, a competition was organised in 1837 to find a more reliable process of coloration. The production of gold ruby glass reached its peak in England during the Victorian era (1837–1901), particularly around Stourbridge (West Midlands) and Bristol. Molineaux Webb & Co. (1826–1928) used gold ruby glass to produce window and vessel glass on quite an extensive scale. Several English companies and Baccarat, a French firm, exhibited pieces of ruby glass at the first world's fair of 1851. The colour came in a variety of less saturated tints, close to pale pink, called cranberry, that was obtained by decreasing the gold concentration in glass. In the US, gold was also used to produce new types of glass, burmese and rose amber glasses. These are opaque glasses, ranging from yellow to pink, obtained from uranium oxide (that gives a soft yellow colour) and from gold (that gives the pink blush).

1.2.7 Pink enamel porcelain: Rose Pompadour and Famille Rose

Although each material was known from 3000 BCE, enamel combination of glass and metal is not found until the twelfth century BCE, when the Myceneans succeeded in making enamels on a gold base.[35] According to Garner's paper, there is no doubt that practical means of making ruby enamels were known from the time of Benvenuto Cellini (1500–1571), i.e. earlier than Cassius and Glauber (see Section 1.2.3), and possibly even earlier still. However, painted enamels with a good deal of pink did not appear before 1667 in south Germany and before the end of the seventeenth century in France.

By 1719, pink enamel was prepared and used in the first porcelain factory to be established in Europe, at Meissen (see Section 1.2.5), using the *Purple of Cassius*.[36] In 1738, the Vincennes porcelain workshop in France, which became the well-known *Manufacture Royale de Sèvres* in 1756 under King Louis XV, produced colours from bright red to violet from gold-based purples.[37] The chemist Jean Hellot succeeded in producing the so-called *Rose Pompadour* in 1757 also based on the use of *Purple of Cassius* (Fig. 1.4). The recipe was based on the preparation of a colloidal sol of gold, a slightly different recipe to the *Purple of Cassius*. It was added to a powdered flux, which was then dried and ground to fine powder. Once suspended in turpentine, the enamel was then used in the decoration of

Fig. 1.4. *Rose Pompadour*; eighteenth century, sugar pot of the Calabre tableware in soft paste porcelain, base cover and bird cartel, from Sèvres-Cité de la céramique, France. Photograph: Gérard Jonca, Sèvres-Cité de la céramique.

porcelains and then fired at temperatures up to 880 °C. These pink enamels were soon introduced in England at Worcester and Chelsea among other early porcelain factories.[36]

Meanwhile, by 1723, the recipe of the *Purple of Cassius* had reached China, probably through Jesuits, and was used successfully for the production of enamel on metal, followed on porcelain, which is designated as *Famille Rose* porcelain (the 'rose family').[14] Around 1735, the Chinese mastered the technique and probably a little before 1740, *Famille Rose* enamels were applied to both porcelain and copper in the Peking area.[35]

1.3 Scientific Approach of the Preparation of the Gold Ruby Colour

1.3.1 *Elucidation of the constitution of the Purple of Cassius in the nineteenth century*

The elucidation of the nature and constitution of the *Purple of Cassius* remained puzzling throughout the whole of the nineteenth century in spite of many studies performed by the most famous scientists of this period.[36]

In 1866, J.C. Fischer of Munich was able to draw up a list of twelve distinguished chemists who held that gold was present as an oxide and of six other ones who believed that it was in metallic form. Among the partisans of the gold oxide, there were Louis-Nicolas Vauquelin in 1811, Jöns Jacob Berzelius in 1831 and Louis Gay-Lussac in 1832. Among those of the metallic form, there were Alphonse Buisson, the assistant of Alexandre Brongniart, the director of Sèvres factory in 1830, and Michael Faraday who conjectured in 1857 that gold was present in solution in a 'finely divided metallic state'.[38] The closest to the truth was Henry Debray, lecturer at the École Polytechnique in Paris who proposed in 1872 that the *Purple of Cassius* consisted of finely divided gold adsorbed on stannic acid. However, it is only at the turn of the century in 1905 that the true nature of the *Purple of Cassius* was finally elucidated by Richard Adolf Zsigmondy (1865–1929). Thanks to the development of a slit ultramicroscope (based on light scattering) that he developed with an optical physicist, Heinrich Siedentopf, he was able to observe finely divided gold particles on colloidal stannic acid. For these investigations, he was awarded the Nobel Prize in Chemistry in 1925. More details and references can be found in Carbert's paper.[36] Zsigmondy also confirmed the presence of colloidal particles in the ruby glass.[39] In 1908 Gustav Mie predicted the optical properties of homogeneous spherical particles.[40] For a spherical nanoparticle much smaller than the wavelength of light (diameter $d << \lambda$), an electromagnetic field at a certain frequency (v) induces a resonant, coherent oscillation of the metal free electrons across the nanoparticle. This oscillation is known as the localised surface plasmon resonance (LSPR).[41,42] The plasmon oscillation of the free electrons of the metal nanoparticles Au, Ag and Cu, results in a strong enhancement of absorption and scattering of electromagnetic radiation in the visible range (around 520 nm for gold) in resonance with the plasmon frequency, giving them intense colours and interesting optical properties. The ratio of scattering to absorption increases with nanoparticle volume. This is more extensively developed in Chapter 3. The dichroism of the Lycurgus cup can be briefly explained as follows. When it is illuminated from outside (Fig. 1.1.a), the green colour of the cup is due to the non-negligible contribution of the scattering contribution of the 50–100 nm particles contained in the cup. The red colour given by the cup when it is illuminated from inside (Fig. 1.1.b) results from the absorption contribution:

the green wavelength is absorbed and the light that goes through the glass appears with the complementary colour, which is red.

1.3.2 *Chemical approach to the formation of the Purple of Cassius*

Two stages are involved in the preparation of the *Purple of Cassius*. The first stage is the formation of a gold sol, and the second one, its stabilisation. The first stage involves a redox reaction between stannous chloride and auric chloride and the formation of metallic gold:

$$2\,Au^{3+} + 3\,Sn^{2+} \rightarrow 2\,Au^{0} + 3\,Sn^{4+}$$

The solution of stannous chloride (Sn^{2+}) also contains stannic ions (Sn^{4+}). In the past, this solution was obtained from the dissolution of tin in *aqua regia*, and the resultant stannic chloride was reduced to produce the required stannous to stannic ratio by a further addition of tin metal. The second stage is the hydrolysis of the stannic chloride into tin hydroxide that flocculates and precipitates. According to Weyl,[39] both processes, the precipitation of tin hydroxide and the formation of metallic gold, occur simultaneously. When tin flocculates and precipitates, the adsorbed gold particles precipitate with it and remain in dispersed form. The gold sol is therefore stabilised by adsorption onto a colloidal tin hydroxide. Further operations are filtration, milling in wet conditions and drying to recover the precipitate. More details can be found in Ref. 36. The particle growth must be controlled, and 'good quality' *Purple of Cassius* requires gold particles of 10–15 nm, but a large number of factors can cause variations in the particle size and therefore in the hue and strength of the colour. The result is that despite the many years over which it has been known and studied, its preparation remains difficult to control. However, this preparation is still in use nowadays to colour glasses and enamels. According to Weyl,[39] the *Purple of Cassius* did not surprise its discoverers by its colour but rather by the stability of its colour at high temperatures. It offered a new red pigment, which could be introduced into glazes and into glass. Gold has such a strong colouring ability that only a minute amount is required even for the deepest colours: 100 to 1000 ppm is sufficient to produce deep pink colour glass whereas the red *sang de boeuf* colour provided by copper requires a concentration a hundred times as high as gold.

1.3.3 *Chemical approach to the preparation of gold ruby glass*

If one keeps in mind all that has been told about the preparation of gold ruby glass so far, it appears that the addition of *Purple of Cassius* to melt glass is not the only way to prepare gold ruby glass. Calcination of gold is proposed in Darduin's manuscript (1585–1654) (Section 1.2.4): layers of gold leaves and sodium chloride are calcined several times in a furnace until the gold leaves become crumbly. The same procedure is reported by Orschall in his treatise of 1684, in which he noticed that after a calcination of eight hours, the salt turns purple.[23] Neri in *L'Arte Vetraria* (1612) reports the calcination of gold with *aqua regia* followed by the calcination of the powder, which turns red and is then added to melt glass.[23] Several experimental studies performed during the nineteenth century were gathered by Weyl, together with his own experiments in another outstanding book after *L'Arte Vetraria*, entitled *Coloured Glasses* and published in 1951.[39] Weyl confirms that there are different ways to prepare gold ruby glass. He writes that metallic gold can be directly added to molten glass, and dissolves with reasonable speed, but that it is more effective to introduce gold in the form of *Purple of Cassius* or of gold chloride prepared by dissolving gold in *aqua regia*.

Practically, the preparation of gold ruby glass is based on three consecutive steps: (i) the addition of gold (several hundred ppm), most often as gold chloride or in the form of *Purple of Cassius*, in melt glass around 1400 °C; at this stage, glass is colourless; (ii) a step of rapid quenching to room temperature and; (iii) usually a step of reheating or annealing (striking step) around 500–650 °C during which the red colour appears.

A simplified scheme of the principle of gold ruby glass formation gathering the different steps associated to physico-chemical phenomena and colours (Fig. 1.5) is drawn from the scheme proposed by Weyl.[39] During the cooling step, there is oversaturation of the 'atomic gold solution' and the formation of gold nuclei. At this point, two extreme cases can be distinguished depending on the initial composition of glass: (i) glasses with a steep solubility curve such as sodium silicate or borax glass, lead to the growth of the nuclei and the formation of large gold particles and a brownish colour, i.e. to bad/spoiled glass (Type II then Type III); (ii) the presence of

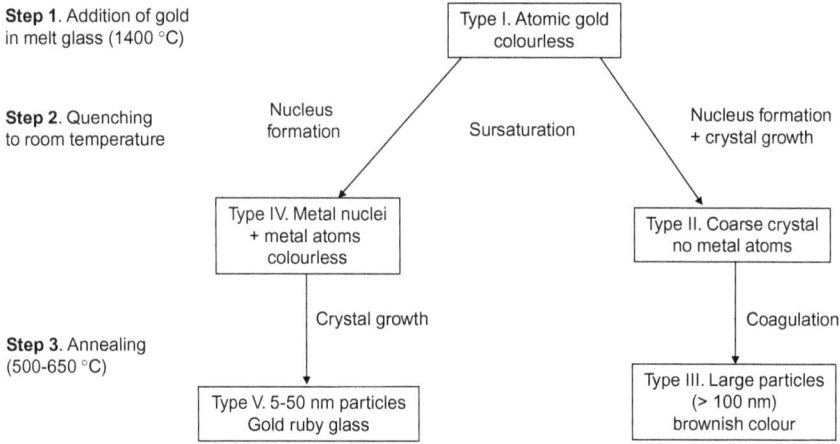

Fig. 1.5. Simplified scheme of the principle of gold ruby glass formation gathering the different steps associated to physico-chemical phenomena and colours; drawn from the scheme proposed by Weyl.[39]

lead, tin or bismuth ions in glass enhances gold solubility, so cooling produces the nuclei (Type IV), but a substantial part of gold still remains in atomic dispersion available for the nourishment of the nuclei; the glass is still colourless at this stage. Reheating to the striking temperature causes the nuclei to grow and brings about the red colour (Type V). By adjusting the gold content, the heat treatments (temperature and duration of cooling and reheating) and the temperature coefficient solubility (controlled by the addition of lead, antimony or tin oxide), it is possible to produce nuclei in sufficient quantity at relatively high temperature, and achieve the desirable hue. One can see how complex the preparation of gold ruby glass is and better understand the difficulties encountered by the glassmakers in the past to produce gold ruby glass and achieve reproducible preparations. Note that some glasses strike on cooling, so the nuclei have the chance to grow during the initial cooling. In such a case, a gold ruby glass is directly obtained; this may happen for instance when tin oxide has been added. Also note that nucleation can be produced with several other agents such as antimony oxides or using ultraviolet, X or gamma radiation.[43]

There are disagreements in the literature whether gold dissolves in glass as gold atoms or ions (Type I).[44] According to a ^{197}Au Mössbauer study performed on quenched colourless glasses (Type IV) by Wagner *et al.*,[45] most of the gold is in the oxidation state I at this stage whether the gold precursor introduced in the glass was $HAu^{III}Cl_4$ or $KAu^{I}(CN)_2$. After annealing when glasses are coloured (Type V), the main species is Au^0. In another study of the same group,[46] ^{119}Sn and ^{197}Au Mössbauer spectroscopy gives an indication on the role that tin plays in the formation of a larger number of gold particles much smaller than in the absence of tin: tin provides condensation nuclei for gold nanoparticles, tin acts as a surfactant at the surface of the gold nanoparticles, which stabilises the small gold metal particles and accelerates the kinetics of formation of gold metal through a redox mechanism similar to that occurring during the formation of *Purple of Cassius* (see Section 1.3.2):

$$Sn^{2+} + 2\,Au^+ \rightarrow Sn^{4+} + 2\,Au^0$$

Weyl[39] states that the particles in glass which strike to purple are between 5 and 50 nm and those which lead to livery ruby are larger than 100 nm. The development of the gold ruby colour is enhanced by the presence of some ions in the base glass. Lead-based glasses produce the best ruby-coloured glasses because the higher the lead content, the higher the gold solubility. The deepest colours can be obtained by adding 1000 ppm of gold chloride in heavy lead glass. In soda-lime-silica glass, deep red colours can be obtained with 100 to 300 ppm of gold.[20]

A fragment of seventeenth century ruby red glass found in the remains of Kunckel's factory at Peacock island was studied by Fredrickx *et al.*[47] Gold concentration was 160 ppm, that of tin oxide (SnO_2) was 525 ppm, and the gold particles displayed a cubo-octahedral morphology and had the right sizes (~40 nm) to provoke the proper red colour through the phenomenon of surface plasmon resonance (see Chapter 3). Iron-containing particles, mostly α-Fe_2O_3, were abundantly found in the glass matrix, and were supposed to have an influence on the colour.

1.4 Conclusion

The history of gold nanoparticles, already covering centuries through their use for the coloration of glass and ceramic, is far from over. Gold nanoparticles are still used to make ruby glass, even though other red colorants based on copper or selenium are also used. For instance, in France,

Fig. 1.6. '*Cristal Rubis*', ruby crystal, wine glass, (from the Baccarat Vega Martini collection. Copyright ©Baccarat).

Saint Gobain produces decorative pink glasses as stain glasses, and Baccarat has developed a series of gold ruby glasses (Vega collection) (Fig. 1.6), and edits glass artefacts containing ruby red glass designed by artists.

In the case of red enamel for porcelain, there are alternative and less expensive colours based on chrome-tin, but they do not offer the same range of hues and give a more opaque finish compared to the translucent effect obtained with the gold-based enamels. Gold-based enamels also withstand a higher temperature during the firing of the colours than the cheaper base metal enamels.[36] Nowadays, Sèvres Manufacture still produces gold-based red enamels, and they are still conducting research for improving the quality and the reproducibility of the colour.

The history of gold nanoparticles, that has paralleled that of gold ruby glass for centuries, is now expanding with the advent of nanosciences and nanotechnologies. Nowadays researches on synthesis, properties and applications of gold nanoparticles involve the many fields of chemistry, biology and physics that are described in this book.

Acknowledgments

The author deeply thanks Marco Verita, Paolo Zecchin and Ian Freestone, for contributing very recent papers or papers in press. The author also thanks Jeannine Geyssant and Dedo von Kerssenbrock-Krosigk for information, and Gérald Dujardin, Richard Holliday and Rachel Doherty for having read the chapter.

References

1. J.A. Harrell and V.M. Brown, *J. Am. Res. Center Egypt* **29** (1992) 81.
2. L.B. Hunt, *Gold Bull.* **9** (1976) 24.
3. G.J. Higby, *Gold Bull.* **15** (1982) 130.
4. R.V. Parish and S.M. Cotrill, *Gold Bull.* **20** (1987) 3.
5. S.P. Fricker, *Gold Bull.* **29** (1996) 53.
6. E.R.T. Tiekink, *Gold Bull.* **36** (2003) 117.
7. H. Knosp, R.J. Holliday and C.W. Corti, *Gold Bull.* **36** (2003) 93.
8. C.L. Brown, G. Bushell, M.W. Whitehouse, D.S. Agrawal, S.G. Tupe, K.M. Paknikar and E.R.T. Tiekink, *Gold Bull.* **40** (2007) 246.
9. J. Pérez-Arantegui, J.t. Molera, A. Larrea, T. Pradell and M. Vendrell-Saz, *J. Am. Ceram. Soc.* **84** (2001) 442.
10. E. Darque-Ceretti, D. Hélary, A. Bouquillon and M. Aucouturier, *Surf. Eng.* **21** (2005) 352.
11. P. Colomban, T. Calligaro, C. Vibert-Guigue, N.Q. Liem and H.G.M. Edwards, *ArcheoSciences* **29** (2005) 7.
12. R.H. Brill and N.D. Cahill, *J. Glass Study* **30** (1988) 16.
13. D.J. Barber, I.C. Freestone and K.M. Moulding, in *From Mine to Microscope — Advances in the Study of Ancient Technology* (eds I.C. Freestone, A.J. Shortland and T. Rehren), Oxbow Books, Oxford, 2009, p. 115.
14. L.B. Hunt, *Gold Bull.* **9** (1976) 134.
15. BritishMuseum, http://www.britishmuseum.org/explore/online_tours/museum_and_exhibition/the_art_of_glass/the_lycurgus_cup.aspx. (Accessed 1 February 2012.)
16. R.H. Brill, *Proc 7th Internat. Cong. Glass, Bruxelles, Section B, Paper 223*, (1965) 1.
17. D.J. Barber and I.C. Freestone, *Archaeometry* **32** (1990) 33.
18. G.L. Hornyak, C.J. Patrissi, E.B. Oberhauser, C.R. Martin, J.-C. Valmalette, L. Lemaire, J. Dutta and H. Hofmann, *NanoStruct. Mater.* **9** (1997) 571.
19. I. Freestone, N. Meeks, M. Sax and C. Higgitt, *Gold Bull.* **40** (2007) 270.
20. M. Verità and P. Santopadre, *J. Glass Studies* **52** (2010) 11.
21. W. Ganzenmuller, *Glastechnische Berichte* **15** (1937) 379.
22. H.A. Sheybany, *Glastechnische Berichte* **40** (1967) 481.
23. P. Zecchin, *J. Glass Studies* **52** (2010) 25–33.
24. S. Franck, *Glass Technol.* **25** (1984) 47.

25. D. von Kerssenbrock-Krosigk, *Rubinglas des ausgehenden 17. und des 18. Jahrhunderts*, Philipp von Zabern, Mainz am Rhein, 2001.
26. M. Bimson and I.C. Freestone, in *Annales du 9ème congrès de l'Association Internationale pour l'Histoire du Verre*, Nancy, France, 1983; Liège, 1985, p. 209.
27. A. Polak, *Glass: Its Maker and its Public*, Weidenfeld & Nicolson, London, 1975.
28. D. von Kerssenbrock-Krosigk, in *Glass of the Alchemists*, The Corning Museum of Glass, New York, 2008, p. 123.
29. P.H. Smith, in *Glass of the Alchemists*, The Corning Museum of Glass, New York, 2008, p. 23.
30. W. Loibl, in *Glass of the Alchemists*, The Corning Museum of Glass, New York, 2008, p. 63.
31. J. Geyssant, in *Bernard Perrot 1640–1709, secrets et chefs d'oeuvre des verreries royales d'Orléans, Somogy Edition d'art, Paris and Musée des beaux-arts d'Orléans, Orléans* (2010) p. 51.
32. I. Biron, B. Gratuze and S. Pistre, in *Bernard Perrot 1640–1709, secrets et chefs d'oeuvre des verreries royales d'Orléans, Somogy Edition d'art, Paris and Musée des beaux-arts d'Orléans, Orléans* (2010) p. 87.
33. C.D. Valence, *Bull. Soc. Archeol. Hist. Orléanais* **20** (2010) 3.
34. O. Drahotova, *Glass Rev.* **28** (1973) 8.
35. H. Garner, *Trans. Orient. Ceram. Soc.* **37** (1967–69) 1.
36. J. Carbert, *Gold Bull.* **13** (1980) 144.
37. O. Dargaud, L. Stievano and X. Faurel, *Gold Bull.* **40** (2007) 283.
38. M. Faraday, *Trans. R. Soc. London* **147** (1857) 145.
39. W.A. Weyl, "Coloured glasses", *The Society of Glass Technology* (1951).
40. G. Mie, *Annalen der Physik* **25** (1908) 377.
41. K.L. Kelly, E. Coronado, L.L. Zhao and G.C. Schatz, *J. Phys. Chem. B* **107** (2003) 668.
42. P.K. Jain, I.H. El-Sayed and M.A. El-Sayed, *Nanotoday* **2** (2007) 18.
43. A. Ruivo, C. Gomes, A. Lima, M.L. Botelho, R. Melo, A. Belchior and A.P.D. Matos, *J. Cult. Heritage* **9** (2008) e134.
44. J.A. Williams, G.E. Rindone and H.A. McKinstry, *J. Am. Ceram. Soc.* **641** (1981) 709.
45. F.E. Wagner, S. Haslbeck, L. Stievano, S. Calogero, Q.A. Pankhurst and K.-P. Martinek, *Nature* **407** (2000) 691.
46. S. Haslbeck, K.-P. Martinek, L. Stievano and F.E. Wagner, *Hyperfine Interact* **165** (2005) 89.
47. P. Fredrickx, D. Schryvers and K. Janssens, *Phys. Chem. Glasses* **43** (2002) 176.

Chapter 2
Introduction to the Physical and Chemical Properties of Gold

Geoffrey C. Bond

Emeritus Professor, Brunel University, Uxbridge UB8 3PH, UK.
Email: geoffrey10bond@aol.com

2.1 Introduction

Gold possesses a unique combination of physical and chemical properties in both the macroscopic and microscopic states; on the macroscopic scale gold is known for its unique yellow colour, for its chemical stability and high redox potential. They are the consequence of an electronic structure, the understanding of which originates with quantum chemistry coupled to Einstein's Theory of Relativity. On the nanoscale, the unusual electronic configuration combines with other effects due to the extremely small dimensions and: (i) high ratio of surface atoms to bulk atoms, so that overall properties are dictated by the surface atoms, (ii) electromagnetic confinement when an optical wave interacts with a gold nanoparticle giving rise to their specific colour through a localised plasmon resonance, and (iii) quantum effects that explain the change from metallic to semiconducting character of very small particles.

Gold is the third member of Group 11 of the Periodic Classification, lying below copper and silver, but its physical and chemical properties are not predictable on the basis of trends observed in other groups; this is at once revealed by its bright metallic yellow colour, which resembles that of copper, but not that of silver. It occupies a position at one extreme of a range of metallic properties, having excellent resistance to corrosion, considerable malleability and high density; these properties ensure its natural occurrence as metallic nuggets and powders, and its suitability for making jewellery and objects of devotion (as discussed in Chapter 1). Its lack of reactivity is demonstrated by its inability to interact with components of the atmosphere and to corrode with the formation of oxides and sulfides in the manner of copper and silver, or to dissolve in common acids. Gold artefacts are recovered unchanged from burial after many centuries and this 'nobility' also gives gold its preferred place as a coinage metal and as a form for securing wealth against all risks except theft. The corollary of its inertness is the thermal instability and ease of reduction of compounds such as Au_2O_3, $AuO(OH)$, $Au(OH)_3$ and Au_2S_3; this follows from the small difference in electronegativity of the component atoms.

These unique physical and chemical properties are responsible for its widespread applications in both the macroscopic state[1] and in the micro-scopic or nanoparticulate state, as described in the following chapters of this book.

2.2 Physical Properties of Massive Gold

2.2.1 Crystal structure

Gold crystallises in the face-centred cubic (fcc) structure, its metallic radius being fractionally smaller than that of silver[2] (Table 2.1); this structure is responsible for its malleability: 1 g can be beaten into a foil of area $\sim$1 m^2, the thickness of which is less than 250 atomic diameters. The same amount can also be drawn into 165 m of wire that is 20 μm in diameter.[4] These characteristics, together with many others, were discussed in detail in a lengthy but fascinating paper by Michael Faraday in 1857.[5] Gold also forms alloys and intermetallic compounds with many other elements,[6] but it has no apparent ability to dissolve or occlude simple gases, although there is indirect evidence that hydrogen atoms can diffuse through it if formed on its surface by dissociation of molecules.[7] In jewellery, pure gold is very malleable and has to be alloyed to make it harder. The gold content is measured in carats, 24 carat gold being pure gold.

Table 2.1. Physical properties of gold compared to those of copper and silver (column 11 elements).

Property	Cu	Ag	Au
Atomic number	29	47	79
Atomic mass/amu	63.55	107.868	196.9665
Electronic configuration	$[Ar]3d^{10}4s^1$	$[Kr]4d^{10}5s^1$	$[Xe]4f^{14}5d^{10}6s^1$
Structure	fcc	fcc	fcc
Metallic radius/nm	0.128	0.14447	0.14420
Density/g cm^{-3}	8.95	10.49	19.32
Melting temp/K	1356	1234	1337
Sublimation enthalpy/kJ mol^{-1}	337 ± 6	285 ± 4	343 ± 11
1st ionisation energy/kJ mol^{-1}	745	731	890
Electrical resistivity at 293 K/micro-ohm cm	1.67	1.59	2.35
Interband transition threshold	$3d \rightarrow 4s$	$4d \rightarrow 5s$	$5d \rightarrow 6s$
/eV	2.1	3.9	1.84
/nm	590	318	674

[a] In 12-coordination.

2.2.2 Density

While in some respects its properties reflect its greater atomic mass compared to copper and silver (e.g. density), in many cases the trend is reversed; thus its melting point and heat of sublimation are almost the same as that of copper (see Table 2.1), the greater strength of the Au–Au bond being a consequence of its shorter than expected length. There is only one naturally occurring isotope of gold, so the atomic mass is known very precisely (196.9665 amu).

2.2.3 Magnetic and electrical properties

It has a non-zero nuclear spin quantum number ($I = 3/2$) and its nucleus is therefore 'magnetic', but its receptivity relative to the proton is only 2.77×10^{-5}, so it is a hard nucleus to study.[3] It also has a large quadrupole moment, which leads to line broadening, so very refined equipment is

31

needed for its study, and the consequential absence of hyperfine structure means that the Nuclear Magnetic Resonance of gold is of limited diagnostic use to chemists.

Its optoelectronic properties are also unpredictable by extrapolation from its antecedents in Group 11. Its electrical resistivity is greater than that of silver (see Table 2.1).

2.2.4 *Colour*

The yellow colour of gold has the same origin as that of copper (see Section 2.3). The explanation of this peculiar appearance is to be found in the electronic properties of the gold or copper atoms themselves, and is linked to the relativistic effects developed in the following section. The colour of gold can be changed by alloying: the addition of copper produces the 'rose gold', which is a soft pink; addition of aluminium leads to purple coloration, of indium to blue, and cobalt to black.

2.3 Relativistic Effects on the Properties of Gold

2.3.1 *The relativistic contraction of the radius of gold atoms*

The trends revealed in Table 2.1 as discussed above, and more especially those in the chemistry of gold in relation to its neighbours, to be considered below, require us to seek a deeper understanding of the electronic structure of gold and of the elements close to it. Inorganic chemists have long realised that the differences between the elements in the First and Second Transition Series are not continued into the Third, and the cause is found to be a consequence of Einstein's Theory of Relativity.[8–10] Modern chemistry relies on quantum physics and on the Schrödinger equation which provides insight for many chemical problems: unfortunately its formulation is non-relativistic, and this limitation becomes significant for elements with atomic numbers $Z > 50$. A relativistic analogue was therefore devised by P.A.M. Dirac and published in 1928;[12] the Dutch physicist, H.A. Kramers, also developed an equivalent treatment at about the same time.[13] Dirac's relativistic wave equation is remarkable in many ways and it provides an explanation of spin in a way that the Schrödinger equation, coupled to the Pauli theory of spin, cannot do. In fact it has been said that *Spin is nature's way of signalling the correctness of Einstein's Theory of Special Relativity.*

Since this introduction cannot provide a detailed exegesis of the Dirac equation, a semi-quantum approach can be used for qualitatively understanding the relativistic effect in gold atoms. In this approach, the electrons are described as orbiting charges characterised by their speed v. Dirac stated that changing to a relativistic wave equation would not significantly affect the calculated properties of the hydrogen atom, but with heavier elements the inner electrons feel the large nuclear charge, and to maintain balance with the strong electrostatic field they must acquire speeds that are comparable with that of light:[8,9] according to Einstein's Theory of Special Relativity this causes their mass (M) to increase and the length (L) to decrease, according to the equations

$$M = M_0/(1 - v^2/c^2)^{1/2}$$

and

$$L = L_0(1 - v^2/c^2)^{1/2}$$

where M_0 and L_0 are their rest mass and length, v their speed and c the speed of light. This is the basis of the *relativistic effect*, which exists in all atoms, but only becomes significant when the atomic number Z exceeds about 50 (Sn). It increases roughly as Z^2, and for gold ($Z = 79$) and mercury ($Z = 80$) the $1s$ electrons have speeds of about 58% of that of light, and their mass is thereby increased by about 20%. The $1s$ orbital therefore shrinks, and the s orbitals of higher quantum number have to contract in sympathy; in fact the $6s$ orbital shrinks relatively more than the $1s$. The same effect also operates to a lesser extent on the p electrons, but d and f electrons are hardly affected, never coming close to the nucleus due to the centrifugal potential $l(l+1)/r^2$, l being the azimuthal quantum number and r the radius. In addition their effective potential is more efficiently screened because of the relative contractions of the s and p shells: they therefore increase in energy and move outwards, this effect being called the *indirect relativistic orbital expansion*. Therefore in the case of gold, the $6s$ outer electronic shell shrinks whereas the $5d$ electrons expand. By evaluating the shell radius of the $6s$ electrons, theoreticians can calculate the radius with or without the relativistic correction.[9] Figure 2.1 plots the ratio of these two radii: if the relativistic effect were insignificant, the ratio would be 1.

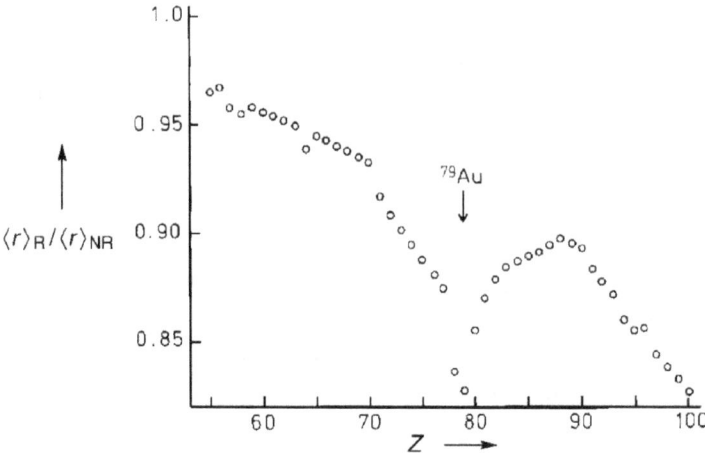

Fig. 2.1. Illustration of the relativistic effect on the radius of the 6s shell. The plot of the ratio of the value calculated taking into account relativity (Dirac equation) over the non-relativistic value is always smaller than 1 (contraction). This contraction is dramatic in the case of gold. Reprinted with permission from Ref. 9.

Among the different elements of the periodic table (at least until $Z = 100$), the relativistic contraction of the 6s shell has a very sharp extremum at platinum and gold, where the contraction is about 0.83 (Figure 2.1); this explains the surprising fact that the gold atom is a little smaller than the silver atom although it has one more electronic shell (Table 2.1). The main chemical consequence is the greater ease with which d electrons can be engaged in chemical activity as atomic number increases.

The greater similarity between the elements in the Second and Third Transition Series compared to those in the First Series (Table 2.1) was formerly ascribed solely to the Lanthanide Contraction, caused by the failure of the 5d and 6s shells to occupy the expected orbitals, because the 5f electrons do not adequately shield them from the increasing nuclear charge, by reason of the disposition of their orbitals: 5d and 6s electrons are therefore drawn towards the nucleus. It is now thought however that the Lanthanide Contraction and the relativistic effect have approximately equal importance, but the latter leads to *selective* effects on the sizes and energies of the various electron shells, these accounting for chemical behaviour that is not otherwise explicable.

2.3.2 *Optical properties, interband transitions and relativistic effect*

The optical absorption of gold in the visible region of the spectrum is due to the relativistic lowering of the gap between the centre of the $5d$ band and the Fermi level (see Fig. 2.2b). A good indication of this relativistic effect is the very low value of the interband transition energy for gold compared to silver (see Fig. 2.2b). The interband threshold as given in Table 2.1 is the energy required to excite electrons from the top of the $5d$ band into the $6sp$ conduction band. In the case of gold its value is 1.84 eV, which means that an optical wave corresponding to a red wavelength is able to excite this transition. In the case of silver, the interband transition is in the UV range so that the visible light is almost not affected after reflecting on a silver surface.[1,4,10,14]

The electronic structure of gold atoms in the massive state is not however exactly that of the free atom, because a weak white line on the leading edge of the L_{III} X-ray absorption edge signifies a small number of vacancies in the d-band caused by d-s hybridisation.[15]

2.4 Chemical Properties of Gold in Relation to Its Neighbours

The chemistry of gold ($5d^{10}6s^1$) is determined by (i) the easy activation of the $5d$ electrons and (ii) its propensity for acquiring a further electron to complete the $6s^2$ level and not to lose the one it has.[16–18] This latter effect awards it a much greater electron affinity and higher first ionisation potential than those of copper or silver (see Table 2.1), and accounts for the ready formation of the Au^{-1} state (see below).[8] The former effect obviously explains the predominance of the Au^{III} state, which has the $5d^8$ configuration (even the Au^V state ($5d^6$) is accessible as in AuF_5^{29}), the Au^I state being of lesser importance and the Au^{II} state being unknown except in a few unusual complexes.[19] Gold's electronegativity (2.4 in Pauling units) equals that of selenium and approaches that of sulfur and iodine (2.5);[4] therefore it is frequently said to have some of the properties of a halogen. Its electrode potential ($E^0 = +1.691$ V) is also extremely high for a metal. Its electronic structure determines its nobility, and its inability in the massive form to

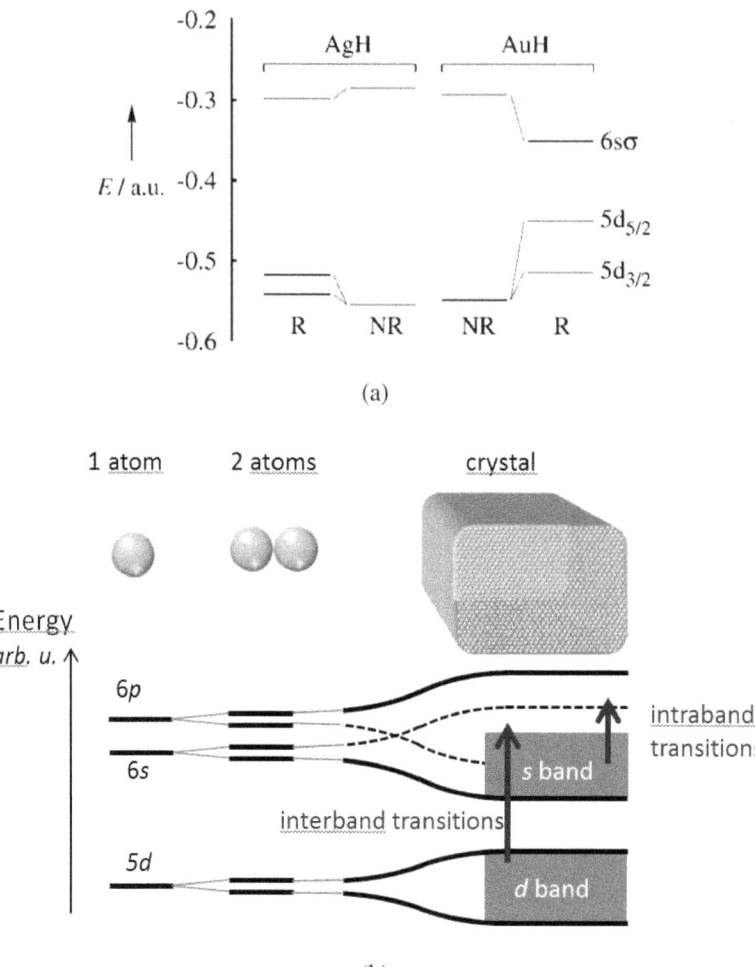

(a)

(b)

Fig. 2.2. (a) Comparison of the influence of the relativistic effect between gold and silver by calculating the electronic transitions of Ag-H and Au-H. For Au-H, the introduction of the relativistic
correction into the calculations considerably lowers the energy of the transition from the $5d$ electrons
to the $6s$ level. Reprinted with permission from Ref. 9. (b) Sketch of the evolution of the electronic
structure of gold: for a single atom, the electronic levels are discrete, as shown above. For a gold dimer
Au–Au, the levels tend to split. For a crystal, this lifting of degeneracy widens and forms a continuum
of levels: the d band emerges from the d electrons of all the gold atoms and is completely filled with
electrons. The conduction band is formed from the $6s$ and $6p$ orbitals and is partially filled (conduction
band). With this structure, light can excite two kinds of transitions: intraband transitions and interband
transitions.

interact with oxygen or sulfur compounds, i.e. to tarnish as silver and copper do, is in line with the instability of its oxide Au_2O_3, which decomposes at about 433 K and has a positive heat of formation. The sulfides Au_2S and Au_2S_3 are known, but are of limited stability and importance.[4]

Gold dissolves in solutions of the heavier alkali metals in liquid ammonia,[20] and the auride ion Au^- is formed; the electrical conductivity of cesium-gold alloys at 873 K shows a very sharp minimum at the 1:1 ratio, and the solid CsAu is regarded as a semiconductor;[8] it has the NaCl structure. Tetramethylammonium auride is isostructural with the bromide, and the deep blue addition compound $CsAu.NH_3$ has recently been prepared and characterised.[8]

The dissolution of gold requires both an oxidant and a ligand to stabilise the resulting cation. Thus it dissolves in *aqua regia* (conc HCl: Conc $HNO_3 = 3:1$) to form $AuCl_4$, and in the presence of oxygen in aqueous CN^- to form the $[Au(CN)_2]^-$ anion.[4]

Finally we may note the existence of compounds of gold which cannot be prepared and stored in a bottle, but whose ephemeral character may imitate transient species formed in catalytic processes. These include the hydrides AuH_3 (i.e. $HAu(H_2)$) and AuH_5 (i.e. $H_3Au(H_2)$) which have been seen in low-temperature matrices, and $AuXe^+$ and $AuXe_2^+$ which have been detected by mass-spectrometry.[8] The compound $[AuXe_4][Sb_2F_{11}]_2$ has however actually been made.

2.5 The Aurophilic Bond

In numerous complexes containing two or more Au^I ions, it is generally observed that distances between pairs of such ions are unusually short (275 to 350 pm), and that some form of bonding must therefore exist between them[8,9,18−20] (see Chapter 9). The effect is termed *aurophilic attraction* or *aurophilicity*. It is also observed when the Au^I ions are in different molecules that pack closely together in the solid state, and when they are located at opposite sides of a ring formed with bidendate ligands (transannular attraction, see Fig. 2.2). There is an enormous literature on gold complexes in which the effect occurs, and this has been reviewed.[21]

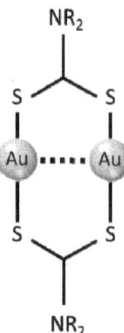

Fig. 2.3. Aurophilic bond between two gold atoms in a dimeric Au^I dithiocarbamate.

Aurophilicity has also been extensively studied by theoretical methods. It appears that the bond is due to dispersion forces of the type that hold molecules together in a liquid or solid, but very much stronger than normal van der Waals forces; it has the same kind of strength as the hydrogen bond in water and alcohols, and takes values between 10 and $100\,kJ\,mol^{-1}$, depending on the separation between the atoms.

2.6 Dependence of Physical and Chemical Properties of Gold on Particle Size

When the size of a gold particle is progressively decreased, significant changes in physical properties and chemical reactivity are observed; they become especially noticeable when size falls below about 10 nm. Such particles, often named *nanoparticles* (although the term lacks quantitative definition), may be obtained by controlled growth of atomic species in various ways: (i) by condensation of vaporised metal atoms under UHV conditions to form gaseous 'clusters' having relatively few atoms (<20);[22] (ii) by reduction of a solution of a gold compound, usually $HAuCl_4$, to form a colloidal dispersion (Chapter 5); and (iii) by deposition of a gold compound followed by decomposition or reduction to the metallic state (Chapter 6) or vaporised gold atoms (Chapter 8) onto a support. The numerous actual and potential applications of nanoparticulate gold, which are the subject of this book, have given rise to numerous studies of their physical properties by a great variety of techniques (Chapters 3 and 4),[10] together with many theoretical studies (Chapter 9). The following paragraphs provide a very brief account of what has been observed.

The most significant consequence of decreasing the particle size is the increase in the surface/volume ratio; the rise in the fraction of surface atoms is responsible for some changes in the structural character. As a rough guide, surface atoms of a 2 nm particle constitute about 60% of the total. Notable changes include: (i) a decrease in the melting temperature (2 nm particles melt at $\sim$500 K compared to 1337 K for massive gold); and (ii) a lowering of the interatomic separation from 0.288 nm to 0.245 nm.[23] These effects arise because surface atoms experience a resulting force acting inwards which is not compensated by atoms above them; it causes a surface energy akin to the surface tension of liquids, and thus to a decrease of interatomic spacing throughout the particle. At the same time the decrease in the mean number of near neighbours allows atoms a greater freedom to vibrate around their normal locations, thus accounting for a lowering of melting temperature. Theoretical studies indicate that surface atoms having low coordination number, such as are found at edges and corners of all particle shapes, are the seat of increased chemical reactivity (see Fig. 2.4), and are therefore the preferred

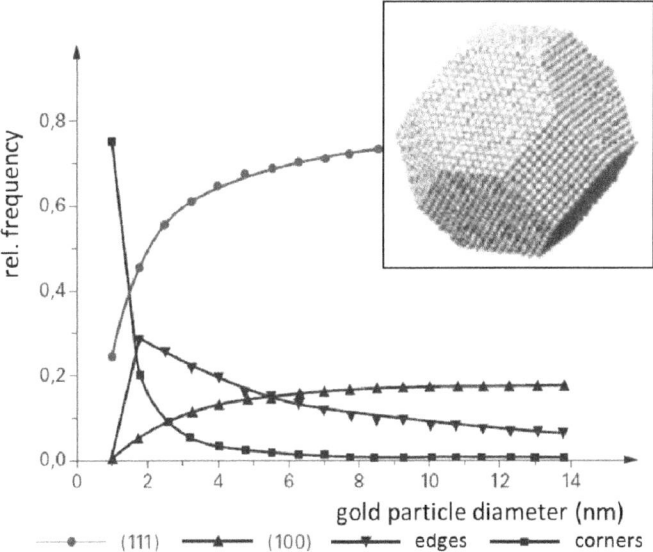

Fig. 2.4. Dependence of relative amounts of surface sites on particle diameter of gold particles, based on cuboctahedron model. Reprinted with permission from Ref. 24. Copyright 2001 Science Reviews Ltd.

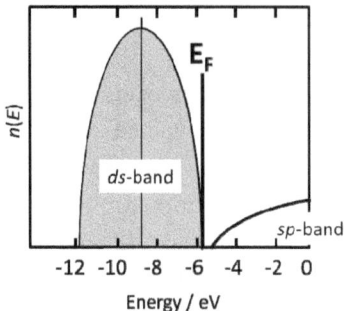

Fig. 2.5. Distribution of energy levels for a 2 nm particle; there is a small band gap above the Fermi energy E_F.[14]

locus for the chemisorption of molecules such as CO and H_2 (Chapter 7). For example, for nanoparticles larger than 2 nm, the majority of surface atoms belong to a (111) facet. For a 10 nm cubo-octahedral nanoparticle, a negligible amount of atoms are corner atoms, whereas 8% populate the edges, 17% the (100) facets and the other ones are part of the (111) surfaces.

Changes in the electronic structure also affect the optical response of particles and the reduction of size down to the nanometre scale account for the various colours exhibited by colloidal dispersions of gold (Chapter 3) and by gold nanoparticles supported on surfaces[25] or in matrixes. Changes to the X-ray photoelectron spectra (XPS) of gold nanoparticles have often been seen; the binding energy of the $4f_{7/2}$ core level typically increases by 0.8–1.0 eV when size falls below 5 nm. Now metallic character is due to the broadening of electron energy levels into overlapping bands, but with very small particles (<2 nm) the bands become narrower, and a gap appears between them (Fig. 2.4);[26] metallic behaviour is then lost, and this is thought to cause the marked rise in catalytic activity for CO oxidation and other reactions (Chapter 7). This may be because then nearly all the surface atoms have low coordination numbers characteristic of edges and corners on larger particles.

2.7 Conclusion

Gold has many chemical and physical properties that are unique and not predictable by extrapolation of those of copper and silver; these include its high electronegativity, its aurophilicity and its variable colours when highly dispersed. All of these characteristics find their origin in the important role that relativistic effects play in the electronic structure of the element and its compounds, and they in turn determine its suitability for the many applications that form the subject of this book.

Acknowledgement

Grateful thanks to Dr O. Pluchery for helpful comments on a draft of this chapter.

References

1. C. Corti and R. Holliday (eds), *Gold: Science and Applications*, CRC Press, Boca Raton, Florida, 2010.
2. A. Bayler, A. Schier, G.A. Bowmaker and H. Schmidbaur, *J. Am. Chem. Soc.* **118** (1996) 7006.
3. M. Tokita and E. Haga, *J. Phys. Soc. Japan* **50** (1981) 482.
4. N.N. Greenwood and A. Earnshaw, *Chemistry of the Elements*, 2nd edn, Butterworth-Heinemann, Oxford, 1997.
5. M. Faraday, *Phil. Trans.* **147** (1857) 145; see also W.D. Mogerman, *Gold Bull.* **7** (1974) 22.
6. W. Rapson, *Gold Bull.* **29** (1996) 141.
7. R.S. Yolles, B.J. Wood and H. Wise, *J. Catal.* **21** (1971) 66.
8. P. Pyykkö, *Angew. Chem. Int. Edn.* **41** (2002) 3573.
9. P. Pyykkö, *Angew. Chem. Int. Edn.* **43** (2004) 4412.
10. G.C. Bond, C. Louis and D.T. Thompson, *Catalysis by Gold*, Imperial College Press, London, 2006.
11. G.C. Bond and E.L. Short, *Chem. and Ind.* **11** (2002) 12.
12. P.A.M. Dirac, *Proc. Roy. Soc. A* **117** (1928) 210; **118** (1928) 351; **123** (1929) 714.
13. T. Der Haar, *Masters of Modern Physics: The Scientific Contributions of H.A. Kramers*, Princeton University Press, Princeton, New Jersey, 1998.
14. G.C. Bond, *Faraday Discussions* **152** (2011) 277.
15. J. Chevrier, L. Huang, P. Zeppenfeld and G. Comsa, *Surf. Sci.* **355** (1996) 1.
16. G.C. Bond and D.T. Thompson, *Catal. Rev. Sci. Eng.* **42** (1999) 319.
17. N. Bartlett, *Gold Bull.* **31** (1998) 22.
18. P. Pyykkö, *Gold Bull.* **37** (2004) 136.
19. M.C. Gimeno and A. Laguna, *Gold Bull.* **36** (2003) 83.
20. F. Mendizabal and P. Pyykkö, *Phys. Chem. Chem. Phys.* **6** (2004) 900.
21. H. Schmidbaur, *Gold Bull.* **33** (2000) 3; **23** (1990) 11.
22. R. Meyer, C. Lemire, S. K. Shaikhutdinov and H.-J. Freund, *Gold Bull.* **37** (2004) 72.

23. J.T. Miller, A.J. Kropf, Y. Zha, J.R. Regalbuto, L. Delannoy, C. Louis, E. Bus and J.A. van Bokhoven, *J. Catal.* **240** (2006) 222.

24. C. Mohr and P. Claus, *Sci. Pro.* **84**, 4 (2001) 311.

25. G.C. Bond and P.A. Sermon, *Gold Bull.* **6** (1973) 102.

26. K. Okazaki, S. Ichikawa, Y. Maeda, M. Haruta and M. Kohyama, *Appl. Catal A: Gen.* **291** (2005) 45.

Chapter 3
Optical Properties of Gold Nanoparticles

Olivier Pluchery

Université Pierre et Marie Curie, Institut des NanoSciences de Paris,
4 place Jussieu, 75005, Paris, France.
Email: olivier.pluchery@insp.jussieu.fr

3.1 Introduction

Among the remarkable properties of gold nanoparticles (AuNPs), the collective behaviour of their conduction electrons that gives rise to the plasmon resonance is a dramatic one. As a consequence, plasmonics is a field that emerged in the 1990s and is now well studied. The number of scientific articles referring to plasmon in their title or abstract climbed from 125 in 1990 to 1480 in 2002 and has reached the number of 5,570 in 2010.[i] This demonstrates that plasmonics is now a well-established research field, with devoted journals and conferences. A major application field of plasmonics is found in biophysics (see Chapter 10). The word plasmon corresponds to the quantum of energy associated with an eigenfrequency of a plasma oscillation. In general, a plasma is a gas where electric charges are free to move under the influence of electromagnetic or gravitational forces. In the case of metals, the conduction electrons play the role of free charges since they are detached from their ionic core and can be excited by an electromagnetic

---

[i] Data obtained through the database provided by ISI Web of Knowledge, searching «Plasmon*» in the topics of the published articles (July 2011).

wave such as an optical beam. The oscillation of the electrons and the oscillation of the electromagnetic field are intrinsically linked. We will describe some cases when they give rise to resonant modes. These modes are sometimes termed by their corresponding quasiparticles: polaritons, when the phenomenon is considered from the point of view of electrical charges, and plasmon when it is studied from the electromagnetic point of view. Therefore plasmon waves correspond to the coupling of two waves: the mechanical oscillations of charges and the electromagnetic oscillations of the electric field.

This phenomenon occurring at the nanoscale shows up immediately at our scale with very specific colours: a solution containing spherical nanoparticles has a red-purple colour. This is called the Localized Surface Plasmon Resonance (LSPR) and is the consequence of the confinement of the electric field within a small metallic sphere. In other words, LSPR results from the Maxwell equations and the boundary conditions imposed to the electric field in spheres whose radius is much smaller than the wavelength. This optical resonance is the origin of many other properties that have made the AuNPs famous: it explains the red-purple colour of spherical nanoparticles, and the slight change of colour when their shape or the surrounding medium are modified; it is responsible for a strong electric field enhancement within a distance of a few times the particle diameter when this resonance is excited by light (near-field exaltation), which explains the peculiar nonlinear optical properties of AuNPs and their strong Raman activities; it also results in the absorption of part of the energy of the impinging light beam and a heating of the nanoparticle (see Chapter 4). The applications of these different properties are discussed in other chapters of the present book (Chapters 10, 11 and 13).

The present chapter focuses on the optical properties associated with plasmons and begins by clarifying the distinction between propagating plasmon waves called SPR (Surface Plasmon Resonance) and the localized plasmon of nanoparticles traditionally called LSPR (Local Surface Plasmon Resonance). We will derive an analytical model for describing the response of spherical metallic particles with the so-called electrostatic model and show that it gives a reasonable agreement with experiments in many cases. Then we will discuss the improvements needed to treat more complex cases when particles are not spherical or have a diameter larger than 60 nm.

3.2 What is the ambition of the present chapter?

Researchers entering the field of gold nanoparticles may find numerous review articles on how the plasmon resonance is used to enhance sensitivity of spectroscopic techniques or how the shape of nanoparticles influences the optical spectra, but it is more difficult to find an accessible electromagnetic description of the optical properties of ordinary nanoparticles. Ordinary nanoparticles are spherical, diluted gold nanoparticles with sizes between 5 and 60 nm and their optical response can be described by the so-called electrostatic model. Such a model is less sophisticated than the very popular Mie theory which relies on a somewhat arduous mathematical formalism. The electrostatic model, however, is a good start to understanding the various effects that are in play when an electromagnetic wave interacts with the electron clouds of the metallic sphere. This chapter offers a derivation of this model as well as some simple improvements for describing the optical response of AuNP.

3.3 Distinction Between Localized Surface Plasmon Resonance (LSPR) and Surface Plasmon Resonance (SPR)

In dealing with plasmon resonance, a certain confusion reigns in the terms used. Two distinct phenomena are called plasmon resonances: the surface plasmon resonance where an evanescent wave crawls along a metallic surface, and the localized plasmon which is an oscillation of the conduction electrons within a metallic nano-object. We have to make this distinction clear.

3.3.1 *Optical properties of metals*

Metals are characterized by their quasi-free electrons in the ground state which are not bound to a single atom anymore but can freely move through the crystalline structure of the metal. These free electrons are responsible for the main properties of metals: high conductivity and high optical reflectivity. When an electromagnetic wave characterized by the electric field $\vec{E}(\vec{r}, \omega)$ interacts with a metal, it tends to make the electron cloud oscillate and this creates a dynamic polarization $\vec{P}(\vec{r}, \omega)$. This quantity expresses how far the electric field succeeds in displacing the electrons relative to the core atoms. The polarized atoms are the sources of a depolarizing field and all these effects are combined into the electric displacement $\vec{D}(\vec{r}, \omega)$ linked to the excitation electric field by the relationship[ii]: (see ref. 1 for a good introduction to the optical properties of condensed matter).

$$\vec{D}\left(\vec{r}, \omega\right) = \varepsilon_0 \varepsilon\left(\omega\right) \vec{E}\left(\vec{r}, \omega\right) \tag{1}$$

$\varepsilon(\omega)$ is the dielectric function of the metal and it captures the entire response of a metal to the electromagnetic excitation wave for the whole frequency spectrum, starting from radio frequencies up to X-ray including of course optical frequencies. If the medium is vacuum, $\varepsilon(\omega) = 1$ and the displacement $\vec{D}$ is simply proportional to $\vec{E}$, but ε accounts for any kind of materials (metals, insulators, transparent or opaque media) and in the general case, it is a complex function expressed as:

$$\varepsilon(\omega) = \varepsilon_1(\omega) + i\varepsilon_2(\omega) \tag{2}$$

For metals, ε is dominated by its real part which is a negative function. ε is linked to the optical complex index $\tilde{n}$ by

$$\varepsilon = \tilde{n}^2 = [n + ik]^2 = n^2 - k^2 + 2nki \tag{3}$$

$\varepsilon, \tilde{n}$ are functions of the angular frequency ω (linked to the wavelength λ) and this dependence will not be expressly indicated in the followings to simplify notations, but it should be kept in mind since the plasmon resonance is directly linked to it.

The oscillator model is the simplest analytical way to describe the polarization and gives rise to the Drude dielectric function[1-3] expressed as follows:

$$\varepsilon_{\text{Drude}} = 1 - \frac{\omega_p^2}{\omega^2 + i\Gamma\omega} \approx 1 - \frac{\omega_p^2}{\omega^2}$$

ω_p is the plasma frequency and Γ the damping constant.

[ii]The SI system of units is used for all the equations of this chapter.

3.3.2 *The dielectric function of gold*

As explained in the previous section, the Drude model is a good start for describing the dielectric functions of metals. The formula given above is slightly modified by replacing the factor 1 by a constant ε_∞ that accounts for transitions that do not need to be explicitly expressed in the visible range.

$$\varepsilon_{\text{Drude}} = \varepsilon_\infty - \frac{\omega_p^2}{\omega^2 + i\Gamma\omega} \tag{4}$$

In the case of gold, the plasma frequency and the damping constant are given by[4]

$$\begin{cases} \hbar\omega_p = 8.95\,\text{eV} & \text{i.e. } \omega_p = 1.36 \times 10^{16}\,\text{rad.s}^{-1} \\ \hbar\Gamma = 65.8\,\text{meV} & \text{i.e. } \Gamma = 1.0 \times 10^{14}\,\text{rad.s}^{-1} \\ \varepsilon_\infty = 9.5 \end{cases} \tag{5}$$

These values are obtained by fitting experimental values with equation (4) in an energy range where the free electrons are the major contributions to the dielectric function[5,6] and these values fluctuate from one author to another.

However when a precise model is needed, the Drude model is far too simplified because it only takes into account the free electrons (intraband transitions) and completely dismisses the contribution from the bound electrons (interband transitions). The latter plays an important role for gold. This is illustrated in Fig. 3.1 where the complex dielectric function is plotted. The accurate measurement of this function is crucial for reasonably modelling the optical response of AuNP. One of the preferred measurements was performed by Johnson and Christy in 1972[5] and an approximated analytical model has been published by Etchegoin.[7] The measures provided by the *Handbook of Optical Constants* have some problems near the interband threshold for gold and must be used with great care.[8] In the present chapter we are using values measured by H. Arwin from the University of Linköping (Sweden)[9,10] because they cover a wide range of wavelength.

3.3.3 *Plasmon resonance at surfaces, SPR*

An electromagnetic wave in the visible range is rapidly screened by a metal. It does not penetrate into metal much farther than the so-called skin depth

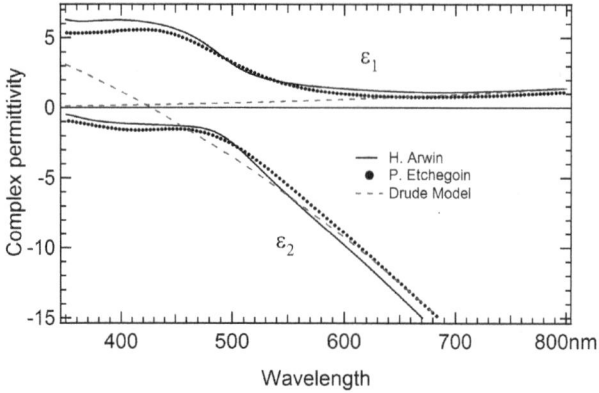

Fig. 3.1. The complex dielectric function of gold $\varepsilon = \varepsilon_1 + i\varepsilon_2$ obtained from an ellipsometric measurement[9,10] (thin line) and from an analytical model[7] (large dots) based on Johnson and Christy's data.[5] The dotted line shows the plot of the Drude model discussed in the text. This model accounts well for the free electron contribution but not for the bound electrons (intraband transitions). This discrepancy is clearly visible in the range of 300 nm to 530 nm.

given by $\delta = \sqrt{2\rho/\mu_0\mu_r\omega}$.[1,3] In the case of light beam impinging on gold with a wavelength of 500 nm the skin depth is $\delta = 20$ nm and strongly depends on the wavelength. Therefore, the conduction electrons are excited by the electromagnetic field within an ultrathin sheet of metal close to the surface. This occurs in two cases: in the case of a flat and infinite interface and in the case of structures whose size is of the order of magnitude of the skin depth. We will quickly consider the first case and then focus our attention on the second one.

In the case of a flat interface between a metal and an insulating medium whose dielectric functions are respectively ε and $\varepsilon_{\mathrm{diel}}$ it is possible to launch a surface wave that stays confined very close to the interface. This wave is a charge density wave with longitudinal structure (unlike light waves that propagate in vacuum with a transverse structure). The charge wave is called a polariton wave and is coupled to an electromagnetic wave, which is the surface plasmon wave. This is the reason why this SPR is sometimes called surface plasmon polariton (SPP).[4] This propagation is characterized by the following dispersion relation:

$$k_x = \frac{\omega}{c} \cdot \sqrt{\frac{\varepsilon(\omega) \cdot \varepsilon_{\mathrm{diel}}(\omega)}{\varepsilon(\omega) + \varepsilon_{\mathrm{diel}}(\omega)}} \qquad (6)$$

The dispersion relation plays a key role in electromagnetism because it controls the propagation of the wave. It is an expression of the link between fundamental quantities in physics: energy ($\hbar\omega$) and momentum ($\hbar\vec{k}$). For a monochromatic light beam of photon energy ω, only waves whose wavevector has a component parallel to the interface k_x given by relation (6) can be excited. In this relationship the number in the square root is greater than one for metals, meaning that the wave vector of the plasmon wave is greater than that of any wave travelling in free space. As a consequence, surface plasmons can only be launched with special setups. The most usual way is to excite the surface plasmon with an evanescent wave resulting from a total internal reflection from a prism in the so-called Kretschmann configuration depicted in Fig. 3.2.[11]

When the geometrical conditions are fulfilled to excite the surface plasmon, this is the surface plasmon resonance (SPR). It shows up as a dip in the reflection spectrum when the incident beam impinges on the gold film with a given angle (see Fig. 3.2-b).

Since SPR only occurs on flat surfaces approximated to planes of infinite extension, we will not discuss this kind of plasmon resonance further and restrict ourselves to charge oscillations within particles of nanometer dimensions in three-dimensional space.

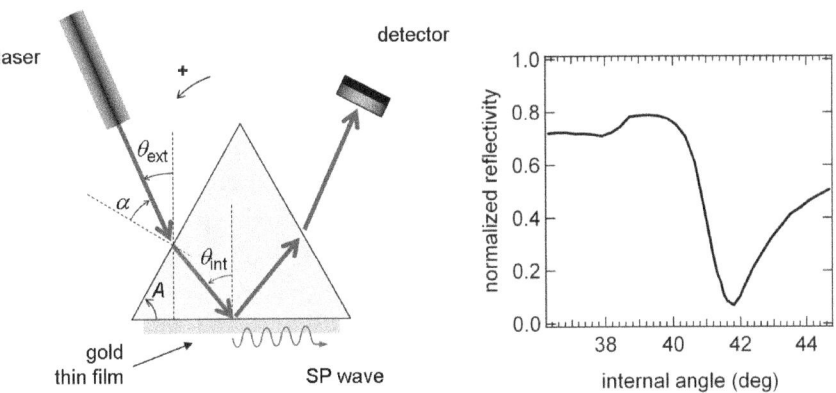

Fig. 3.2. Surface Plasmon Resonance: (a) sketch of the prism for coupling an excitation wave of a laser to the surface plasmon (SP) wave. The coupling is controlled by the incidence angle. (b) when the coupling is achieved the reflected beam undergoes a strong intensity drop measured at $\theta_{\text{int}} = 41.8°$ in the present case.

3.3.4 *Localized surface plasmon resonance in nanoparticles, LSPR*

In nanoparticles, the charge oscillation is different from the SPR described above. Qualitatively, if particles have sizes much smaller than the wavelength of light and smaller than the penetration depth of the field (i.e. particle size around 20 nm), the electron cloud of the particle is entirely probed by the electric field. The whole assembly of electrons is polarized, and this creates surface charges that accumulate alternately on opposite ends of the particle. This oscillating polarization of the particles creates an electric field opposed to the excitation field and results in a restoring force. This oscillation is partially damped. The damping occurs through two channels: creation of heat and light scattering. All this can be described as a dipolar oscillator characterized by a resonance frequency $\omega_{plasmon}$ that will be discussed and used throughout this book.

In the following chapter, this resonance will be called Localized Surface Plasmon Resonance (LSPR). This denomination is generally accepted although this localized plasmon oscillation is not primarily a surface effect, but a bulk effect taking place in the very small and confined volume of metallic nano-objects. Thus, some authors prefer using other denominations such as nanoparticle plasmon (NPP[12]), surface plasmon on metal nanoparticles (SPN[13]).

The main properties of LSPR can be understood within the dipolar model and can be summarized as follows:

- Spectrally, the plasmon resonance appears in the visible or near-infrared range for gold or silver nanoparticles. A light beam going through an assembly of homogeneous nanoparticles is partially absorbed at the plasmon resonance frequency so that the emerging beam displays a spectrum with a sharp absorption at $\omega_{plasmon}$. At the same time the nanoparticles exhibit light scattering with a cross section much larger than conventional dye.[14,15]
- The LSPR strongly depends on the environment close to the particle surface. For example, if the layer of molecules adsorbed on the NP changes, the SPR shifts. The AuNP typical shift for a protein interaction is of the order of magnitude of 10 nm (see Chapter 11). This effect is the basis of plasmonic biosensing.

- When excited at the resonance, the dipole radiates a near-field electro-magnetic wave, whose amplitude can be enhanced by a factor up to 10. This plasmon amplification is widely used for enhancing the sensitivity of biosensors (Chapter 10).

3.4 Theoretical Description of the Localized Plasmon Resonance

3.4.1 *About Mie theory*

An exact theory in case of spherical particles of any size is provided by Mie theory[16] (Gustav Mie, 1908). It gives the exact solution of a plane wave inter-acting with a metallic sphere. The electromagnetic fields are expanded in multipole contributions and the expansion coefficients are found by applying the correct boundary conditions for electromagnetic fields at the interface between the metallic nanoparticle and its surroundings. For small particles (<60 nm), it is sufficient to restrain the multipole expansion to its first term, which is dipolar. This approximation is the dipolar approximation, also called the quasistatic or Rayleigh limit. We will develop this approximation which is sufficient to grasp the working principles of plasmonics; readers interested in the full Mie theory should refer to textbooks such as Born and Wolf[3] or Bohren and Huffmann.[17]

3.4.2 *The quasistatic approximation for describing the localized plasmon resonance*

Let us consider a metallic spherical particle of radius R which is submitted to an excitation electric field aligned along the x-axis: $\vec{E} = E(\vec{r}, t)\vec{e}_x$ (see Fig. 3.3). In the quasistatic limit, the electronic polarization is exactly in phase with the excitation field (no retardation effect) and the electrons are displaced as a whole. Therefore the charge distribution in the particle can be treated as if it were a static distribution.[4,17] The electric field should obey the Laplace equation:

$$\Delta V = 0 \tag{7}$$

Where V is the electric potential linked to the electric field by $\vec{E} = -\vec{\nabla}V$.

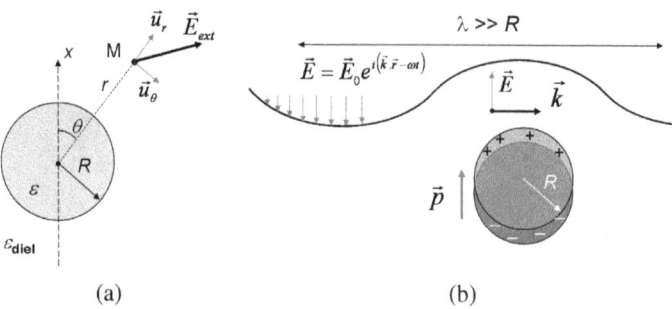

Fig. 3.3. (a) Sketch of the metallic sphere and the coordinates used in the electrostatic model to calculate the external electric field. (b) Under the influence of an excitation wave with a wavelength greater than the dimension of the sphere, the electrons oscillate as a whole and the system can be handled as an oscillating dipole.

Due to the symmetry of the problem, spherical coordinates are used. Moreover the x-axis is an axis of symmetry so that the third coordinate usually required for spherical coordinates becomes useless. Therefore equation (7) written in spherical coordinates becomes:

$$\frac{1}{r}\frac{\partial}{\partial r^2}(rV) + \frac{1}{r^2 \sin\theta}\frac{\partial}{\partial\theta}\left(\sin\theta\frac{\partial V}{\partial\theta}\right) = 0 \tag{8}$$

The solutions of this differential equation are the spherical harmonics with the following form:

$$V(r,\theta) = \sum_{n=0}^{\infty}\left(A_n r^n + \frac{B_n}{r^{n+1}}\right)P_n(\cos\theta) \tag{9}$$

where A_n and B_n are coefficients to be determined. P_n are the Legendre polynomials which often appear in physics problems expressed in spherical coordinates.[iii]

This electric potential has two different forms $V_{int}(r,\theta)$ and $V_{ext}(r,\theta)$ inside and outside the metallic particle. Moreover it should obey the

---

[iii] The Legendre Polynomials are obtained as solutions of the Legendre equations or the nth element of the recurrence relation of Bonnet. The first four terms are: $P_0(X) = 1$; $P_1(X) = X$; $P_2(X) = 3/2X^2 - 1/2$; $P_3(X) = 5/2X^3 - 3/2X$.

following boundary conditions:

- The electric field should be properly defined at $r = 0$: $\frac{\partial V_{int}}{\partial r}\Big|_{r=0}$ exists.
- The electric field far away from the particle should match the excitation field, which is constant in the electrostatic approximation, so that $\lim_{r \to \infty} V_{ext} = 0$.
- At the interface, the electric fields obey the continuity equation at the particle surface: $\varepsilon(\omega)E_{int}(r = R) = \varepsilon_{diel}(\omega)E_{ext}(r = R)$.
- Finally, the potential should be continuous at the particle surface: $V_{int}(r = R) = V_{ext}(r = R)$.

Applying these conditions allows us to determine the coefficients for the electric potentials inside and outside the particle. For example, for the external potential one determines that all the A_n coefficients are null except one: $A_1 = -E_0$ and similarly for the B_n coefficients, the only non-zero coefficient is given by:

$$B_1 = E_0 4\pi\varepsilon_0 R^3 \frac{\varepsilon - \varepsilon_{diel}}{\varepsilon + 2\varepsilon_{diel}} \tag{10}$$

Once the electric potential is obtained, the electric field is deduced by using the gradient operator: $\vec{E} = -\vec{\nabla}V$.

After some calculations one obtains the following expression for the electric field outside of the nanoparticle, which is a sum of the incident field and a second field produced by the particle:

$$\vec{E}_{ext} = \vec{E}_0 - \alpha E_0 \left[-2\frac{\cos\theta}{r^3}\vec{u}_r - \frac{\sin\theta}{r^3}\vec{u}_\theta \right] \tag{11}$$

Where α is the sphere polarizability given by

$$\alpha = 4\pi\varepsilon_0 R^3 \frac{\varepsilon - \varepsilon_{diel}}{\varepsilon + 2\varepsilon_{diel}} \tag{12}$$

The electromagnetic response of the particle is captured in the polarizability. It is clear that the external field will go through a maximum when the polarizability is maximized. The parameter that depends on frequency is $\varepsilon = \varepsilon(\omega)$. Therefore $|\alpha|$ is maximized when the following relationship is fulfilled:

$$|\varepsilon + 2\varepsilon_{diel}| \quad \text{is a minimum} \tag{13}$$

As a good approximation the dielectric permittivity of the surrounding medium, ε_{diel} is a constant and real parameter. Therefore equation (13) leads to the following condition applied to the real part of ε: $\varepsilon_1 + 2\varepsilon_{\text{diel}} = 0$. In case of a nanoparticle in water, $\varepsilon_{\text{diel}} = 1.77$, and the condition becomes $\varepsilon_1 = -2 \times 1.77 = -3.54$. From the plot of the dielectric function of gold in Fig. 3.1, it is easy to check that this condition leads to a plasmon resonance at 520 nm. If the particle was in air, the plasmon resonance would be at 504 nm. These calculated values for the plasmon resonance in water or in air correspond closely to experimental values of spherical gold nanoparticles.

3.4.3 Extinction and scattering cross sections

However, in order to proceed one step further and calculate the intensity of light being scattered or absorbed, the electrostatic model has to be completed to take into account that light is a wave and not a static electric field. Upon interaction with the particle, light is absorbed (absorption cross section σ_{abs}) and scattered (scattering cross section σ_{scatt}). As a result, the beam going through a particle undergoes an extinction characterized by the extinction cross section (σ_{ext}). The three cross sections are linked by the simple relation

$$\sigma_{\text{abs}} = \sigma_{\text{ext}} - \sigma_{\text{scatt}} \tag{14}$$

These phenomena are strongly frequency dependent and, therefore, time dependent. The model of the radiating dipole allows for calculating the extinction and scattering cross sections and links them to the polarizability found in equation (12):

$$\sigma_{\text{ext}} = 3\frac{2\pi}{\lambda}\sqrt{\varepsilon_{\text{diel}}}\text{Im}(\alpha)$$

$$= 9\frac{2\pi}{\lambda}\varepsilon_{\text{diel}}^{3/2}V\frac{\text{Im}(\varepsilon)}{|\varepsilon + 2\varepsilon_{\text{diel}}|^2}$$

(15-a & 15-b)

$$\sigma_{\text{scatt}} = 3\frac{(2\pi)^3}{\lambda^4}\varepsilon_{\text{diel}}^2|\alpha|^2$$

$$= 3\frac{(2\pi)^3}{\lambda^4}\varepsilon_{\text{diel}}^2V^2\left|\frac{\varepsilon - \varepsilon_{\text{diel}}}{\varepsilon + 2\varepsilon_{\text{diel}}}\right|^2$$

(16-a & 16-b)

This description is acceptable for homogeneous, metallic spherical nanoparticles whose diameter is roughly between 10 and 60 nm (see the following section).

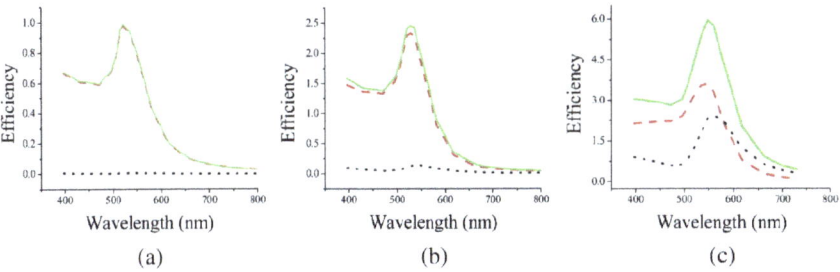

Fig. 3.4. Comparison of the extinction (solid line), absorption (dashed line) and scattering (black dots) efficiencies as a function of particle diameter for gold nanoparticles. In the case of particles of 20 nm diameter (graph a), absorption is largely dominating as well as for 40 nm (graph b). For 80 nm NP, scattering and absorption are of the same order of magnitude (graph c). Calculations are made with the Mie theory. Reprinted with permission from Ref. 15. Copyright 2006, American Chemical Society.

For small particles, only absorption is in play with negligible scattering. A proper calculation within the Discrete Dipole Approximation (DDA) formalism has been conducted by El-Sayed and coworkers[15] and shows that for 20 nm AuNP, no scattering occurs and the particles only absorb radiation. For diameters of 80 nm, both cross sections are equivalent, and for larger particles scattering dominates (see Fig. 3.4). In other words, if one wants to have brilliant particles, they must choose particles with a large diameter, for example larger than 40 nm. For biomedical applications where AuNPs can be used as markers, this will be important. AuNP nanospheres with a diameter of 40 nm have a calculated absorption cross-section of $2.93 \times 10^{-15}\,\mathrm{m}^2$ (thus corresponding to a molar absorption coefficient ε of $7.66 \times 10^9\,\mathrm{M}^{-1}.\mathrm{cm}^{-1}$) at a plasmon resonance wavelength maximum λ_{max} of 528 nm. This value is five orders larger than the molar extinction coefficient for indocyanine green $\varepsilon = 1.08 \times 10^4\,\mathrm{M}^{-1}.\mathrm{cm}^{-1}$ at 778 nm), a NIR dye commonly used in laser photothermal tumor therapy (see Chapter 10).

3.4.4 *Experimental illustrations*

The electrostatic approximation and the analytical model provided by equation (15) accurately describe simple experimental situations such as spherical nanoparticles in suspensions with a concentration such that the particles stay far from each other (a few particle diameters). In this case one usually

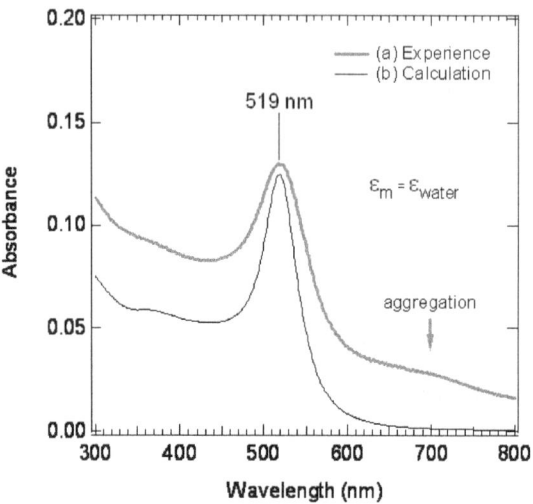

Fig. 3.5. Comparison of measured and calculated extinction spectra of 14.2 nm gold nanoparticles prepared by the Turkevich method. The experimental spectrum is taken in a $b = 1$ mm thick cuvette and the density of AuNP is $n = 1.76 \times 10^{18}$ m^{-3}. A beginning of particle aggregation is visible at 700 nm on the experimental spectrum and indicated on the graph.

measures the absorbance of a solution placed in the cuvette of spectrophotometer. The absorbance is linked to the extinction cross section by:

$$A = -\frac{1}{\ln 10} n b \sigma_{\text{ext}} \tag{17}$$

where n is the number of nanoparticles per unit volume, b the length probed by the optical beam (thickness of the cuvette) and σ_{ext} is the cross section given by relation (15). As an example, Fig. 3.5 shows a comparison of the measured absorbance of 14.2 ± 1.3 nm gold nanoparticles in suspension in water with the calculated absorbance.

Systematic checks have been conducted and show that the plasmon resonance is measured at 519 nm in aqueous solutions and does not shift for diameters from 4 to 35 nm. The extinction coefficients are precisely measured and can be used in a Beer-Lambert law.[18] According to such a rule, the measured absorbance writes:

$$A = \varepsilon b C \tag{18}$$

56

where ε is the molar extinction coefficient, b is the cuvette thickness and C the AuNP concentration. For example, in the case of a 8.6 nm citrate capped AuNP, $\varepsilon = 5.14 \times 10^7 \, \mathrm{M}^{-1}.\mathrm{cm}^{-1}$ and for a 20.6 nm AuNP, $\varepsilon = 8.78 \times 10^8 \, \mathrm{M}^{-1}.\mathrm{cm}^{-1}$.[18]

Aggregation of AuNPs is the most frequent cause of plasmon shift in such colloidal solutions; it causes the plasmon to redshift. An illustration can be found here.[19] The coupling between neighbouring particles is discussed in a following section.

3.4.5 *Local field enhancement and applications*

From equation (11) the electric field radiated by a nanoparticles can be expressed as

$$E_{\mathrm{out}} = E_0 \frac{3\varepsilon}{\varepsilon + 2\varepsilon_{\mathrm{diel}}} \tag{19}$$

At $\lambda = 530$ nm in air this radiated field is five times larger than the excitation field E_0. This enhancement is confined to the close vicinity of the particle and has given birth to the concept of optical nano-antennas.[4,12] This local field amplification is also very important for explaining the high sensitivity obtained in Surface Enhanced Raman Scattering (SERS), or more complex nonlinear optical spectroscopies[20] for which applications are developed in Chapter 10.

3.4.6 *Beyond the quasistatic and dipolar approximations*

For spherical nanoparticles Mie theory is an efficient albeit heavy approach to describe the optical response. However, for nanoparticles of non-spherical shapes, exact solutions cannot be derived except for spheroids[21] or infinite cylinders[22] exclusively.

For other shapes numerical methods have been developed and some of them are quickly reviewed below.

Discrete Dipole Approximation (DDA) is the most popular numerical approach[23] and is used by a considerable number of groups. The nano-object to be analyzed is represented by a tridimensional cubic array of dipoles. The polarization at each point is induced by a local electric field which is produced by the external exciting field and the sum of the fields generated by all the other point dipoles. In case of a mapping of the nano-object with N points, the global polarization vector is the solution of 3N linear equations that are solved numerically. Quantities like extinction or scattering cross sections are deduced for this set of dipoles in the far field as well in the near-field region. DDA can represent an object or multiple objects of arbitrary shape and composition and yields results typically with 10% accuracy.

The **Multiple Multipole Method** (MMP) is a semi-analytical approach to solve Maxwell's equations in multiple connected media that have to be homogeneous, isotropic and linear.[24] The region of interest is divided into connected domains and the field inside each domain is described by a series expansion of known analytical solutions of Maxwell's equations. MMP mainly uses multipolar expansion with different origins as basis functions of the series expansion. The computational advantage of MMP is that only the boundaries, not the domains themselves, need to be discretized. The choice of suitable sets of basis functions is the most difficult task in MMP as no optimum can be defined in a unique way. The use of multipoles and not just dipoles makes this method more powerful for describing near-field effects, but increases its complexity.

With the **Finite Difference Time Domain** (FDTD) method the Maxwell's curl equations are solved explicitly.[25] The equations are dis-cretized both over time and space. The method is based on a time marching algorithm that runs over a carefully defined spatial grid. It can be used for studying both the near- and far-field electromagnetic responses for hetero-geneous materials of arbitrary geometry. The time marching aspect of the method allows one to make direct observations at any position in space (near field and far field) and at any time during the simulation. This last feature often brings new insight into the dynamics of the system under study.

The **T-matrix method** is especially suited to calculate the extinction, absorption and scattering of an ensemble of nanoparticles taking into account their multipole contribution and not just their dipolar response.[26] The transition matrix (T-matrix) establishes the relationship between the different expansion coefficients of the multipolar expansion. The averaging over the different orientations taken by the particle is made in an analytical way. Multiple scattering effects are accounted for with this method.

3.5 Factors Shifting the Plasmon Resonance of Gold Nanoparticles

The position of the plasmon resonance is basically given by relationship (13) with a rather good accuracy in the case of spherical particles of moderate sizes. It is evident that this relationship does not offer any practical "handle" to shift the plasmon resonance. In particular relations (13) as well as (15) show that the nanoparticle size does not influence the LSPR position but only its intensity through the volume of the sphere. Yet when applications are envisioned, it is often necessary to tune this plasmon wavelength. This is the case in the field of biosensors when one wants to excite nanoparticles through the human skin and needs to use a light source compatible with the biological window (650–900 nm). Another case among many others is the use of the local field enhancement at the plasmon resonance: it might be necessary to tune this nano-antenna to the wavelength of the light source.

The LSPR can be adjusted in three ways:

1. by changing the medium surrounding the particle,
2. by changing the shape of the nanoparticles: ellipsoids, cube, triangle, icosahedra etc.,
3. by using core-shell particles.

As a general behaviour, these parameters shift the LSPR to higher wave-lengths compared to the case of a sphere in air ($\lambda_{plasmon} = 504$ nm). Such a redshift pushes the resonance away from the interband transitions of gold and therefore decreases their mutual coupling. Subsequently the LSPR appears as a narrower and more pronounced resonance. These effects are developed in the following three sections.

3.5.1 *Influence of the surrounding medium*

From the discussion of § 3.4.2, it is clear that the surrounding medium plays a role in the plasmon resonance through its optical index n linked to the dielectric permittivity: $\varepsilon_{diel} = n^2$. Within the limits of the electrostatic approximation already mentioned above, the formula (13) provides a good accuracy between calculated and measured wavelengths. It shows that the higher the optical index, the higher the plasmon resonance. Some results are summarized in Table 3.1 below.

Along with the shift of the plasmon resonance to higher wavelengths, the increase of the index of the surrounding medium is accompanied by a sharp increase of the absorption cross section. This is illustrated in Fig. 3.6 where

Table 3.1. Calculated plasmon resonance wavelengths in case of spherical gold nanoparticles in different surrounding media. The electrostatic model is used with the dielectric function of gold provided by H. Arwin.[10]

			Plasmon resonance of spherical particles in various surrounding media			
n	1	1.33	1.47	1.66	1.77	2.79
ε_{diel}	1	1.77	2.16	1.77	3.13	7.78
material	air	water, glycerol	silica glass, DMSO	alumina (Al$_2$O$_3$)	sapphire	titanium dioxide (TiO$_2$)
$\lambda_{plasmon}$ (nm)	504	519	527	541	551	678

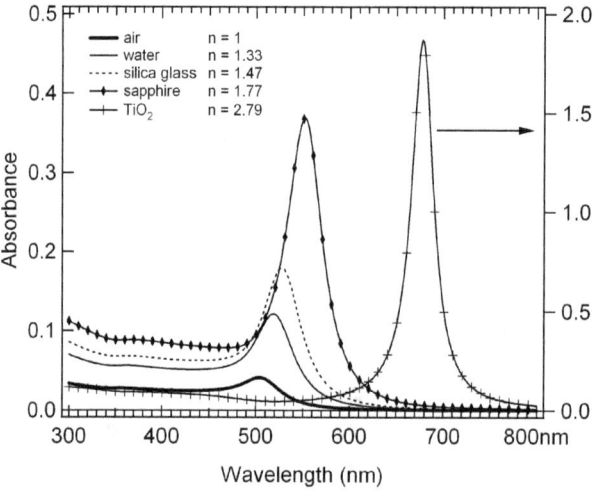

Fig. 3.6. Calculated spectra of the absorbance through an assembly of gold nanoparticles when the surrounding medium is modified. This assembly is made of a sheet of particles of diameter of 14.2 ± 1.3 nm, with a density $n = 1.76 \times 10^{18}$ m^{-3} and a thickness $b = 1$ mm.

the evolution with surrounding medium of the absorbance of an assembly of gold particles of 14 nm diameter is calculated. In the present case this assembly exhibits an absorbance of 0.04 in air, 0.18 in glass and 1.87 in titanium dioxide.

3.5.2 *Plasmon resonance of ellipsoids and other shapes*

For non-spherical particles, the absorption and scattering cross sections can be calculated from relations (16-a) and (17-a) as soon as the polarizability α is known. However, an exact analytical model for calculating this polarizability is only available in the case of ellipsoids[21] and infinite cylinders.[22]

We first focus on the case of ellipsoids.[6,27] They possess three plasmon resonances corresponding to the oscillation of electrons along the three principal semi-axes, denoted a, b and c. By changing the axes lengths the LSPR can be tuned from ca. $\lambda = 500$ nm up to the infrared range. It is obvious that the excitation of one of these plasmon resonances depends on the direction of the impinging electric field. Therefore it depends on the direction of the light beam and its polarization. Moreover a crucial challenge lies in synthesizing such objects (see Chapter 5): they are produced as assemblies and their size and shape are usually not homogeneous. Since they cannot be probed as individual objects except with dedicated experimental setups,[28,29] their optical response is an average over all the sizes and orientations present in the assembly. This may result in a smoothing or vanishing of the plasmon resonances.

For an individual metallic ellipsoid, the polarizability along the axis i ($i = x, y$ or z) is given by:

$$\alpha_i = \frac{V_e}{4\pi} \times \frac{\varepsilon - \varepsilon_m}{\varepsilon_m + L_i(\varepsilon - \varepsilon_m)} \tag{20}$$

ε is the dielectric function of the metal, which is a complex number and a function of the wavelength, and ε_m is the dielectric function of the surrounding medium which can be considered as a real number, independent of the wavelength in the visible range of interest, as long as this medium is transparent with negligible absorption.

V_e is the volume of the ellipsoids given by: $V_e = {}^{4\pi}\!/_3 abc$, with a, b and c being the semiaxes of the ellipsoids.

L_i is the depolarization factor, which is a purely geometric factor.

For simplicity we will consider revolution ellipsoids with two equal axes. These ellipsoids correspond to the majority of the experimental cases and belong to two families: they can be prolate (rugby-ball-like with $a > b = c$) or oblate (pumpkin-like with $a = b > c$). Among the three proper modes, two of them are degenerated and two depolarization factors are equal. They appear as functions of the eccentricity e of the corresponding ellipse along axis i. In case of prolate spheroids the depolarization factors are given by[6,27]:

$$L_x = \frac{1 - e^2}{2e^3} \left(\log \frac{1 + e}{1 - e} - 2e \right)$$

$$L_y = L_z = \tfrac{1}{2}(1 - L_x) \quad \text{and} \quad e = \sqrt{1 - b^2/a^2} \tag{21}$$

And for oblate spheroids:

$$L_z = \frac{1 + e^2}{e^3}(e - \tan^{-1} e)$$

$$L_x = L_y = \tfrac{1}{2}(1 - L_z) \quad \text{and} \quad e = \sqrt{a^2/c^2 - 1} \tag{22}$$

We label the two modes as longitudinal modes (LM) when the electric field is along the symmetry axis and transverse modes (TM) when it is perpendicular to this axis. Notice that this model assumes that the dipolar approximation is acceptable (nanoparticles of moderate size) and that no quadrupolar modes are entering in play. For a sphere ($a = b = c$), the three resonances are degenerated and the depolarization factors are all equal to 1/3. In this case, it is easy to check that equation (20) gives the polarizability of the electrostatic model of relation (12).

This model allows for a fully analytical calculation of the extinction cross section, assuming a dipolar model for the ellipsoid. Some results are presented in Fig. 3.7 in the case of particles in air, and compared to spherical gold particles with a radius of 7 nm whose plasmon wavelength is at 504 nm. This graph plots the extinction efficiency, which is the ratio of the extinction cross section to the geometrical cross section. The ellipsoid is chosen so that its volume is the same as the volume of a sphere with $r = 7$ nm. For example, in the case of the prolate ellipsoid of aspect ratio $a/c = 2$, the dimensions are the following: a = 11.2 nm, b = 5.56 nm and c = 5.56 nm. In this case the LM is shifted to 532 nm and its intensity is enhanced by a factor of almost 5. TM is slightly blueshifted down to 500 nm and partially damped. By increasing the aspect ratio, this trend is amplified.

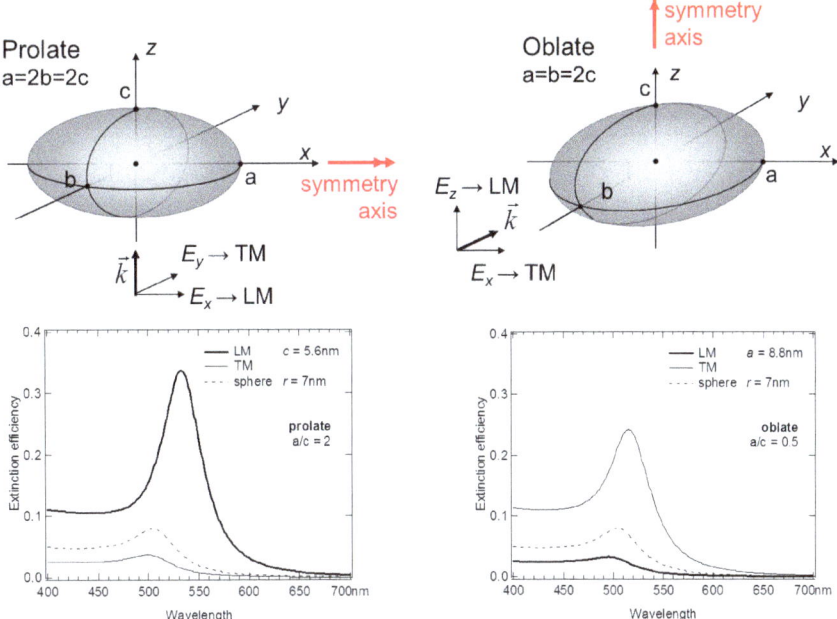

Fig. 3.7. Calculated extinction efficiency of prolate and oblate ellipsoids within the dipolar approximation. Ellipsoids of aspect ratio of 2 are compared to a sphere of equal volume. The longitudinal mode corresponds to an excitation field parallel to the symmetry axis and the transverse mode is perpendicular to this axis. In case of prolate ellipsoids, LM is the most sensitive to the change of shape whereas for oblate particles TM is the most sensitive.

For example, elongated ellipsoids of aspect ratio of 10 exhibit a longitudinal mode in the near infrared at 1037 nm whereas the transversal mode stays at 495 nm. Some other results as well as the values of the depolarization factors are given in Table 3.3. The case of oblate ellipsoids is similar with the slight difference that the TM is now the most sensitive to the change of particle shape and shifts to higher wavelengths when the aspect ratio is increased. For example, an oblate ellipsoid of aspect ratio of 2 exhibits a transverse plasmon at 515 nm. Other values are summarized in Table 3.3.

For an electric field of random orientation relative to the symmetry axis, the extinction spectrum is a combination of the two modes and will exhibit two resonances.[6]

With the help of other calculation methods, the plasmon modes of particles of more complex shapes can be calculated.[6,30] New modes are

63

Table 3.2. Calculated plasmon resonance wavelengths for **prolate** gold spheroids (a > b = c) of different aspect ratios in air.

	Prolate spheroids						
Aspect ratio (a:b)	1:1	2:1	4:1	6:1	8:1	10:1	20:1
L_x	0.3333	0.1735	0.0754	0.0432	0.0284	0.0203	0.0067
$L_y = L_z$	0.3333	0.4132	0.4623	0.4784	0.4858	0.4898	0.4966
λ_{LM} (nm)	504	532	634	760	897	1037	>1700
λ_{TM} (nm)	504	500	498	496	495	495	495

Table 3.3. Calculated plasmon resonance wavelengths for **oblate** gold spheroids (a = b > c) of different aspect ratios in air.

	Oblate spheroids						
Aspect ratio (a:c)	1:1	2:1	4:1	6:1	8:1	10:1	20:1
$L_x = L_y$	0.3333	0.2363	0.1482	0.1077	0.0845	0.0695	0.0369
L_z	0.3333	0.5272	0.7036	0.7846	0.8308	0.8608	0.9262
λ_{TM} (nm)	504	515	545	579	615	650	807
λ_{LM} (nm)	504	495	492	490	490	490	488

showing up. Many calculations have been done to address the case of nanorods which can be at first assimilated to ellipsoids but exhibit more complex plasmonic structures.[31] Plasmon modes of varying shapes have been calculated in the case of silver particles: they are plotted in Fig. 3.8 for a cube, several truncated cubes, a cuboctahedron, an icosahedron and a sphere. Although the plasmon resonances are different for silver and for gold, the trend is similar: the number of plasmon modes increases the more that the shape differs from a sphere. A cube for example exhibits six resonances. It is generally observed that the vertices of the nanoparticles play an important role in the optical response, because the sharper they become, the greater the number of resonances. A main resonance with a dipolar character

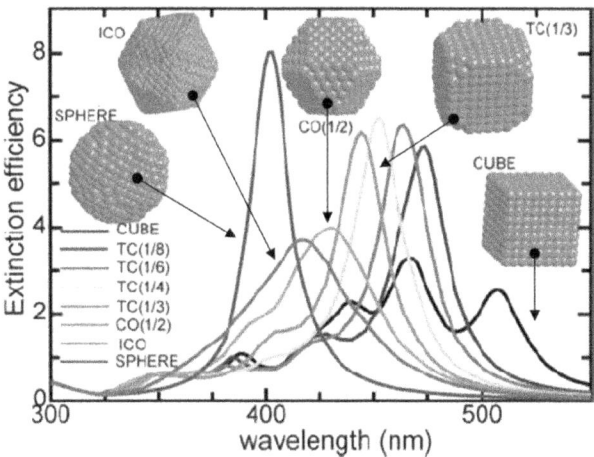

Fig. 3.8. Extinction efficiencies for silver nanoparticles of different shapes. Reprinted with permission from Ref. 6. Copyright 2007 American Chemical Society.

is always present along with other secondary resonances of lower intensity. It is also observed that as the nanoparticle becomes more symmetric, the main resonance is always blueshifted.[30]

3.5.3 The case of very small (less than 5 nm) and very large gold nanoparticles (greater than 60 nm)

In a first approach, size does not affect the SPR position but the balance between scattering and absorption. This is true as long as the approximation that describes the particle as a radiating dipole stands. However, as soon as the particle is larger than 60 nm, multipolar effects come into play[32] and show up as larger bands and tend to shift the dipolar contribution. The displacement of the electron cloud is no longer homogeneous over the entire particle and the oscillating charges can no longer be simply described with two opposite charges at a fixed distance (model of the dipole). The quadrupolar and higher multipolar plasmonic contributions are always located at shorter wavelengths compared to the dipolar ones, which are displaced to higher wavelengths. In the case of spherical nanoparticles, these effects can be analytically predicted using Mie theory and taking into account terms of orders higher than 2 in the multipolar development of the electromagnetic field.[32]

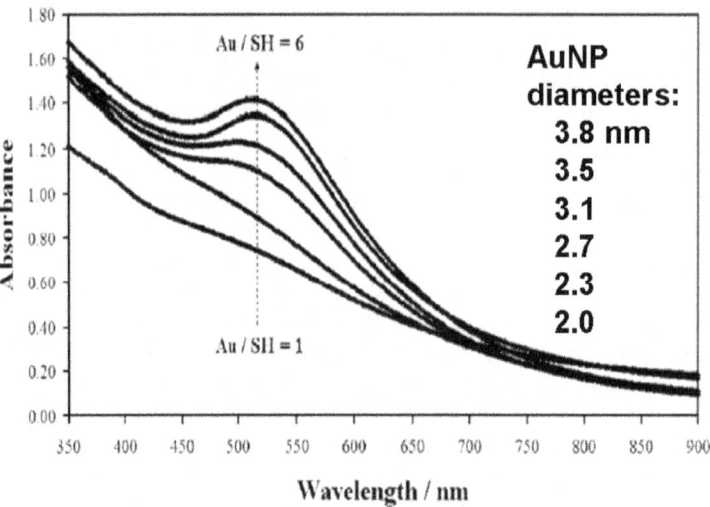

Fig. 3.9. Experimental absorbance spectra of small AuNP showing the uptake of the LSPR when the nanoparticles become greater than 2.5 nm. Reprinted with permission from Ref. 34 Copyright Elsevier 2008.

If the particle size becomes smaller than the mean free path of the free electrons (the conduction band electrons), the collisions of electrons with the particle surfaces becomes important.[6,33] This results in a slight broadening of the plasmon band for AuNPs smaller than 10 nm, and becomes really obvious below 5 nm.[34,35] For gold nanoparticles smaller than 2.5 nm, experimental studies of the UV-visible extinction coefficients show that the plasmon resonance experiences a slight blueshift and eventually vanishes as shown in Fig. 3.9.[34] The theoretical description of this effect needs a semi-quantum description to take into account the interplay between the d electrons and the s electrons of the conduction band.[35] For large particles, this interplay between bound and free electrons also takes place but it is captured by the bulk dielectric function as already discussed above and shown in Fig. 3.1.

For small nanoparticles, a typical quantum phenomenon that influences the optical response is the spill-out of the electrons. It occurs at the surface of a metal and corresponds to the fact that the density of the electrons clouds does not undergo an abrupt transition. For spherical nanoparticles for example the radius of the electron sphere is slightly larger than the radius of the positively charged sphere constituted with the cationic cores. The amount with which the electrons spill out is of the order of magnitude of a few atomic units (0.529 Å) which is not negligible anymore for small nanoparticles and lead to a slight blueshift in the case of gold.[36]

3.6 Optical Response of Assemblies of Nanoparticles

Nanoparticles are always produced in great numbers, in colloidal solutions or with other synthesis methods, and it occurs frequently that they interact with each other, especially when their density becomes high. Interactions affect the LSPR when particles are close to each other and when they are interacting with a supporting substrate. These effects modify the spectral position of the LSPR; however a primary consequence of considering an assembly of nanoparticles is to modify the absorption cross section. A method for taking into account assemblies is to use the Effective Medium Approximation. All these effects will be quickly reviewed but it is clear that the next sections are just rough introductions.

3.6.1 *Supported gold nanoparticles*

The optical response of nanoparticles deposited on a plane substrate still exhibits a plasmon resonance but this is affected by the electromagnetic coupling with the substrate. Therefore the LSPR signature depends on the distance to the substrate and its nature (metallic or dielectric substrate).[33] Briefly, the oscillating charge distribution of the AuNP induces image charges within the substrate and the two charge distribution couple.[33] In a first approximation this can be treated as a dipole-dipole interaction. If the exciting electric field is polarized perpendicular to the interface it generates a charge electric distribution symmetric to the interface plane with the opposite sign (see Fig. 3.10-a). As a result the induced dipole is aligned to the excitation dipole whereas the two dipoles are anti-parallel for an electric field polarized along the interface (see Fig. 3.10-b). The corresponding calculated spectra are presented in Fig. 3.10-c and 10-d in the case of 10 nm spherical silver nanoparticles positioned at a distance d from a Al_2O_3 substrate.[6] The LSPR dependence with particle-surface distance occurs when the electric field is normal to the interface. In this case Fig. 3.10-c shows that no

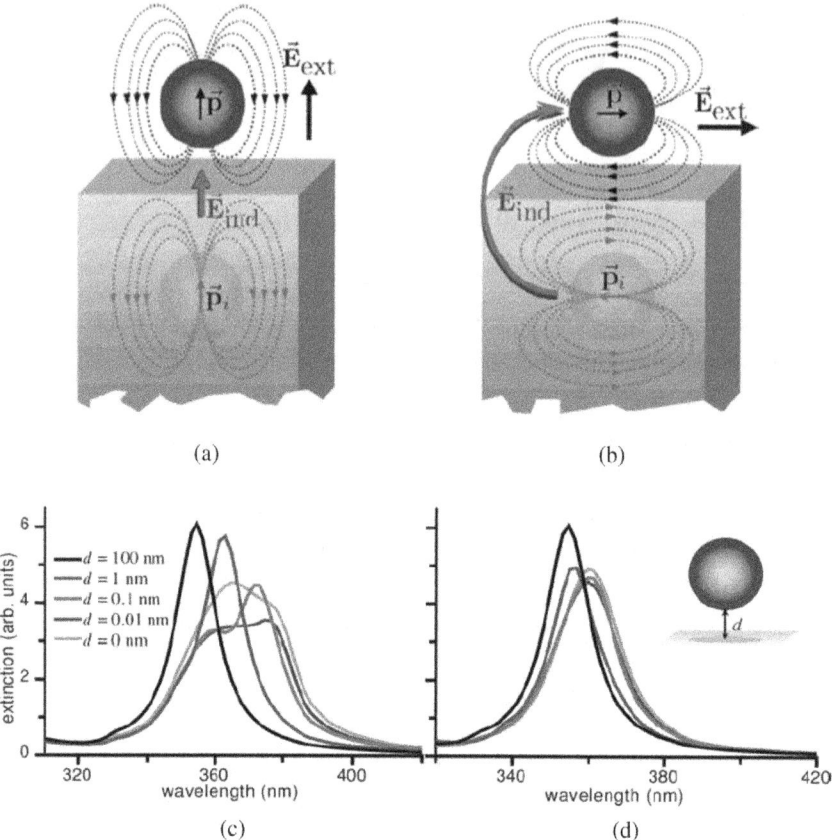

Fig. 3.10. Calculation of the electromagnetic coupling of a silver nanoparticle (10 nm diameter) with an Al_2O_3 substrate. In case (a), the exciting field is normal to the substrate, and the extinction spectrum (c) is strongly modified when the particle-substrate distance decreases below 1 nm. When the exciting field is parallel to the surface (b, and spectra d) the dependence is not as obvious. Reprinted with permission from Ref. 6. Copyright 2007 American Chemical Society.

interaction occurs when the particle is 100 nm away from the surface and the extinction is predicted to be identical to that of isolated nanoparticles (355 nm for silver spheres). The coupling with the surface becomes visible when the particle approaches at distances smaller than 1 nm and it shows up as a weak shoulder at 346 nm due to a quadrupolar contribution. By further approaching the particle to 0.1 nm the quadrupolar contribution becomes almost as important as the dipolar contribution and this latter is shifted to higher wavelengths.

The case of AuNP deposited on a substrate is investigated experimentally by Wang[37] and Kooij.[38]

3.6.2 *Nanoparticle coupling*

Coupling of nanoparticles strongly affects the LSPR but usually it occurs when a colloidal solution starts to precipitate and this coupling shows up first as a shoulder in the infrared region and may grow as a large peak around 700 nm when particles come very close to each other.

Roughly, the optical response of a pair deviates from the mere sum of the individual particle contributions as soon as the surface-to-surface distance (*d*) becomes of the order of their lateral size (radius *R*). For a transverse excitation, the optical spectrum is not very sensitive to *d* (weak blueshift) and the plasmon resonance of the pair remains close to the LSPR of each particle. On the contrary, in the case of a longitudinal excitation, the LSPR is strongly shifted toward long wavelengths as *d* is reduced. Whatever the experimental context, these observations are quite general and are corroborated by several theoretical studies considering the ideal case of spherical particle dimers, or even more complex geometries.[29,39]

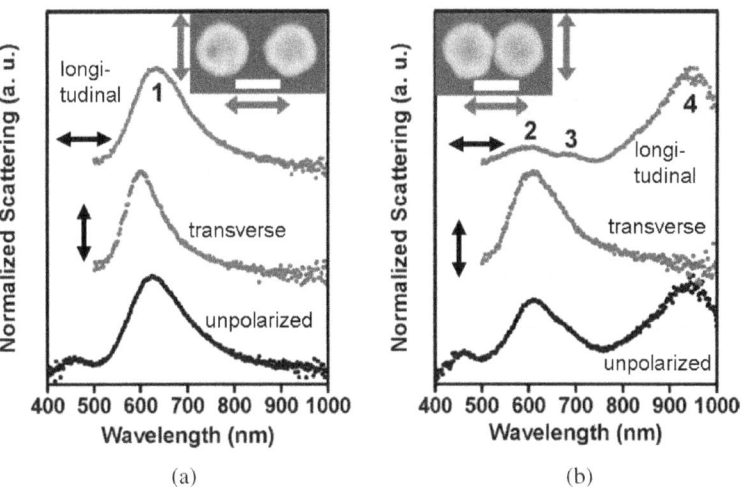

Fig. 3.11. Plot of the scattering cross sections of two interacting nanoshells made of a silica core (radius 40 nm) capped with a gold shell (radius 60 nm). In case (a) the nanoparticles are distant of 20 nm and in case (b) they are linked by a nonanedithiol molecule that keeps them separated of 1.5 nm. Reprinted with permission from Ref. 29.

3.6.3 *Effective medium approximation methods*

A medium constituted by metallic nanoparticles embedded in another material is an example of an inhomogeneous medium that can be treated by the Effective Medium Approximations (EMA) or Effective Medium Theories (EMT). The idea of the EMA is to describe the optical response of this complex medium by a single dielectric function which depends on three parameters: the dielectric function ε of the metallic inclusions, that of the embedding matrix ε_m and the volumic fraction of metal denoted f. The power of these approaches is that they provide with acceptable dielectric functions without specifying neither the relative positions of the particles nor their shape or sizes with a domain of validity that will be shortly discussed below. Once the dielectric function is known, it is easy to access the optical index and predict all sorts of situations such as reflection by a thin film of this complex material deposited on a substrate or its transmission coefficient.

The most commonly used EMA is the Maxwell Garnett (MG) model set up by the physicist of this name in 1904 to explain the coloration of glasses containing metallic inclusions.[40] MG theory is a local-field theory that gives the same result as the Clausius Mossotti formula. It assumes that the metallic inclusions are acting like isolated electric dipoles so that the resulting polarisation of the composite material is the sum of the individual microscopic polarizabilities. This microscopic polarizability can be calculated if the local field acting at an individual dipole is known. This is achieved with the help of the so-called Lorentz sphere which surrounds the dipole of interest. The Lorentz sphere delimits two regions: the outer region where the influence of the dipoles is treated globally with a mean-field theory and the inner region with neighbouring dipoles. The full derivation of this model and the different size scales cannot be discussed in details and could be found elsewhere.[33,41] However, in the case of a metal (ε) included in a matrix (ε_m) with a volumic fraction f, the resulting dielectric function is given by:

$$\frac{\varepsilon_{MG} - \varepsilon_m}{\varepsilon_{MG} + 2\varepsilon_m} = f \times \frac{\varepsilon - \varepsilon_m}{\varepsilon + 2\varepsilon_m}, \tag{23}$$

which can be solved into:

$$\varepsilon_{MG} = \varepsilon_m \frac{\varepsilon(1 + 2f) + 2\varepsilon_m(1 - f)}{\varepsilon(1 - f) + \varepsilon_m(2 + f)} \tag{24}$$

This approach holds as long as the individual dipoles are independent. If one considers that this condition is fulfilled when one dipole feels only 1% of the electric field created by its neighbour, then the maximum acceptable volumic fraction is $f_{max} = 0.1$.[41] Moreover the calculations summarized above assume that the inclusions are spherical which explains the similarity of equation (23) with equation (12) and follows the choice of spherical Lorentz cavities.

Many other models of EMA exist such as the Bruggeman approximation which considers a composite medium made by quasi equivalent mixing of two components.[33,42]

3.7 Conclusion

This chapter has provided an overview of the basic principle of plasmonics and cannot pretend to be comprehensive given that plasmonics is a rapidly growing research field. However, the key features have been presented, as well as the main conceptual tools, so that the reader can access more specialized review articles. Some other chapters of the present book will open some complementary perspectives. Chapter 4 deals with thermo-optical properties that emerge from the plasmon resonance; Chapters 10 and 11 will focus on some possible applications in biology, drug delivery and therapy.

Among the very various fields of modern optics, plasmonics is perhaps not the most innovative, but it plays a special role in a revival of optics. Plasmon amplification, enhanced sensitivity, or plasmonic coupling are phenomena occurring in the near-field range. And everybody knows that the behaviour of an electromagnetic wave in the near-field range strongly differs from the usual far-field optics. Plasmonics has opened a door into subwavelength optics that can be overstepped even by non-specialists. Yet many non-physicists handle near-field effects, trying to engineer nanoparticles geometry in order to enhance bio-sensor sensitivity. However, controlling these near-field phenomena in far field is still a very difficult game. Much is still to be understood.

Finally a key for controlling the plasmonic effects probably relies on close work between the physicists who seek to better understand how the nanoparticle shape influences its optical response and chemists who can produce more and more exotically shaped nanoparticles. This connection between chemistry and optics is occurring and some promising illustrations will be found in Chapters 10 and 13 of this book.

References

1. M. Fox, *Optical Properties of Solids*, Oxford University Press, Oxford, 2001.
2. N. Ashcroft and D. Mermin, *Solid State Physics*, 1st edn, Brooks Cole, New York; London, 1976.
3. M. Born and E. Wolf, *Principles of Optics: Electromagnetic Theory of Propagation, Interference and Diffraction of Light*, 7th edn, Cambridge University Press, Cambridge, 1999.
4. L. Novotny and B. Hecht, *Principles of Nano-Optics*, Cambridge University Press, Cambridge, 2006.
5. P.B. Johnson and R.W. Christy, *Phys. Rev. B* **6** (12) (1972) 4370.
6. C. Noguez, *J. Phys. Chem. C* **111** (10) (2007) 3806–3819.
7. P.G. Etchegoin, E.C. Le Ru and M. Meyer, *J. Chem. Phys.* **125** (16) (2006).
8. D.W. Lynch and W.R. Hunter, *Handbook of Optical Constants*, Academic Press, New York, 1985.
9. K. Johansen, H. Arwin, I. Lundstrom and B. Liedberg, *Review of Scientific Instruments* **71** (9) (2000) 3530–3538.
10. Dielectric function measured by H. Arwin on a 200 nm thick gold film in vacuum and provided to the author. Unpublished data.
11. H. Raether, *Surface Plasmons on Smooth and Rough Surfaces and on Gratings*, Springer Verlag, New York, 1986; O. Pluchery, R. Vayron and K.-M. Van, *Eur. J. Phys.* **32** (2011) 585.
12. V.M. Shalaev and S. Kawata, *Nanophotonics with Surface Plasmons*, Elsevier, Amsterdam; Oxford, 2007.
13. M.L. Brongersma and P.G. Kik, *Surface Plasmon Nanophotonics*, Springer-Verlag, New York, 2006.
14. P.K. Jain, I.H. El-Sayed and M.A. El-Sayed, *Nano Today* **2** (1) (2007) 18–29.
15. P.K. Jain, K.S. Lee, I.H. El-Sayed and M.A. El-Sayed, *J. Phys. Chem. B* **110** (14) (2006) 7238–7248.
16. G. Mie, *Annal. Physik* **25** (1908) 377–445.
17. C.F. Bohren and D.R. Huffman, *Absorption and Scattering of Light by Small Particles*, Wiley-Interscience, New York, 1998.
18. X. Liu, M. Atwater, J. Wang and Q. Huo, *Colloid Surface B* **58** (1) (2007) 3–7.
19. S. Basu, S. K. Ghosh, S. Kundu, S. Panigrahi, S. Praharaj, S. Pande, S. Jana and T. Pal, *J. Colloid & Interf. Sci.* **313** (2) (2007) 724–734.
20. O. Pluchery, C. Humbert, M. Valamanesh *et al.*, *Phys. Chem. Chem. Phys.* **11** (35) (2009) 7729–37.
21. S. Asano and G. Yamamoto, *Appl. Opt.* **14** (1), (1975) 29–49.
22. A. C. Lind and J. M. Greenberg, *J. Appl. Phys.* **37** (8) (1966) 3195–203.

23. G.C. Schatz, *Journal of Molecular Structure: THEOCHEM*, **573** (1–3) (2001) 73–80.

24. L.Novotny, R.X. Bian and X.S. Xie, *Phys. Rev. Lett.* **79** (4) (1997) 645–648.

25. C. Oubre and P. Nordlander, *J. Phys. Chem. B* **108** (46) (2004) 17740–17747.

26. M.I. Mishchenko, D.W. Mackowski and L.D. Travis, *Appl. Opt.* **34** (21) (1995) 4589–4599; B.N. Khlebtsov and N.G. Khlebtsov, *J. Phys. Chem. C* **111** (31) (2007) 11516–11527.

27. E. Lifshitz, M. Landau and L.D. Pitaevskii, *Electrodynamics of Continuous Media*, 2nd edn, Butterworth-Heinemann, Burlington, MA, 1984.

28. S. Berciaud, L. Cognet, P. Tamarat and B. Lounis, *Nano Lett.* **5** (3) (2005) 515–518; C. Novo, A.M. Funston and P. Mulvaney, *Nature Nanotechnology* **3** (10) (2008) 598–602.

29. J.B. Lassiter, J. Aizpurua, L.I. Hernandez, D. W. Brandl, I. Romero, S. Lal, J.H. Hafner, P. Nordlander and N.J. Halas, *Nano Lett.* **8** (4) (2008) 1212–1218.

30. A.L. Gonzalez and C. Noguez, *J. Comput. Theor. Nanosci.* **4** (2) (2007) 231–238.

31. J. Pérez-Juste, I. Pastoriza-Santos, L.M. Liz-Marzán and P. Mulvaney, *Coordination Chemistry Reviews* **249** (17–18) (2005) 1870–1901.

32. K.L. Kelly, E. Coronado, L.L. Zhao and G.C. Schatz, *J. Phys. Chem. B* **107** (3) (2003) 668–677.

33. U. Kreibig and M. Vollmer, *Optical Properties of Metal Clusters*, Springer Verlag, Berlin, 1995.

34. G.A. Rance, D.H. Marsh and A.N. Khlobystov, *Chem. Phys. Lett.* **460** (1–3) (2008) 230–236.

35. E. Cottancin, G. Celep, J. Lerme, M. Pellarin, J.R. Huntzinger, J.L Vialle and M. Broyer, *Theor. Chem. Acc.* **116** (4–5) (2006) 514–523.

36. W. Eckardt, *Phys. Rev. B* **29** (1984) 1558.

37. D.-S. Wang and C.-W. Lin, *Opt. Lett.* **32** (15) (2007) 2128–2130.

38. E.S. Kooij, H. Wormeester, E.A.M. Brouwer, E.v. Vroonhoven, A.v. Silfhout and B. Poelsema, *Langmuir* **18** (11) (2002) 4401–4413.

39. S. Marhaba, G. Bachelier, C. Bonnet, M. Broyer, E. Cottancin, N. Grillet, J. Lerme, J.-L. Vialle and M. Pellarin, *J. Phys. Chem. C* **113** (11) (2009) 4349–4356.

40. J.C. Maxwell Garnett, *Philosophical Transactions of the Royal Society* **203** (1904) 385.

41. S. Berthier, *Optique des Milieux Continus*, Polytechnica, Paris, 1993.

42. D.A.G. Bruggeman, *Annal. Physik* **24** (1935) 636.

Chapter 4
Photothermal Properties of Gold Nanoparticles

Bruno Palpant

Ecole Centrale Paris, Laboratoire de Photonique Quantique et
Moléculaire, UMR 8537- CNRS, Ecole Normale Supérieure
de Cachan, Grande Voie des Vignes, 92295 Châtenay-Malabry
cedex, France. Email: bruno.palpant@ecp.fr

4.1 Introduction

An outstanding number of different developments and applications have
recently arisen from the ability of noble metal nanoparticles to act as
nanoscale heat sources under light irradiation. The localized surface plasmon resonance (LSPR) is an efficient way to input energy in such small
metallic objects, as seen in the preceding chapter, thus enhancing the yield
of the nanoscale light-to-heat conversion. It has also been seen that the
characteristics of the LSPR are affected by different intrinsic or extrinsic
parameters, which allows us to detect different kinds of local modifications through optical monitoring. The main part of this chapter will present
the photo-induced generation of thermal energy within and around gold
nanoparticles, by providing the basic principles and considering cases useful for some applications. The temperature (or, more rigorously, the local
thermal state) dependence of the optical properties of a medium containing gold nanoparticles will then be briefly examined. Finally, the variation
of the melting temperature with Au-NP size (known as the melting point
depression) will be presented.

4.2 Light Induced Heating of Gold Nanoparticles

4.2.1 *Introduction: the key role of electron-phonon scattering*

It is more than likely that the optical response of metals confined at the nanoscale shares many common characteristics with the bulk metal one. Hence, as the dielectric function of gold in the visible range is dominated by the contributions of both interband and intraband transitions,[1] the optical properties of Au-NPs (provided their size is sufficiently large, see Chapter 3, § 3.3) are ruled by the microscopic coupling and exchange mechanisms involved in such transitions. The plasmon resonance phenomenon — which arises from the dipolar nature of the wave-driven collective electron excitation allowed by the small NP size — is essentially a coherent set of in-phase intraband transitions. It is then also closely dependent on the energy dissipation processes involved in individual intraband transitions, which are classically accounted for in the bulk metal response by the phenomenological scattering rate Γ of the Drude model. If the Drude model is assumed to depict the NP dielectric function, the spectral width of the SPR is directly given by $\hbar\Gamma$ (see Chapter 3, §3.3).

It may certainly be interesting to remind ourselves of the origin of the implication of collisions in metal intraband transitions. The conduction electrons being quasi-free in noble metals,[2] their dispersion law is quasi-parabolic (apart from the gap opening at the edge point L of the Brillouin zone stemming from the weak periodic ionic potential). Let us consider an initial (final) state with energy E_i (E_f) and wave vector $\mathbf{k}_i$ ($\mathbf{k}_f$). As the photon momentum is negligible against the electron one in the visible spectral domain, an electron transition induced by simple photon absorption imposes $\mathbf{k}_f \approx \mathbf{k}_i$ due to momentum conservation (the transition is said to be vertical). This means that a final state in the conduction band with $E_f \neq E_i$ cannot be reached by such an inelastic electron-photon scattering. Thus, an optical transition within the conduction band is impossible without the help of a collision with a third particle or quasi-particle which provides the momentum difference between the initial and final states. Let us underline that such an interband transition can occur through either emission or absorption of a third particle. The significant contributions to the collision-assisted intraband transitions at room temperature are the electron-electron and electron-phonon scattering ones.[3] Indeed, electron scattering by defects and impurities can be neglected, except at very low temperature. Furthermore, the contribution of electron-electron collisions[4] does not exceed 15% of the total scattering rate.[5] Even at higher electron temperatures reached with ultrashort laser excitation (see below) the electron-phonon scattering contribution remains by far the dominant one. This reveals that the plasmon resonance is tightly linked with the scattering of electrons with phonons.

Heat can be seen as a broadband incoherent statistical set of vibrations and is then supposed to be supported, in a solid, by phonons. Heat transport is ensured by carriers which are elementary particles, as electrons, phonons, or even photons. In the former two cases the mechanism involved is conduction, whereas in the latter the transfer occurs by radiation. When exposed to an incident light in (or in the vicinity of) the visible spectral domain, a gold nanoparticle can gain energy by absorbing photons through electron transitions. The main relaxation process is then the electron-phonon scattering, as stated above. Of course, an individual electron-phonon collision can result either in an energy gain or an energy loss for the electron; but as the initial energy supplied to the NP by photon absorption is input in the form of electron excitations the subsequent overall energy transfer progresses from electrons to lattice vibrations. This mechanism is then the cause of the photo-induced heating of a metal NP.

To end this introduction, let us notice that before going deeper into any calculation we can foresee the great interest of the plasmon resonance for realizing nanosize heat sources. Indeed, the SPR absorption is an efficient and fast way of inputting energy into a NP by macroscopic light excitation, the inner exchange mechanisms then allowing this energy to be mainly converted into heat at the nanoscale. This explains the spectacular recent rise in studies and applications of Au-NPs for their photothermal ability (see Chapter 10).

4.2.2 *A series of energy exchanges*

By simply depicting, in the preceding section, the link between the absorption of photons in gold NPs (at or off the plasmon resonance) and the energy dissipation through electron-phonon collisions, we have implicitly introduced the notion of dynamics. The optical properties of such nano-objects

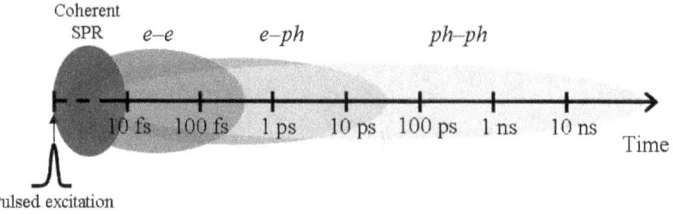

Fig. 4.1. Schematic illustration of the series of energy exchanges involved in the optical impulse response of a gold nanoparticle on a logarithmic time scale.

are indeed driven by a series of interlinked internal and external energy exchanges, each of them characterized by a typical time scale. In order to gain a deeper insight into these properties, it appears useful to study the NPs impulse response, as is typical for dynamic systems in physics. Furthermore, regarding the ability of Au-NPs to behave as thermal nanosources, such a study may allow us (i) to understand the stationary regime of photo-induced heating, and (ii) to propose new ways of controlling the heating.

Let us then imagine that a "very short"[i] light pulse is sent onto a gold nanoparticle (see Fig. 4.1). Part of the incoming light is then absorbed for inducing electron transitions. Let us stress that it is *a priori* possible to induce, whether interband transitions, or intraband ones. In the former case, photons generate individual electron-hole pairs, while in the latter they promote electrons up to higher levels within the conduction band. We have seen in a previous section that in spherical Au-NPs both the plasmon resonance and interband transitions can be excited in the same spectral range. In this case the SPR is damped and broadened due to Landau damping. Beyond this, for the sake of simplicity, we will disregard the possible excitation of interband transitions in the description of the dynamics mechanisms. Provided the ultrashort wave packet spectrum matches the SPR energy, the plasmon resonance is excited, that is, a resonant coupling with the electromagnetic wave induces a coherent set of in-phase electron excitations in the conduction band. It has been shown through optical experiments using second-harmonic generation autocorrelation, spectral hole burning or measurements of the SPR bandwidth of single NPs that the dephasing time (T_2)

---

[i]By very short, we mean the duration of which is about a few electromagnetic field oscillation cycles.

of the SPR is a few femtoseconds.[6] The NP is then left with an extra energy stored in the electron gas.

As a matter of fact, a few electrons have gained photon energy by absorption, the other ones remaining in the non-excited states. This puts the electron distribution out of equilibrium. Energy is then redistributed among the whole quasi-free electron gas by electron-electron collisions, leading to the recovery of an internal thermal equilibrium within the conduction band. This process occurs on a time scale which ranges from a few tens to several hundreds of femtoseconds, depending mainly on the initial energy input (the higher the proportion of excited electrons in the gas the faster the energy redistribution by collisions). At the same time, electrons scatter with phonons. Actually, there are no real collisions with such quasi-particles; rather, a quantized vibration mode of the crystal lattice (i.e. a phonon) induces a modification of the periodic potential experienced by the electrons, and then a modification of the wave function of the latter. The typical time scale of this process is about one picosecond. As for the electron-electron (e-e) scattering rate, the actual electron-phonon (e-ph) relaxation time depends on several factors such as particle size and input excess energy.

Finally, as the NP is not isolated but is embedded in a medium, there is a thermal energy exchange at the interface through phonon-phonon collisions, leading to the cooling down of the NP. The dynamics of this process, as will be seen later, are very sensitive to the heat transfer properties in the host medium. It may range from a few picoseconds to nanoseconds. Let us notice that, as Au-NPs are most of the time dispersed in a liquid or solid insulating medium, the heat carriers in the latter are mainly phonons (convection may be neglected at these small time scales). Of course, the NP cooling goes together with the transient heating of its close environment, which may be exploited to use gold NP as photo-induced heat nanosources. If the neighbouring NPs are sufficiently close in the medium (that is, if the NP density is high enough), then the thermal energy released by these neighbours will affect the cooling dynamics. Let us stress that in the case of very close NPs a thermal exchange through radiative transfer is possible.[7] We won't address this particular effect in this chapter.

As a consequence of this series of energy exchanges, the internal energy of the electron gas subsequent to light pulse absorption undergoes (i) a sudden and strong rise, (ii) an inner redistribution within

79

the electron gas (athermal regime), (iii) a fast decrease (*e-ph* scattering) and (iv) a slow return back to equilibrium (thermal transfer to the host medium).

The particle temperature is ruled by the balance between the gain of energy from *e-ph* collisions and the heat release towards the host medium. It then presents an increase on a picosecond time scale followed by a slow decrease. In the subsequent sections, the different steps described above will be examined in deeper detail.

4.2.3 *Athermal regime*

Monovalent bulk metals like alkali and noble metals exhibit quasi-free electron behaviour in the conduction band.[ii] Indeed, their Fermi surface in the reciprocal space is very close to a sphere, which reveals a parabolic dispersion law:

$$E(\mathbf{k}) = \frac{\hbar^2 \mathbf{k}^2}{2m^*}, \tag{1}$$

where E, $\mathbf{k}$ and m^* are the electron energy, wave vector and effective mass, respectively, and $\hbar$ denotes the reduced Planck constant.[2] This property bestows on us the right to use the quasi-free electron model for describing the conduction electron gas. Hence, at thermal equilibrium, the latter obeys a Fermi–Dirac distribution:

$$f(E) = \left[1 + \exp\left(\frac{E - E_F}{k_B T_e}\right)\right]^{-1}. \tag{2}$$

T_e is the electron temperature, defined as a means to characterize the electron internal energy at equilibrium, E_F is the Fermi energy and k_B the Boltzmann constant. Before light excitation (Fig. 4.2, left), T_e equals the lattice temperature T_l, and $T_l = T_0$ (initial temperature). When sending the light pulse, part of it is absorbed to induce electron transitions the nature of which is, either intraband only if the photon energy $\hbar\omega$ is lower than the interband transition threshold, E_{ib}, or otherwise both intra- and interband. For the sake of simplicity, we will restrict ourselves to the first case only. The electron distribution is then dug just below E_F and those electrons

---

[ii] For noble metals, nevertheless, the lattice periodic potential opens a gap at the point L of the edge of the Brillouin zone.

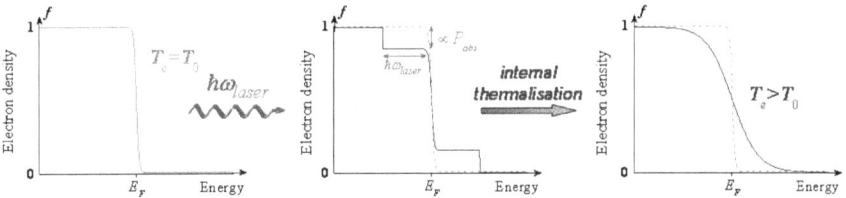

Fig. 4.2. Evolution of the conduction electron occupation rate as a function of electron energy subsequent to a light pulse absorption. Left: distribution at initial room temperature; middle: just after photon absorption; right: after internal thermalization.

which have gained energy by photon absorption are promoted to energy levels just above E_F (Fig. 4.2, middle). One is then left with a distribution out of thermal equilibrium; f does not follow a Fermi–Dirac distribution anymore and no electron temperature can be defined. This is the athermal regime. Note that this regime can be neglected when the excitation pulse width is long relative to the typical *e-ph* scattering time, as in this case the electron and the phonon gas are at every instant at quasi-thermal equilibrium.

By electron-electron collisions the energy is internally redistributed within the electron gas. This process is as efficient as the number of excited electrons is high, which explains that the duration of the athermal regime decreases with increasing laser power. The distribution then recovers a Fermi–Dirac statistics at a temperature $T_e > T_0$. This is known as the *Fermi smearing*.

To describe the electron properties in the athermal regime the relevant parameter then appears to be the electron distribution $f(E, t)$ which depends on energy and time, the dynamics of which is governed by the Boltzmann equation[8-10]:

$$\frac{\partial f(E, t)}{\partial t} = \frac{\partial f(E, t)}{\partial t}\bigg|_{source} + \frac{\partial f(E, t)}{\partial t}\bigg|_{e-e} + \frac{\partial f(E, t)}{\partial t}\bigg|_{e-ph}. \qquad (3)$$

Here we neglect electron diffusion (assuming that the particle size is smaller than the wave penetration depth, the excitation can be considered as homogeneous) as well as the environment (the influence of which will be significant at longer times). The source term refers to the instantaneous modification of f under photon absorption. Its value at a given electron energy then depends on photon energy $\hbar\omega$ and is proportional to the number of photons absorbed per time unit, this number being proportional to the instantaneous power absorbed from the laser, $P_{abs}(t)$, as illustrated on Fig. 4.2 (middle graph). $P_{abs}(t)$ can be evaluated from the particle absorption cross section (or absorption coefficient of the medium knowing the NP density) and the laser pulse intensity and time profile. The second and third terms in Eq. 2 denote the contributions of electron-electron and electron-phonon scattering, respectively, to the variation rate of f. Several approaches have been proposed to treat these contributions.[8-10] The *a priori* most rigorous one consists in including explicitly all the detailed scattering processes.[10] For the *e-e* term, it amounts to integrate over all the wave vectors of the 2nd electron and all the momentums exchanged in the elastic collision. For the *e-ph* term, both the absorption and emission of phonons have to be accounted for and the integration runs over all the wave vectors of the phonons exchanged. A significant simplification can be introduced by considering the weak perturbation regime, which validates the use of the relaxation time approximation for the two scattering contributions separately:

$$\frac{\partial f(E,t)}{\partial t} = \frac{f(E,t) - f_0(E)}{\tau(E)}. \tag{4}$$

f_0 is the initial equilibrium distribution and τ the typical collision time. In the case of the *e-ph* term, one has to distinguish the contribution of the spontaneous emission of phonons from the ones of phonon stimulated emission and absorption, the two latter depending on the number of states available in the phonon bath.[9] For the *e-e* scattering term, the Landau theory of Fermi liquids allows us to express the *e-e* mean collision time: $\tau_{e-e}(E) = \tau_0 E_F^2/(E - E_F)^2$.[11] This accounts for the fact that the scattering probability decreases as E gets closer to the Fermi level, which stems from the Pauli principle. τ_0 is a few tenths of a femtosecond. In the limit of very weak perturbations, some authors have split the distribution into a major part still at thermal equilibrium and a small athermal part: $f = f_{thermal} + f_{athermal}$, with

$$\int_{-\infty}^{+\infty} f_{athermal}(E)dE << \int_{-\infty}^{+\infty} f_{thermal}(E)dE.^{10}$$

82

The model can be further refined, for instance by including the competition between transitions within the conduction band and the creation of electron-hole pairs in the valence and conduction bands.[12]

Let us notice that the only difference between the approaches for a bulk metal and a NP has up to now rested on the disappearance of the electron diffusion term in the second case of Eq. 2 (provided that the excitation can be considered as homogeneous in the NP). In fact, some authors have shown that in the relaxation time approximation the *e-e* and *e-ph* mean collision times exhibit a size dependence.[13] Indeed, they both decrease with decreasing particle radius R due to the decrease of the Coulomb interaction screening. Moreover, the appearance of low-frequency acoustic vibration modes in finite-size NP induces a new relaxation channel for the electrons, which increases the *e-ph* scattering rate.[14]

4.2.4 *Thermal regime*

Once the thermal equilibrium is recovered within the conduction electron gas, the energetic couplings within the nanoparticles and with their environment can be treated by more classical thermodynamics approaches as the electron temperature can be defined. Let us stress that, while the model described in the preceding section enables us to account for the off-equilibrium situation for the conduction electrons, it is restricted to short times after excitation as the heat exchange with the host medium is not accounted for. Moreover, in many cases the athermal regime can be disregarded, as in pump-probe time-resolved experiments carried out with a high pump pulse intensity which shortens the athermal regime duration, or for which the time scale under consideration is much larger than this duration. It can of course be also neglected for long pulse excitation (pulse width larger than the *e-ph* collision time) as in this case the electron gas and the crystal lattice are permanently at equilibrium.

4.2.4.1 *Two-temperature model*

For describing the coupling between electrons and phonons, the two-temperature model (TTM) developed for bulk metals has been adapted to NPs. It consists in writing the two coupled differential equations ruling the time evolution of the electron and lattice internal energies:

$$C_e \frac{\partial T_e}{\partial t} = -G(T_e - T_l) + P_{abs}(t), \qquad (5)$$

$$C_l \frac{\partial T_l}{\partial t} = G(T_e - T_l). \qquad (6)$$

Here C_e and C_l denote the electron gas and lattice specific heats, respectively. As in noble metals the conduction electrons exhibit a quasi-free electron behaviour, as stated in §4.2.3, C_e can be deduced from free electron quantum statistics: $C_e = \gamma_e T_e$, where γ_e is a constant the value of which depends on the metal (for gold, $\gamma_e = 66\,\mathrm{J\,m^{-3}\,K^{-2}}$). G is the *e-ph* coupling constant (for gold, $G = 3 \times 10^{16}\,\mathrm{W\,m^{-3}\,K^{-1}}$). As for the athermal regime, $P_{abs}(t)$ represents the instantaneous power absorbed per metal volume unit (source term) and has the profile of the incident light pulse. Again, as the NP size is assumed to be smaller than the light penetration depth, the excitation is homogeneous and electron diffusion can be neglected.

The parameters involved in the TTM are usually taken as the bulk phase ones, but it may be easy to replace some of them in a phenomenological manner to account for finite size effects in NP. This has been done, for instance, to deduce the size-dependent G value from pump-probe experiments using Eqs. 5 and 6.[15] Some other authors have shown that G may vary with temperature which itself depends on laser power.[16]

4.2.4.2 Three-temperature model

The TTM completely neglects the thermal influence of the surrounding host medium. This might be valid as long as the heat exchange at the interface remains negligible, that is, when photo-heating with an ultrashort pulse and considering the electron temperature only during the first few picoseconds (see Fig. 4.3). Of course, if the contact between the particle and the matrix is poor, if the thermal resistance at the interface is high, or if the matrix has a low thermal conductivity, its influence can be neglected over a larger time scale. In the general case, it has to be taken into account. This is the purpose of the three-temperature model (3TM). For this, Eq. 6 has to be modified as to add the contribution of the instantaneous heat released through the interface, $H(t)$:

$$C_l \frac{\partial T_l}{\partial t} = G(T_e - T_l) - \frac{H(t)}{V}, \qquad (7)$$

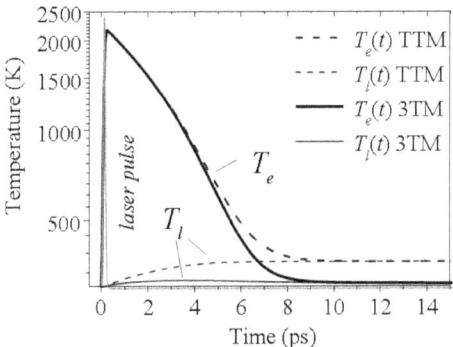

Fig. 4.3. Time evolution of the electron (T_e, thick lines) and lattice (T_l, thin lines) temperatures of a Au-NP in silica after light pulse absorption [$P_{abs}(t)$, grey line] calculated without (TTM, dashed line) and with (3TM, solid line) considering the heat release to the host medium and a purely diffusive heat transfer in the latter. Pulse duration and peak power absorbed are worth 110 fs and 1.4×10^{21} W m^{-3}, respectively. After Ref. 16.

where V is the NP volume. The time evolution of the factor H (and subsequently of T_e and T_l) then strongly depends on the characteristics of the thermal transport in the surrounding medium, namely, the ability of the latter to evacuate heat and then to cool down the particle. This point will be given a special attention in the next section. Let us underline beforehand the role that the quality of the contact may play at the interface. It has been recently shown that an interface thermal resistance (known as Kapitza resistance) may modify the NP cooling down dynamics.[17] The value of this resistance has been extracted from time-resolved pump-probe experiments. The authors ascribe this phenomenon to the acoustic impedance mismatch between the NP and its surrounding medium. It is also likely in some other situations, depending on the synthesis technique, that the medium close to the interface presents a partial porosity or that the mismatch induces lattice defects, which may also affect the thermal resistance.

4.2.5 *Heat transfer to the host medium*

4.2.5.1 *Different approaches depending on the heat transfer characteristics*

As we have seen, the cooling down of a Au-NP after ultrashort light pulse absorption or, by extension, the topography of the temperature field around a Au-NP under light irradiation — whatever the time profile of the latter — strongly depends on the characteristics of the heat transfer in the NP host medium. It is easily understandable that, for instance, the higher the thermal conductivity of this medium, the faster the relaxation and then the lower the temperature at a given distance from the NP. This has been experimentally shown to influence the dynamics even at short times,[18] especially by comparing the optical relaxation of silver or gold NPs in silica (or glass) and alumina, the conductivities of which are roughly in the ratio of one to thirty.[19,20] Beyond the only conductivity, the detailed mechanisms involved in the heat transport through the surrounding medium play a crucial role in the Au-NP photo-induced thermal response. This involvement depends on the excitation conditions as well as on the observation ones as we will now show. The different approaches which are used to describe heat transport in a medium can be split into two categories. In the first one, known as molecular dynamics (MD), matter is described by its constituents (atoms) and the modelling consists in determining the motion of each atom by (usually) solving the classical second law of Newton once the suited analytical description of the interatomic forces is chosen.[21] Once all the motions are calculated, statistical physics allows us to determine relevant quantities of the system thermodynamics as their mean values and fluctuations. Whereas, as we have seen above, molecular dynamics has allowed us to address the problem of photo-induced phase transform and partial melting of Au-NPs, this powerful method has up to now been used very little to model the thermal transport in the medium surrounding a metal NP subsequent to the photo-induced heating of the latter.[22] Rather, continuous-media approaches are employed. They consider that all the media can be described by continuously varying quantities such as phonon density, energy and flux. In this category, the most general theory appropriate for this problem is the Boltzmann transport equation (BTE). It allows us to describe the time evolution of the local phonon density and is particularly suited for off-equilibrium situations. If the spectral composition of the heat transport is disregarded (namely, if a spectrally-integrated phonon mean free path, Λ_{ph}, and lifetime, τ_{ph}, can be defined in a phenomenological manner) then they may be used as relevant parameters to validate successive simplifications of the BTE. First, the definition of a characteristic phonon lifetime itself may result in the use of the time relaxation approximation that we have already addressed in the section devoted to the athermal regime for the electrons. This approximation enables us to simplify the BTE. Further, it has been shown by Chen that an approach much simpler to bring into play than the BTE can nevertheless provide very similar results in the case of transient nanoscale heat transfer. This is the *ballistic diffusive equations* (BDE) method.[23] This takes into account the fact that at short time and space scales both ballistic and diffusive transport mechanisms can contribute to the overall heat propagation in a medium. This point will be given a more detailed development below. If the ballistic mechanism is not significant, we are then left with the Cattaneo-Vernotte model which is a hyperbolic extension of the parabolic Fourier law for the diffusion that includes the finite lifetime of thermal phonons. Finally, the simplest approach is the Fourier law which assumes an infinite speed for the heat transport and is then based on the assumption of a high number of anterior diffusive events to drive the thermal energy evolution at a given point of the medium and a given time. The validity of the Fourier law is then ensured when the observation timescale is much larger than τ_{ph} and the typical observation distance is much larger than Λ_{ph}. As usual examples, at room temperature $\tau_{ph} \sim 1040$ fs and $\Lambda_{ph} \sim 1.6$ nm in water, $\tau_{ph} \sim 130$ fs and $\Lambda_{ph} \sim 0.5$ nm in silica (amorphous SiO_2) and $\tau_{ph} \sim 850$ fs and $\Lambda_{ph} \sim 5.4$ nm in alumina (amorphous Al_2O_3). Hence, for a Au-NP in silica the Fourier law can be used to describe thermal transport in the matrix in order to determine on a few picosecond timescale the NP transient thermal response (see Fig. 4.3) or the temperature field in its vicinity — but not closer than a few nanometres to the NP surface. Note that, on the contrary, this would not be valid for an Au-NP in alumina.

In that case, the interface heat exchange factor $H(t)$ of Eq. 7 is simply worth

$$H(t) = S\kappa_m \left.\frac{\partial T_m}{\partial r}\right|_{r=R} \tag{8}$$

if a perfect contact is assumed, i.e., $T_l(t) = T_m(t, R)$ where $T_m(t, r)$ is the matrix temperature at distance r from the NP centre and time t. S is the particle surface and κ_m is the matrix conductivity (W m^{-1} K^{-1}). The temperature field in the matrix is then ruled by the classical diffusion law

$$\frac{\partial T_m}{\partial t} = \frac{\kappa_m}{C_m}\Delta T_m \tag{9}$$

where C_m denotes the matrix volume heat capacity ($J\,m^{-3}\,K^{-1}$). This equation together with Eqs. 5, 7 and 8 then constitute the usual form of the 3TM.[24]

The Fourier law becomes invalid at small space and time scales, and then in the cases of transient regimes at very short times (of the order of τ_{ph}), small distances (of the order of Λ_{ph}) and then at high metal concentration in a nanocomposite medium, as well as for NPs in a host medium having a high thermal conductivity. Let us imagine that an Au-NP is suddenly heated by photo-absorption. At the very first instants after this excitation, the matrix in the close vicinity will be heated by thermal transfer through the interface. On the one hand there exists no anterior diffusive event to ensure this heating. This "phonon rarefaction" slows down the NP cooling as compared with what would be deduced from the classical diffusion law. On the other hand, if the observation time is lower than τ_{ph} and the distance from the NP smaller than Λ_{ph}, phonons emitted by the NP wall can reach the point under consideration in the surrounding medium in a ballistic (direct) way. The BDE approach introduced above is then suited to describe such phenomena. We now supply some basic elements to present the physical principle of this method, and we refer the reader to Ref. 25 and references therein for more details. As the medium is out of thermal equilibrium, temperature is no longer a good parameter to describe thermal transport. Rather, the BDE uses both internal energy $u(\mathbf{r}, t)$ and heat flux $\mathbf{q}(\mathbf{r}, t)$ which are defined locally in time and space. Then the contributions of both diffusive and ballistic phonons are added as $\mathbf{q} = \mathbf{q}_b + \mathbf{q}_d$ and $u = u_b + u_d$.

As for photons, it is possible to define a ballistic phonon intensity $I_{b\omega}(t, \mathbf{r}, \mathbf{\Omega})$ at point M($\mathbf{r}$) and time t, in a given direction $\mathbf{\Omega}$ and at a given phonon frequency ω (see Fig. 4.4):

$$I_{b\omega}(t, \mathbf{r}, \mathbf{\Omega}) = I_{wall,\omega}(t - r'/|\mathbf{v}|, \mathbf{r} - r'\mathbf{\Omega}, \mathbf{\Omega}) \exp(-r'/\Lambda_\omega), \qquad (10)$$

where r' is the distance from M to the NP wall in direction $\mathbf{\Omega}$ and $\mathbf{v}$ is the heat carrier speed. Λ_ω is the mean free path of phonons with angular frequency ω and corresponds to the characteristic attenuation length of the phonon intensity as can be seen in the exponential of Eq. 10. $t - r'/|\mathbf{v}|$ represents the carrier travel delay from its emission by the NP wall to point M. By definition, the NP wall emits ballistic phonons only, with a frequency-dependent intensity $I_{wall,\omega}$. The ballistic flux at a given point and in a given direction

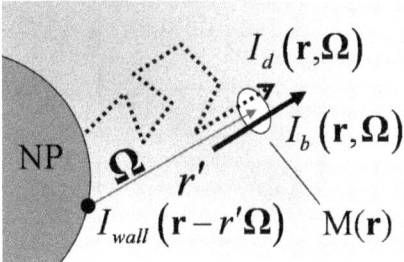

Fig. 4.4. Schematic representation of the principle of the BDE. The ballistic phonon intensity I_b at point M(**r**) for a direction Ω results from the emission of a ballistic phonon by the NP wall at point $\mathbf{r} - r'\Omega$. I_d denotes the diffusive intensity.

is calculated by integrating the phonon intensity over all frequencies and summing over all directions. The frequency integration of the wall intensity $I_{wall,\omega}$ is proportional to the temperature of the NP if any interface thermal resistance is disregarded. After some developments regarding the diffusive contribution,[25] the following equation is obtained:

$$\tau_{ph}\frac{\partial^2 u_d}{\partial t^2} + \frac{\partial u_d}{\partial t} = \frac{\kappa_m}{C_m}\Delta u_d - \nabla \cdot \mathbf{q}_b. \tag{11}$$

The contribution of the ballistic transport is then added to the hyperbolic Cattaneo-Vernotte expression in the form of the divergence of the ballistic flux. After imposing the suited initial and boundary conditions, the heat exchanged at the interface in the 3TM (Eq. 7) is now evaluated through

$$H(t) = \int_S \mathbf{q}(\mathbf{r}, t) \cdot \mathbf{n} ds \tag{12}$$

where the integral runs over the whole NP surface and $\mathbf{n}$ is the unit vector normal to the surface.

The effect of phonon rarefaction is illustrated in Fig. 4.5 where a 100-fs laser pulse heats an Au-NP surrounded by an alumina shell, the thickness of which is set as the phonon mean free path (5.4 nm). The calculation is performed for two extreme boundary conditions: adiabatic external wall (isolated system) and external wall thermalized at room temperature. Whatever the boundary conditions imposed, the results show a slowing down of the NP cooling (and then a higher temperature reached) after energy injection by ultrashort laser pulse absorption as compared to the predictions of the classical Fourier law.[26]

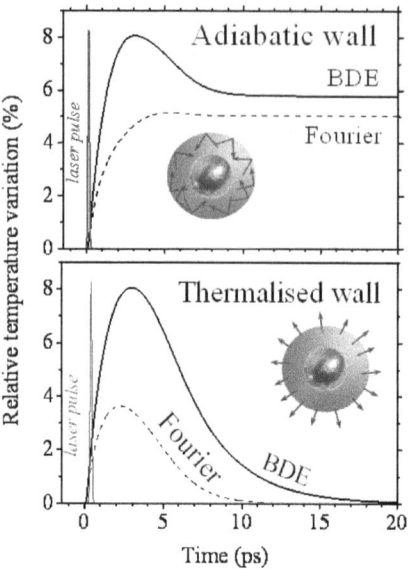

Fig. 4.5. Time evolution of the relative temperature excess of a Au-NP (10 nm radius) surrounded by an alumina shell after absorption of a 100-fs laser pulse. Shell thickness is of the order of the thermal phonon mean free path. The calculation is performed in the framework of the Ballistic-Diffusive Equations. Top: adiabatic outer wall; bottom: thermalized outer wall. The result of the classical Fourier diffusion law is added for comparison.

Heating the nanoparticle surroundings

For many applications Au-NPs are used as heat nanosources. It is then interesting to study not only the heating and subsequent cooling of such a nanoparticle under light excitation, but also the ones of the surrounding medium. We will distinguish the transient regime and the steady state for the excitation.

Transient regime. As an example Fig. 4.6 shows the mapping of the temperature field around a core-shell nanoparticle consisting of a 40 nm silica bead coated with a 0.5 nm thick gold layer (allowing to have a SPR spectral location optimized for biomedical purposes, see Chapter 10). The nanoshell is irradiated with a 7 ns Gaussian light pulse with a peak power absorbed of 5×10^{17} W m^{-3} (typical for nanosecond lasers). The surrounding medium is silica, but the method could be used for any medium. Such mapping can be useful for biomedical applications, for selecting the irradiation power or

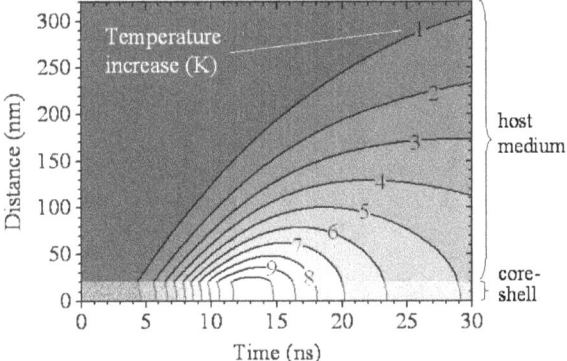

Fig. 4.6. Isothermal plot of the medium (silica) surrounding a core-shell NP consisting of a 40 nm silica bead coated with a 0.5 nm thick gold layer, irradiated with a nanosecond laser pulse. The calculation is carried out in the framework of the 3TM by using Eq. 9 for the thermal transport in silica. Pulse duration and peak power absorbed are worth 7 ns and 5×10^{17} W m^{-3}, respectively. After Ref. 16.

the molecular spacer length between the NP heat source and the biological object to be targeted.

Whereas the method presented above can be easily applied as far as the geometry of the system remains spherical, it becomes tougher for any particle shape or for assemblies of NPs. One may then use three dimensional finite elements methods to solve the set of coupled equations of the 3TM.

Another interesting feature for applications lies in the dependence of the temperature field generated on the light irradiation duration. It is well known from heat physicists that the shorter the heat impulse the smaller the spatial heated range. This is now used in the metal micromachining industry as laser cutting with femtosecond pulses is much more precise and creates many fewer defects than by using continuous lasers. Here, changing the pulsewidth can allow us to tune the effective thermal spatial range, as illustrated on Fig. 4.7. This figure presents the space and time evolution of the temperature in the matrix (silica again) in perfect contact with an Au-NP heated by photo-absorption of a Gaussian light pulse. In (a) the pulsewidth is worth about 100 fs, whereas in (b) it is 30 ps and in (c) 7 ns. The respective input energies correspond to realistic experimental situations with lasers and Au-NPs (peak intensity I_{00} of 11×10^9, 81×10^7 and 3.25×10^6 W cm^{-2}, respectively). It can be seen that despite the similar temperatures reached in

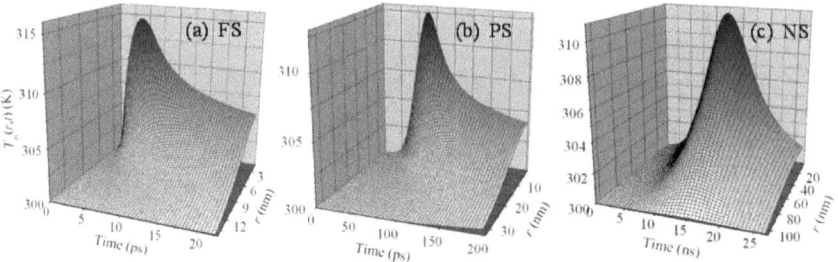

Fig. 4.7. Temperature in the matrix (silica) around an Au-NP (radius 1.3 nm) as a function of time *t* and distance from the NP centre, *r* (the back wall corresponds to the NP surface) as calculated by using the 3TM with Fourier law and perfect contact at the interface. The NP is irradiated by a Gaussian laser pulse in three different temporal regimes: (a) femtosecond, in which the conditions are those of Fig. 4.3; (b) picosecond, with 30 ps pulsewidth and 1×10^{19} W m^{-3} peak power absorbed, and (c) nanosecond, the conditions being those of Fig. 4.6. After Refs. 16, 24.

the three cases the typical spatial heating range in the ultrashort regime is about ten nanometres while it is about 40 nm in the picosecond regime and reaches about 100 nm in the nanosecond one. This pulsewidth dependence of the spatial range of the photo-induced heating around a metal NP has also been reported very recently in Ref. 27 where the effect of a femtosecond laser pulse illumination (ultrashort case) with the one of a CW laser (steady state case) are compared.

Steady-state heating. The approaches described above are well suited to depict the NP short-time transient thermal response, as they include the distinction between electrons and phonons in the metal and may also account for the phonon rarefaction at short time and space scales. However, when the photo-absorbed energy input is slow against the typical lifetimes of these off-equilibrium states it is possible to simplify the problem. This may be the case, for instance, for long-lasting laser pulses and *a fortiori* for CW beams. This steady-state analysis has been carried out by some authors.[28,29] For this, the energy absorbed is considered as entirely con-verted into heat. The electron excitation and the electron-phonon coupling can then be disregarded. The heat source term $Q(\mathbf{r}, t)$ is thus calculated by evaluating the power density of the electromagnetic field absorbed in the NP. This first step can be performed either through a simple approach when the field in the NP can be analytically calculated (as for instance in the case of an isolated sphere in the stationary regime, see Chapter 3, § 3.4.2)[28] by calculating the absorption cross section otherwise if the field can be assumed as homogeneous in the NP (using, for instance, the Discrete Dipole Approximation[30]), or by numerically solving Maxwell's equations in the most general case (Finite Difference Time Domain approach or Boundary Element Method[29]). Two interesting alternative approaches have been recently proposed by G. Baffou and co-workers, in which a thermal extension of the Green Dyadic Method, already evoked above, or the DDA (see Chapter 3, § 3.4.6) is used.[31] The authors use these approaches to determine the temperature field topography around different kinds of gold nanostructures. Moreover, they show that the optical near-field distribution at the SPR does not always superimpose on the thermal one. Even if the absorption power density is not uniform in the NP, the usual high contrast between metal and surrounding medium thermal conductivities induces a fast homogenisation of the metal temperature in the NP at the typical interface exchange timescale.[29] By using the Green dyadic method to determine the field distribution in gold nanostructures under light absorption, and then the heat power density, the authors of Ref. 32 have shown that the NP shape has a strong influence on the heat density distribution. The temperature increase is, however, almost uniform as the thermal conductivity of gold is high. The equation which now has to be solved is the classical Fourier law for heat diffusion with a source term and a spatial dependence of the parameters:

$$C(\mathbf{r})\frac{\partial T(\mathbf{r}, t)}{\partial t} = \nabla[\kappa(\mathbf{r})\nabla T(\mathbf{r}, t)] + Q(\mathbf{r}, t). \tag{13}$$

C and κ are still the volume specific heat and the thermal conductivity, respectively. It is then possible to extract, by simple analysis, some characteristics of the heating of the NP and its environment. If the continuous light source is switched on at time $t = 0$, the typical time needed to reach the steady state is $\tau_{st} \approx R^2/D_m$,[28] where R is still the NP radius and $D_m = \kappa_m/C_m$ is the host medium (matrix) diffusivity. Hence $\tau_{st} \approx 6$ ns for a 30-nm Au nanosphere in water. Beyond, the temperature excess reached by the NP and the temperature excess profile in the host medium in the stationary regime can be evaluated as[28, 29]:

$$\Delta T(r) = \frac{\sigma_{abs}I_0}{4\pi\kappa_m r}. \tag{14}$$

I_0 denotes the incident light intensity in the host medium (W m^{-2}), σ_{abs} is the NP absorption cross section (m^2), κ_m is the matrix conductivity (W m^{-1} K^{-1}).

Size-dependence of the cooling down

From a very simple and classical consideration, it can easily be deduced that the heat release rate from a NP to its environment subsequent to light energy absorption is ruled by the competition between the energy stored in the NP, proportional to its volume, and the thermal exchange through the interface, proportional to the NP surface (see Eqs. 7 and 8). Consequently, the bigger the NP the slower the cooling down, and then the higher the temperature reached by the NP.[20,24] Beyond, due to confinement some parameters are likely to be modified by finite size effects. Let us merely summarize them. The increasing influence of the NP surface when decreasing particle size results in the acceleration of the global relaxation dynamics. Indeed, the *e-e* collision time decreases due to (i) quantum size effects (electron *spill-out* and skin of reduced polarisability at the surface) inducing the decrease of the screening of the Coulomb interaction,[33] and (ii) electron recombination with *d*-holes, resonant with the SPR; this process is all the more efficient as R is small.[12] Furthermore, the *e-ph* collision time also decreases, subsequent to (i) the decrease of the Coulomb interaction screening, and (ii) the appearance of new vibration modes (capillary and acoustic) in small NPs.[34] The characteristic relaxation times τ_{e-e} and τ_{e-ph} are then reduced in a ratio of 1 to 0.4 and 1 to 0.6, respectively, from bulk gold to a Au-NP with $R = 2$ nm.[33]

Influence of the NP density

Until now, we have considered isolated particles only. However, thermal energy exchange between neighbouring particles through the interstitial medium may affect their thermal response, either in the stationary regime or in the transient one. It can be easily understood in a qualitative manner by examining Eq. 8. Indeed, it can be seen that the heat exchange rate crucially depends on the thermal gradient at the interface, which is a very well-known result of classical thermal transfer physics. Hence, the colder the host medium, the faster interface heat exchange (that is, the faster the NP cooling in the transient regime). Now, if a NP *A* has a neighbouring one

B, the heat released by B induces the reduction of the thermal gradient in the vicinity of A, the cooling of which consequently slows down. Provided A is still irradiated and goes on absorbing light energy, its temperature reaches a higher value than in the absence of B. An outcome of this is that in a nanocomposite medium the temperature of both the NPs and their close vicinity increases with NP density for a fixed absorbed power per particle.[24] Similarly, it has been recently shown by using the thermal extension of the DDA method that in a two-dimensional array of $N \times N$ Au-NPs irradiated in the steady state the temperature at the centre increases linearly with N.[31]

4.3 Thermo-Optical Properties of Gold Nanoparticles

We have seen that the specific optical properties of Au-NPs can lead to make them efficient nanoscale thermal sources, thanks to an internal series of energy exchanges leading to the conversion of light into heat. Inversely, one may be interested in the way a temperature variation leads to a change in the nanoparticle optical properties. This is generally called the *thermo-optical response* of materials. This point may be considered either in the transient case or in the stationary one. By transient we design the case of ultrafast variations where the optical response, which is mainly governed in metals by the behaviour of electrons, cannot be directly linked with the particle temperature as long as the electrons and the crystal lattice are not at equilibrium. This essentially concerns the domain of ultrafast relaxation studied by pump-probe spectroscopy using femtosecond lasers. This has been largely studied for more than two decades and this is out of the scope of this chapter.

In the "stationary" regime a variation of temperature induces a modification of the material complex optical index. Provided the change is sufficiently weak around a given temperature, this variation can be linearised as:

$$\Delta \tilde{n}(t) = (\partial n/\partial T + i\partial \kappa/\partial T)\Delta T(t) = (d_T n + id_T \kappa)\Delta T(t). \quad (15)$$

$d_T n$ and $d_T \kappa$ are the thermo-optical coefficients for refraction and extinction, respectively.

4.3.1 *Bulk gold*

Thanks to several experimental and theoretical investigations carried out in the 1970s on the electronic properties and band structure of noble metals, the mechanisms responsible for the optical and thermo-optical response of bulk gold are well known. Let us summarize the latter. When increasing gold temperature, the contribution of both interband and intraband electron transitions to the optical properties are modified due to (i) the increase in the *e-e* and *e-ph* collision rates. This affects the quasi-free conduction electron contribution, and is significant at low photon energy, well below the interband transition threshold; (ii) the modification of the electron distribution around the Fermi level. This mainly modifies the contribution of interband transitions to the complex dielectric function around their onset (from ~2 to 3 eV), as can be calculated through the Rosei model for the band structure;[35] (iii) the thermal expansion of the crystal lattice, which shifts the electron band energies non uniformly and lowers the Fermi level.[36] This third effect results in the modification of the absorption spectrum above the interband transition threshold, especially around the maximum of the contribution of the L point of the Brillouin zone (~3.5–4.0 eV).[37]

The accurate analysis of different works published a few decades ago has then allowed us to extract the mean values of the thermo-optical coefficients of bulk gold in the range from 295 to 670 K.[38] Their dispersion curve reveals a monotonous decrease of the thermo-optical effect with increasing photon energy, stemming from the contribution of intraband transitions, on which superimpose strong features ascribed to the temperature dependence of the interband transitions, especially in the spectral domain of their threshold.

4.3.2 *Gold nanoparticles*

As gold dielectric function evolves with temperature, the optical properties of Au-NPs are likely to do so. In fact, the thermo-optical response of gold in a nanoparticle is strongly affected by the local electromagnetic field enhancement at the SPR.[38] This has been put in evidence experimentally on the absorption part of the optical response.[39] The consequences of heating a medium containing Au-NPs on its SPR band are a quenching, a broadening

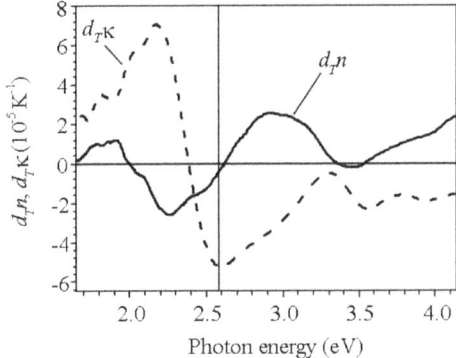

Fig. 4.8. Spectral dependence of the thermo-optical coefficients $d_T n$ and $d_T \kappa$ of a thin film containing Au-NPs (volumic fraction $f = 6.6\%$, particle radius $R = 1.3$ nm, layer thickness: 150 nm). The vertical line denotes the experimental spectral location of the SPR maximum.

and a slight redshift. Furthermore, the refractive part of the thermo-optical response has also been shown to be significant, responsible for instance for a thermal lensing effect when using lasers.[40] As an illustration, Fig. 4.8 shows the spectral variations of both $d_T n$ and $d_T \kappa$ in a Au:SiO$_2$ nanocomposite film (metal volume fraction: 6.6%; mean NP radius: 1.3 nm) as deduced from ellipsometry measurements carried out at 20 °C and 190 °C. It is noteworthy that the profile of the curves is fully correlated to both the characteristics of interband transitions (onset around 2.0 eV, absorption at higher photon energy) and the effect of the SPR around its maximum at about 2.6 eV (denoted by the vertical line). As can be seen, in the visible spectral domain, the thermo-refractive coefficient may be either positive or negative depending on the wavelength, resulting in a convergent or divergent thermal lens.[41] By (analytically or numerically) differentiating relative to temperature any model describing the optical response of a nanoparticle or a nanocomposite medium from the knowledge of the metal dielectric function, it is possible to obtain the temperature dependence of this response. Of course, all the optical and morphological parameters of the model which are likely to exhibit a temperature dependence, beyond the metal index itself, may be preliminarily identified and included in the differentiation. For instance, it has been shown that the thermo-optical properties of a medium matching the conditions of validity of the Maxwell Garnett theory (MGT, see Chapter 3, § 3.6.3) can be quite well reproduced by calculating $\partial \tilde{\varepsilon}_{eff} / \partial T$.[40]

4.4 Melting Point Depression in Gold Nanoparticles

As many properties of NPs evolve with their size, the thermal properties of Au-NPs are likely to present size dependence as well. We have already briefly introduced some of these effects in the preceding sections; it appears interesting here to report such a size dependence for the melting point of Au-NPs, T_{melt}. Indeed, a strong decrease of T_{melt} with decreasing NP size has been both observed and predicted. While this effect was demonstrated in some metallic nanoparticles almost six decades ago thanks to electron diffraction analysis,[42] its experimental evidence in the case of Au-NPs was first reported in the 1970s.[43,44] Actually, this effect concerns all kinds of nanoparticles and is known as the *melting point depression*. Its physical origin probably stems from the increase of the contribution of surface atoms to the total number of atoms when decreasing particle size. As the cohesion energy of surface atoms is lower than the one of volume atoms, partial melting is favoured at the surface[45] and the NP surface melting temperature is lower than the bulk material one.[46] The classical thermodynamics approach accounting for such an effect is known as the Gibbs–Thomson equation, involving the liquid-solid interface curvature dependence of T_{melt}.[47] For a sphere of radius R, it writes[48–53]:

$$T_{melt}(R) = T_{melt}(\infty) \left(1 - \frac{2\sigma_{sl}}{H_f \rho_s R} \right) \qquad (16)$$

where $T_{melt}(\infty)$ denotes the bulk melting point (1337 K for gold), σ_{sl} is the solid-liquid interface energy, H_f is the bulk latent heat of fusion and ρ_s is the solid state density. The $1/R$ divergence from the bulk value has been rather well evidenced by experiments (electron diffraction analysis, calorimetric or field emission measurements), as can be observed on Fig. 4.9 where the variation of T_{melt} with $1/R$ is reported from four different studies.[43,44,48,49] As several works have been devoted to the determination of the melting mechanisms and temperature of Au-NPs by molecular dynamics (MD) calculations,[50–53] we have also reported some of these works on Fig. 4.9. The discussion about the differences in the methods and results of these studies is out of the scope of this chapter. It can be seen that the approach of Shim *et al.*[52] meets particularly well some of the experimental results within a large size range.

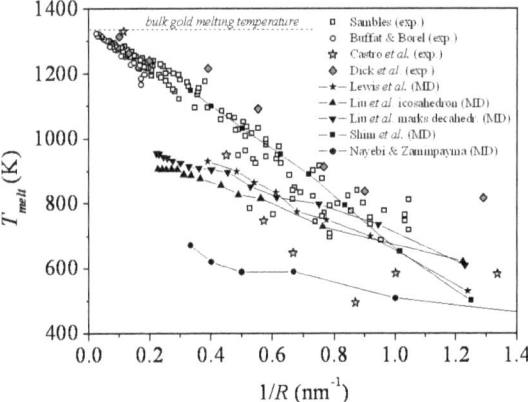

Fig. 4.9. Variation of the melting temperature of Au-NPs as a function of the inverse of their radius. Open symbols: experimental results from Sambles,[43] Buffat and Borel,[44] Castro *et al.*[48] and Dick *et al.*[49] Closed symbols: molecular dynamics calculation from Lewis *et al.*,[50] Liu *et al.*,[51] Shim *et al.*,[52] Nayebi and Zaminpayma.[53] The dashed line denotes the asymptotic value of T_{melt} for bulk gold.

The NP melting mechanisms have also been addressed through optical experiments on gold nanorods, by investigating their thermally- or photo-induced transition into the more stable spherical shape by both transmission electron microscopy and conventional optical spectrometry monitoring their SPR bands.[54,55] The role of the rigidity of the host medium has been emphasized.[54] There has been some attempt to determine the temperature of laser-induced melting of spherical Au-NPs by monitoring the change in the vibration period of their low-frequency breathing acoustic mode by time-resolved pump-probe experiments; however, they have failed to reach a clear conclusion due to the absence of any abrupt discontinuity at the phase transition.[55,56] Time-resolved X-ray scattering experiments have shown that laser-induced heating of Au-NPs can result in the lattice expansion and disappearance of long-range order for sufficient laser pulse power, signing the melting phase transition.[57]

Acknowledgements

B.P. warmly thanks Majid Rashidi-Huyeh, Yannick Guillet, Bruno Gallas, Yann Chalopin and Sebastian Volz for their contribution and enriching discussions.

References

1. M. Fox, *Optical Properties of Solids*, Oxford University Press, Oxford, 2001.
2. N.W. Ashcroft and N.D. Mermin, *Solid State Physics*, 1st edn, Brooks Cole, New York; London, 1976.
3. J.M. Ziman, *Electrons and Phonons*, Clarendon Press, Oxford, 1960.
4. R.N. Gurzhi, *Soviet Phys. JETP* **8** (1959) 673.
5. J.B. Smith and H. Ehrenreich, *Phys. Rev. B* **25** (1982) 923.
6. B. Lamprecht, A. Leitner and F.R. Aussenegg, *Appl. Phys. B* **64** (1997) 269; T. Ziegler, C. Hendrich, F. Hubenthal, T. Vartanyan and F. Träger, *Chem. Phys. Lett.* **386** (2004) 319; C. Sönnichsen, T. Franzl, T. Wilk, G.v. Plessen, J. Feldmann, O. Wilson and P. Mulvaney, *Phys. Rev. Lett.* **88** (2002) 077402–1.
7. P.-O. Chapuis, M. Laroche, S. Volz and J.-J. Greffet, *Appl. Phys. Lett.* **92** (2008) 201906.
8. W.S. Fann, R. Storz, H.W.K. Tom and J. Bokor, *Phys. Rev.* **46** (1992) 13592–13595; R.H.M. Groeneveld, R. Sprik and A. Lagendijk, *Phys. Rev. B* **51** (1995) 11433–11445; G. Tas and H.J. Maris, *Phys. Rev. B* **49** (1994) 15046–15054; C. Suárez, W.E. Bron and T. Juhasz, *Phys. Rev. Lett.* **75** (1995) 4536–4539; V.E. Gusev and O.B. Wright, *Phys. Rev. B* **57** (1998) 2878–2888; Y. Guillet, E. Charron and B. Palpant, *Phys. Rev. B* **79** (2009) 195432.
9. C.-K. Sun, F. Vallée, L. Acioli, E.P. Ippen and J.G. Fujimoto, *Phys. Rev. B* **48** (1993) 12365–12368; C.-K. Sun, F. Vallée, L. Acioli, E.P. Ippen and J.G. Fujimoto, *Phys. Rev. B* **50**, (1994) 15337–15348.
10. N. Del Fatti, R. Bouffanais, F. Vallée and C. Flytzanis, *Phys. Rev. Lett.* **81** (1998) 922–925.
11. D. Pines and P. Nozières, *The Theory of Quantum Liquids, Vol. I: Normal Fermi Liquids*, W. A. Benjamin Inc., New York, 1966.
12. T.V. Shahbazyan, I.E. Perakis and J.-Y. Bigot, *Phys. Rev. Lett.* **81** (1998) 3120.
13. C. Voisin, D. Christofilos, N. Del Fatti, F. Vallée, B. Prével, E. Cottancin, J. Lermé, M. Pellarin and M. Broyer, *Phys. Rev. Lett.* **85** (2000) 2200–2203; A. Arbouet, C. Voisin, D. Christofilos, P. Langot, N. Del Fatti, F. Vallée, J. Lermé, G. Celep, E. Cottancin, M. Guadry, M. Pellarin, M. Broyer, M. Maillard and M.P. Pileni, *Phys. Rev. Lett.* **90** (2003) 177401.
14. E.D. Belotskii, S.N. Luk'yanets and P.M. Tomchuk, *Sov. Phys. JETP* **74** (1992) 88–94.
15. J.H. Hodak, I. Martini and G.V. Hartland, *J. Phys. Chem. B* **102** (1998) 6958; H. Inouye, K. Tanaka, I. Tanahashi and K. Hirao, *Phys. Rev. B* **57** (1998) 11334; J. Sasai and K. Hirao, *J. Appl. Phys.* **89** (2001) 4548.
16. M. Rashidi-Huyeh, *Influence des effets thermiques sur la réponse optique de matériaux nanocomposites métal-diélectrique*, PhD thesis of Université Pierre et Marie Curie, Paris (2006).
17. V. Juvé, M. Scardamaglia, P. Maioli, A. Crut, S. Merabia, L. Joly, N. Del Fatti and F. Vallée, *Phys. Rev. B* **80** (2009) 195406.
18. M.B. Mohamed, T.S. Ahmadi, S. Link, M. Braun and M.A. El-Sayed, *Chem. Phys. Lett.* **343** (2001) 55.

19. V. Halté, J.Y. Bigot, B. Palpant, M. Broyer, B. Prével and A. Pérez, *Appl. Phys. Lett.* **75** (1999) 3799.

20. Y. Hamanaka, J. Kuwabata, I. Tanahashi, S. Omi and A. Nakamura, *Phys. Rev. B* **63** (2001) 104302.

21. D.C. Rapaport, *The Art of Molecular Dynamics Simulations*, 2nd edn, Cambridge University Press, Cambridge, 2004; M.P. Allen and D.J. Tildesley, *Computer Simulation of Liquids*, Oxford University Press, Oxford, 1989.

22. S. Merabia, S. Shenogin, L. Joly, P. Keblinski and J.-L Barrat, *PNAS* **106** (2009) 15113.

23. G. Chen, *Phys. Rev. Lett.* **86** (2001) 2297–2300; G. Chen, *J. Heat Transfer ASME* **124** (2002) 320–328.

24. M. Rashidi-Huyeh and B. Palpant, *J. Appl. Phys.* **96** (2004) 4475.

25. B. Palpant, *Thermal Nanosystems; Nanomaterials* (ed S. Volz), Series Topics in Applied Physics, vol. 118, Springer, New York, 2009, p. 127.

26. M. Rashidi-Huyeh, S. Volz and B. Palpant, *Phys. Rev. B* **78** (2008) 125408.

27. G. Baffou and H. Rigneault, *Phys. Rev. B* (accepted).

28. A.O. Govorov, W. Zhang, T. Skeini, H. Richardson, J. Lee and N.A. Kotov, *Nanoscale Res. Lett.* **1** (2006) 84.

29. G. Baffou, R. Quidant and F.J. García de Abajo, *ACS Nano* **4** (2010) 709.

30. B.T. Draine and P.J. Flatau, *J. Opt. Soc. Am. A* **11** (1994) 1491.

31. G. Baffou, R. Quidant and C. Girard, *Phys. Rev. B* **82** (2010) 165424.

32. G. Baffou, R. Quidant and C. Girard, *Appl. Phys. Lett.* **94** (2009) 153109.

33. J. Lermé, G. Celep, M. Broyer, E. Cottancin, M. Pellarin, A. Arbouet, D. Christofilos, C. Guillon, P. Langot, N. Del Fatti and F. Vallée, *Eur. Phys. J. D* **34** (2005) 199.

34. E.D. Belotskii and P.M. Tomchuk, *Int. J. Electron.* **73** (1992) 955.

35. R. Rosei, F. Antonangeli and U.M. Grassano, *Surf. Sci.* **37** (1973) 689.

36. P. Winsemius, H.P. Lengkeek and F.F. v. Kampen, *Physica B* **79** (1971) 529.

37. P. Winsemius, F.F. v. Kampen, H.P. Lengkeek, and C.G.V. Went, *J. Phys. F* **6** (1976) 1583.

38. M. Rashidi-Huyeh and B. Palpant, *Phys. Rev. B* **74** (2006) 75405.

39. R.H. Doremus, *J. Chem. Phys.* **40** (1964) 2389; R.H. Doremus, *J. Chem. Phys.* **42** (1965) 414; U. Kreibig, *J. Phys. F: Metal Phys.* **4** (1974) 999; A. Heilmann and U. Kreibig, *Eur. Phys. J. AP* **10** (2000) 193; S. Link and M.A. El-Sayed, *J. Phys. Chem. B* **103** (1999) 4212; D. Dalacu and L. Martinu, *Appl. Phys. Lett.* **77** (2000) 4283; L.M. Liz-Marzán and P. Mulvaney, *New. J. Chem.* **22** (1998) 1285.

40. B. Palpant, M. Rashidi-Huyeh, B. Gallas, S. Chenot and S. Fisson, *Appl. Phys. Lett.* **90** (2007) 223105.

41. Y. Guillet, M. Rashidi-Huyeh, D. Prot and B. Palpant, *Gold Bull.* **41** (2008) 341.

42. M. Takagi, *J. Phys. Soc. Jpn.* **9** (1954) 359.

43. J.R. Sambles, *Proc. Roy. Soc. Lond. A* **324** (1971) 339.

44. P. Buffat and J.-P. Borel, *Phys. Rev. A* **13** (1976) 2287.

45. R. Kofman, P. Cheyssac, A. Aouaj, Y. Lereah, G. Deutscher, T. Ben-David, J.M. Penisson and A. Bourret, *Surf. Sci.* **303** (1994) 231.

46. P.R. Couchman and W.A. Jesser, *Nature* **269** (1977) 481.

47. L. Makkonen, *Langmuir* **16** (2000) 7669.

48. T. Castro, R. Reifenberger, E. Choi and R.P. Andres, *Phys. Rev. B* **42** (1990) 8548.

49. K. Dick, T. Dhanasekaran, Z. Zhang, and D. Meisel, *J. Am. Chem. Soc.* **124** (2002) 2312.
50. L. Lewis, P. Jensen and J.-L. Barrat, *Phys. Rev. B* **56** (1997) 2248.
51. H. Liu, J.A. Ascencio, M. Perez-Alvarez and M.J. Yacaman, *Surf. Sci.* **491** (2001) 88.
52. J. Shim, B.J. Lee and Y.W. Cho, *Surf. Sci.* **512** (2002) 262.
53. P. Nayebi and E. Zaminpayma, *J. Cluster Sci.* **20** (2009) 661.
54. S.-S. Chang, C.-W. Shih, C.-D. Chen, W.-C. Lai and C.R.C. Wang, *Langmuir* **15** (1999) 701.
55. H. Petrova, J.P. Juste, I. Pastoriza-Santos, G.V. Hartland, L.M. Liz-Marzán and P. Mulvaney, *Phys. Chem. Chem. Phys.* **8** (2006) 814.
56. G.V. Hartland, M. Hu and J.E. Sader, *J. Phys. Chem. B* **107** (2003) 7472.
57. A. Plech, V. Kotaidis, S. Grésillon, C. Dahmen and G.v. Plessen, *Phys. Rev. B* **70** (2004) 195423.

Chapter 5
Synthesis of Gold Nanoparticles in Liquid Phase

Daeha Seo and Hyunjoon Song

Department of Chemistry, Korea Advanced Institute of Science and Technology, Daejeon, 305-701, Republic of Korea.
Email: hsong@kaist.ac.kr

5.1 Introduction

Nanometer-sized gold particles are of fundamental and practical interest, as their chemical, electronic, and optical properties can potentially be exploited for diverse applications.[1-5] These properties are extremely dependent upon their size and shape, therefore numerous approaches to synthetically and systematically control morphology and surface compositions have been developed in recent decades. Design of a generic method for the preparation of gold nanostructures with a broad range of well-defined and controllable morphologies is needed in order to fully exploit their unique properties for practical applications. Numerous methods have been developed to prepare gold nanostructures, including electrochemical and gas and liquid phase methods. Among these, liquid phase synthesis offers many advantages such as large-scale synthesis and ready quality control. In particular, the synthesis under well-controlled reaction conditions possibly provides uniform and monodisperse nanoparticles. The liquid phase preparation is also important in biological applications because of their compatibility in environments that living cells can exist.

In this chapter, we briefly introduce chemical properties important for liquid phase synthesis (Section 5.2), and review general strategies for the synthesis, shape control, and surface functionality modification of gold nanostructures in liquid phase. Generally, gold nanoparticles are synthesized via the reduction of gold(I) or gold(III) precursors in aqueous or organic media in the presence of surface stabilizers. However, a certain synthetic condition provides only a narrow-size range of the nanoparticles, and thus appropriate synthetic methods must be chosen in consideration of various factors including final particle size and size distribution, readiness of the reaction procedure, reaction scale, and surface functionality suitable for specific applications (Section 5.3). Shape control of gold nanoparticles is another important and timely issue, because light scattering and surface-dependent chemical activity are known to be very sensitive to the morphology of nanostructures (Section 5.4). Surface functionality is deterministic for the aim of chemical and bio-applications (Section 5.5). Synthesis of gold nanoparticles on solid supports is reviewed in upcoming chapters by chemical (Chapter 6) and physical (Chapter 8) methods.

5.2 Chemical Properties and Characterization of Gold Nanoparticles for Liquid Phase Synthesis

5.2.1 *Structure and size range of gold nanoparticles*

A representative structure of gold nanoparticles synthesized in liquid phase comprises three parts including inner gold atoms with a closed-packed crystal structure (central atoms), outer layers exposed on the surface (surface atoms), and surface protecting organic ligands or surfactants (Fig. 5.1).[6] This structure is generally referred to as a 'monolayer protected cluster'

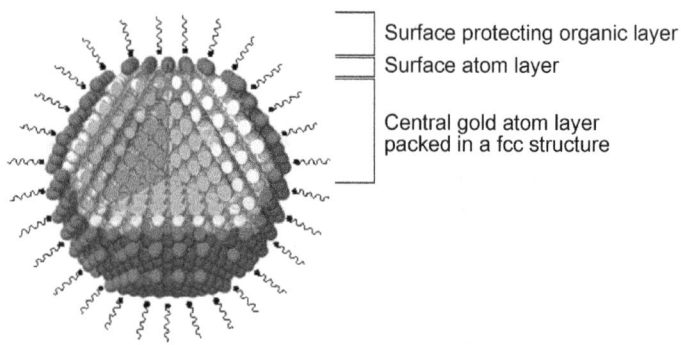

Surface protecting organic layer
Surface atom layer

Central gold atom layer
packed in a fcc structure

Fig. 5.1. A structural model of a single gold nanoparticle.

Table 5.1. Classification of gold nanoparticles by size range.

Size range	Class	Major property	Example
< 2 nm	cluster	molecular- to metal-like	Au_{11}, Au_{28}, Au_{39}, Au_{55}, Au_{102}
2–10 nm	catalytic particle	enhanced catalytic activity	tiny nanoparticles
10–300 nm	plasmonic crystal	surface plasmon resonance	polyhedrons, rods, wires, plates

or briefly MPC. The central gold atoms determine the crystallinity of the structure. The geometry of the surface atoms is different from that of the central atoms, and forms surface facets and edges that dominate catalytic activity. The surface protecting ligands are anchored on the surface atoms, stabilize them, and sometimes provide surface functionality that influences the chemical nature of the particles in solution.

Gold nanoparticles can be classified into three types according to their size and size-dependent properties: size ranges of less than 2 nm, 2–10 nm, and 10–300 nm (Table 5.1). Particles smaller than 2 nm in diameter are called "gold clusters",[7–9] which consist of a few tens to a few hundreds of gold atoms. The cluster surface is normally stabilized by organic molecules such as organothiolates, and physical properties vary from molecular-like to metal-like. Evolution of an electronic structure in gold clusters is observed in this range. Particles with several nanometer diameters are very useful in heterogeneous catalysis.[10] The surface atom fraction versus a total number of atoms (or volume) rapidly increases with reduction of the particle diameter. Surface atoms are generally more active than closed-packed atoms inside the particle, due to their outwardly exposed dangling bonds. When the particle size is less than 5 nm, surface properties exceed intrinsic features of the materials, and dominate total physical properties. For instance, the melting point of bulk gold is 1064°C, but it abruptly decreases to 380°C for 1.5 nm-sized particles due to high surface energy.[11,12] Large surface area and high density of edge and kink atoms are other factors that lead to increased chemical activity compared to bulk materials. Particles with sizes of 10–300 nm strongly scatter light, mostly in the visible range, which is attributed to localized surface plasmon resonance (LSPR).[1–4] The LSPR behaviour is extremely dependent upon the size and shape of metallic objects. Gold nanoparticles with this range of diameter are ideal to adjust LSPR extinctions from visible to near infrared (NIR) ranges (Chapter 3).

5.2.2 *Electrochemical potentials of gold precursors*

Gold(III) halides are the precursors commonly used in gold nanoparticle synthesis. The gold precursors are prepared by dissolution of bulk gold in aqua regia or metal cyanide. Under well-controlled reaction conditions, reducing agents (i.e. $NaBH_4$ by Brust *et al.*,[13] ascorbic acid by Murphy *et al.*,[14] THPC (tetrakis(hydroxymethyl)phosphonium chloride)-NaOH by Baiker *et al.*,[15] alcohols for the polyol process,[16] and citrate for Turkevich *et al.*[17] see Section 5.3.1) are introduced in the precursor solution. However, because the reduction potentials of the gold ions are higher than those of other noble metal ions,[18] and they depend on the nature of the precursors (Table 5.2), proper choice of the reducing agent is very important to synthesize regular nanoparticles with narrow shape and size distributions.

5.2.3 *Surface energy and particle morphology*

The structural total energy (E_t) using macroscopic concepts as a guide is given by

$$E_t(N) = E_B N + E_\sigma N + E_\gamma S$$

Table 5.2. Standard reduction potentials of various gold precursors at 25°C, 1 atm[18].

Reaction	E^o/V
$Au^+ + e^- \leftrightarrow Au$	1.692
$Au^{3+} + 3e^- \leftrightarrow Au$	1.498
$AuBr_2^- + e^- \leftrightarrow Au + 2Br^-$	0.959
$AuBr_4^- + 3e^- \leftrightarrow Au + 4Br^-$	0.854
$AuCl_4^- + 3e^- \leftrightarrow Au + 4Cl^-$	1.002
$Au(OH)_3 + 3H^+ + 3e^- \leftrightarrow Au + 3H_2O$	1.450

where N is the number of atoms, E_B is the bulk energy per atom, E_σ is the structural strain energy per atom, E_γ is the average surface energy per unit area, and S is the surface area of the particles.[19,20] When the particles are anchored with surface-protecting ligands, the interaction between surface atoms and protecting ligands should be added as a separate term. As the particle size decreases, the surface area decreases more slowly than the number of atoms, and the surface energy dominates the total energy. In principle, a sphere is the most thermodynamically stable structure in an isotropic system, because it has the minimum number of surface atoms per volume and the minimum surface energy.[21] Taking into consideration atomic packing in a tiny nanoparticle surface, however, low Miller index facets such as {100}, {111}, and {110} are more stable than other facets due to their high density of atomic packing, and these facets tend to form the surface of polyhedral structures. Among the low index facets, the surface energy is in the order of $\gamma\{111\} < \gamma\{100\} < \gamma\{110\}$, where the {111} facets have the largest surface density and the lowest number of dangling bonds of the surface atoms. Consequently, the particles tend to have {111} facets on the surface without any kinetic adjustment. Among three-dimensional polyhedrons, a decahedron (Fig. 5.2) is one of the most stable structures, having only {111} surfaces and a nearly spherical shape, although it has strain energy originating from twin boundaries.[12,19,22] Theoretical calculations show that decahedral and icosahedral shapes with {111} surface facets are excellent models for particles with a size of less than 5 nm. However, a single-crystalline cuboctahedron is a common structure for the larger particles, because lattice strain rapidly increases with particle size enlargement. Introduction of surface regulating agents and heterometallic species that are selectively bound to the specific facets can precisely alter the relative surface energies and the resulting particle morphology.

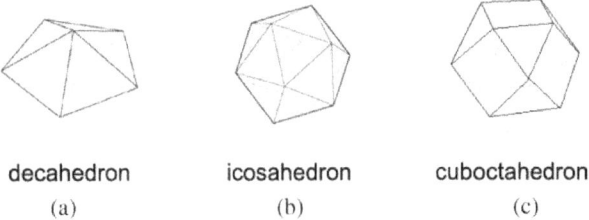

decahedron   icosahedron   cuboctahedron

(a)   (b)   (c)

Fig. 5.2.   Ideal (a) decahedral, (b) icosahedral, and (c) cubo-octahedral structures.

5.2.4 *Characterization of nanoparticles*

5.2.4.1 *Morphology characterization*

Electron microscopy including transmission electron microscopy (TEM) and scanning electron microscopy (SEM) makes use of the wave character of electrons, but is analogous to optical microscopy in that an image of an object is constructed from the scattering pattern of a focused beam of electrons. The technique of high resolution transmission electron microscopy (HRTEM) together with selected electron diffraction (SAED) and energy dispersive X-ray spectroscopy (EDX) can be utilized to analyze the crystal structure and composition of objects. X-ray diffraction (XRD) is meanwhile one of the most important techniques for the determination of structure and composition.

For surface plasmon resonance, UV-visible spectroscopy provides valuable information regarding shape, size, interparticle distance, and aggregation of nanoparticles. Dynamic light scattering (DLS) analysis determines the average hydrodynamic size and size distribution profiles of the particles in solution.

5.2.4.2 *Surface characterization*

X-ray photoelectron spectroscopy (XPS) is an essential tool to analyze surface atomic layers of nanoparticles within several nanometers from the surface. Information of atomic composition and chemical state of the surface atoms can be obtained from XPS spectra. Fourier transform infrared (FT-IR) spectroscopy is meanwhile a powerful tool to characterize functional groups of the surface protecting organic layers. Thermal analysis (TA) including thermogravimetric analysis (TGA) and differential scanning calorimetry (DSC) can be used to analyze the amount of organic residues and surface melting properties.

5.2.4.3 *Theoretical simulation*

The surface plasmon phenomenon is basically a resonance between light and conductive matter, and thus it can be comprehensively analyzed by Maxwell equations.[23] The Mie theory provides an exact solution of the equations, but is limited to analysis of the morphology of spheres and spheroids.

Recent advances in electrodynamic theory make it possible to determine numerically converged solutions for nanoparticles of arbitrary shapes with dimensions up to a few hundred nanometers. The discrete dipole approximation (DDA)[24,25] and finite difference time domain (FDTD)[26,27] methods are generally used for analyzing absorption and scattering and bands of gold nanoparticles.

5.3 Synthetic Methods of Gold Nanoparticles in Liquid Phase

Known since ancient times, the synthesis of colloidal gold was originally used as a method of staining glass (see Chapter 1). According to the demands of given applications such as chemo- and biosensors, catalysts, and electric circuits, numerous synthetic processes including the use of chemical reductants in both aqueous and non-aqueous solvents, as well as photochemical, radiolytic, electrochemical, sonochemical, and microwave assisted methods, have been developed.

5.3.1 *Kinetic consideration for highly monodisperse nanoparticles*

The synthesis of nanoparticles with a narrow size distribution has long been of scientific and technological interest since Faraday's production of gold sols in 1857.[28] For the preparation of monodisperse nanoparticles, the nucleation step must be separated from the growth step in order to prevent simultaneous secondary nucleation and growth. LaMer *et al.* proposed the concept of "burst nucleation", where many nuclei were simultaneously generated, and applied their concept to synthesize a series of monodisperse nanoparticles through a temporally discrete nucleation event followed by controlled growth on the existing nuclei.[29] It is necessary to induce a single nucleation event in the nucleation step and to prevent additional nucleation during the growth step to prepare highly uniform nanoparticles.[30]

There are two major techniques to separate nucleation and growth. The seed-mediated growth is utilizing preformed nanocrystal seeds followed by slow growth on their surface. The precursor concentration is kept at a low level during the growth to prevent homogeneous nucleation. This method is also useful to generate core-shell nanocrystals by introducing hetero-components[31] or to generate anisotropic structures such as nanorods by adding surface regulating surfactants (see Section 5.4.1). The second is so-called "hot-injection", which was originally introduced in the synthesis of semiconductor nanoparticles.[32] This method is rapid injection of a high concentration of the precursor solution into a hot surfactant solution, leading to fast and concomitant formation of nuclei. The nucleation rate rapidly decreases just after the nucleation process, and the growth occurs on the seed surface.[33]

During the growth step, the particle size distribution is generally diminished by "size focusing" through mass-transport processes.[34] Talapin *et al.* simulated two related mechanisms for the control of size distributions.[35] If no additional nucleation occurs and the precursor concentration is relatively high, the particle growth rate is inversely proportional to its radius, and thus the particle size distribution always decreases as long as all particles are continuously growing. On the other hand, under the low precursor concentration, Ostwald ripening occurs.[36] In this process, small crystallites have higher surface free energy, and thus are less stable against the dissolution in solvent than the large crystallites. This stability gradient leads to slow diffusion of the materials from the small to large crystal surfaces, and the relative standard deviation approaches to a constant in the equilibrium state between the dissolution and reprecipitation.

5.3.2 *Chemical reduction of gold precursors*

Reduction of gold salts is a simple process that only requires mixing of the reagents under well-controlled external conditions. These conditions can affect the final morphology of the particles in subtle ways. According to the synthetic route used, different characteristics of final products are obtained (Table 5.3). Besides the strength of the reductant, the action of a stabilizer is critical in liquid phase synthesis. The reaction temperature is one of the main factors to determine particle size, because the oxidation potential and related kinetics of the reductant are normally dependent upon the temperature. Representative examples are: 1) organic phase synthesis involving a two-phase process, with 2–10 nm range (Brust method, Fig. 5.3); 2) the single-phase water based reduction of a gold salt by citrate,

Table 5.3. Various well-known synthetic methods of gold nanoparticles in liquid phase.

Reduction method	Reaction media	Reductant	Surface protecting agent	Particle size range (nm)	Reaction temperature (°C)	Ref.
Brust–Schiffrin method	organic	NaBH$_4$	organothiol	2–10	R.T.	13, 45–52
Turkevich method	aqueous	citrate	citrate	10–20	100	17, 38–42
Murphy method	aqueous	ascorbic acid	CTAB	10–50	R.T.	14
Perrault method	aqueous	hydroquinone	citrate	50–200	R.T.	43
Polyol process	alcohol	diols	PVP	20–200	20–300	37, 64–67

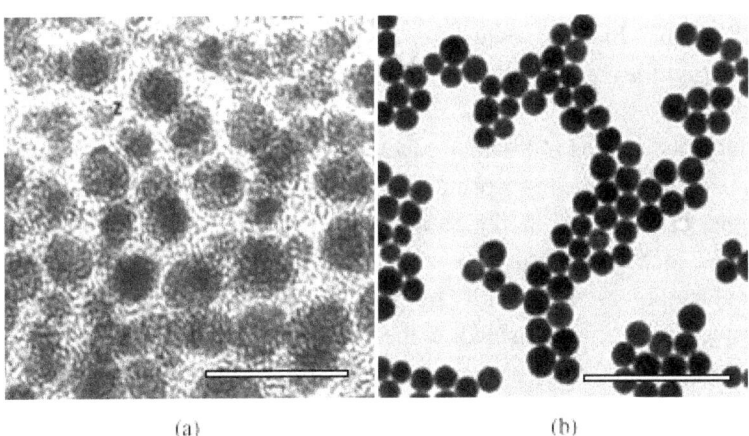

(a)   (b)

Fig. 5.3. TEM images of gold nanoparticles prepared through (a) Brust–Schiffrin method, (b) polyol process. Scale bar: (a) 10 nm and (b) 500 nm. (a) Modified with permission from Ref. 13, Copyright 1995 Royal Society of Chemistry, (b) Ref. 37, Copyright 2006 American Chemical Society.

which produces almost spherical particles over a tunable range of 10–20 nm diameter (Turkevich method, Fig. 5.4a,b); and 3) alcohol phase synthesis over a tunable range of 20–300 nm diameter (polyol process, Fig. 5.3).

5.3.2.1 *Chemical reduction in aqueous media*

Baiker method is the most pertinent one to prepare gold clusters with smaller than 2 nm in aqueous solution.[15] In this method, the gold clusters were synthesized through the reduction of gold(III) ions by tetrakis (hydroxymethyl) phosphonium chloride (THPC).

For producing monodisperse gold nanoparticles of 10–20 nm diameter in an aqueous solution, the citrate-reduction method of $HAuCl_4$ has been widely adopted and improved. This method was pioneered by Turkevich *et al.* in 1951[17] and refined by Frens in the 1970s.[38,39] The reaction kinetics have been addressed in a study from Chow and Zukoski focusing on the stabilization mechanism.[40] In this reaction, sodium citrate behaves both as a reducing agent and as a capping agent that stabilizes the nanoparticles. Generally, this method is useful to produce modestly monodisperse spherical gold nanoparticles suspended in water in diameters of 10–20 nm with a relatively narrow size distribution (Fig. 5.4a,b).[41] Reduction of the sodium citrate concentration diminishes the citrate ion concentration available for particle stabilization, which causes aggregation of small particles into larger ones. During the synthesis, networks of gold nanowires are formed as a transient intermediate.[42]

Other reductants such as inorganic reductants and organic amines have also been successfully utilized. Ascorbic acid is one of the most widely used reducing agents. In this reaction, the gold ion forms a complex with a surfactant molecule, such as cetyltrimethylammonium bromide (CTAB), but is not directly reduced by ascorbic acid. The addition of gold seeds or a small amount of a strong reducing agent leads to the growth of gold nanostructures with different morphologies at room temperature (Fig. 5.4c–f).[14]

For the realization of monodisperse gold nanoparticles of 50–200 nm in diameter in an aqueous solution, Perrault and Chan used hydroquinone to reduce $HAuCl_4$ in an aqueous solution that contains gold nanoparticle seeds (Fig. 5.4g–j).[43] This process is similar to that used in photographic film development, wherein silver grains within the film grow through the

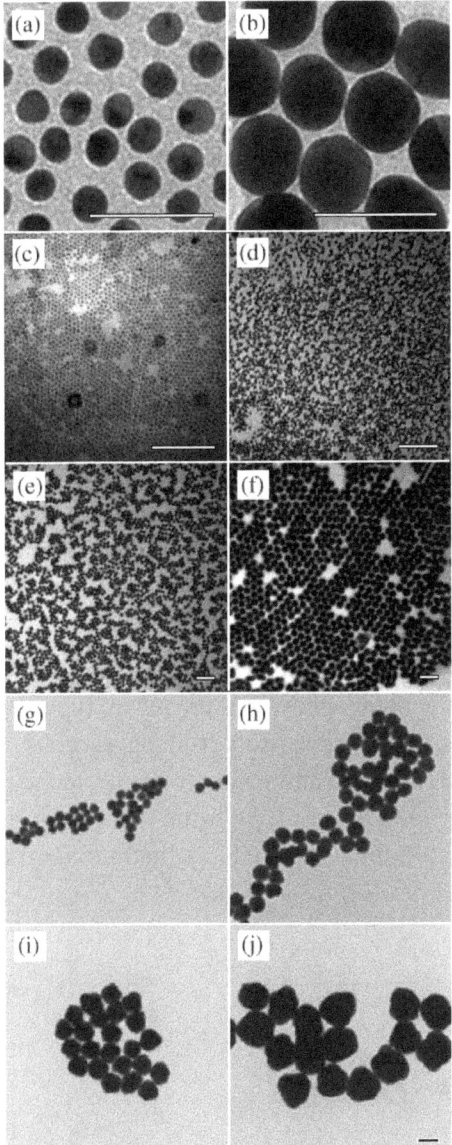

Fig. 5.4. TEM images of colloidal gold particles prepared from a: b) Turkevich method; c–f) Murphy method; and g–j) Perrault method. Average diameter of each particle is : a) 20 nm; b) 40 nm; c) 5.5 nm; d) 8.0 nm; e) 17 nm; f) 37 nm; g) 50.0 nm; h) 74.8 nm; i) 95.0 nm; and j) 175.7 nm. a, b) Modified with permission from Ref. 41, Copyright 2010 American Chemical Society, c–f) Ref. 14, Copyright 2001 American Chemical Society, g–j) Ref. 43, Copyright 2009 American Chemical Society.

addition of reduced silver onto their surface. Similarly, gold nanoparticles can catalyze the reduction of gold ions onto their surface. The presence of a stabilizer such as citrate results in controlled particle growth. Typically, the gold seeds are produced via citrate reduction. The hydroquinone reduction method extends the size range of monodispersed spherical particles. This method can produce particles of at least 50–200 nm, whereas the Frens method is ideal for obtaining particles of 10–20 nm diameter.

5.3.2.2 *Chemical reduction in organic media*

Stabilization of gold nanoparticles with organothiol ligands was first reported by Mulvaney and Giersig in 1993.[44] The Brust–Schiffrin method, for the synthesis of small gold clusters and nanoparticles ranging in diameter between 1.5 and 5.2 nm,[13] has since had a considerable impact on the overall field, as it offers facile synthesis of thermally and air stable gold nanoparticles of reduced dispersity and controlled size (Figure 5.5).[45–50] Indeed, these monolayer protected clusters (MPCs) can be repeatedly isolated and redispersed in common organic solvents without irreversible aggregation or decomposition, and they can be easily handled and functionalized, similar to stable organic and molecular compounds. This procedure typically involves three steps: phase transfer of gold precursors from an aqueous solution to the organic phase using a long chain ammonium salt (e.g. tetraoctylammonium bromide (TOAB)); reduction of Au(III) (from $AuCl_4^-$) to Au(I) by thiol through the formation of gold-thiol intermediates; and further reduction of

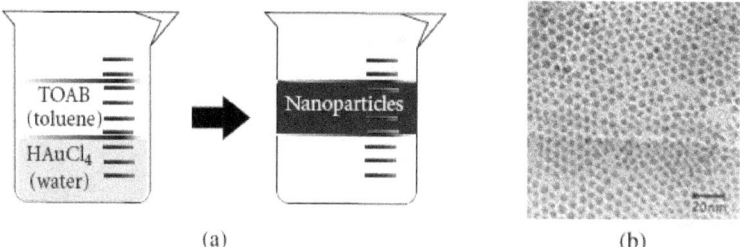

(a)   (b)

Fig. 5.5. (a) Schematic illustration of Brust–Schiffrin method for gold nanoparticles, and (b) TEM image of resulting gold nanoparticles. Modified with permission from Ref. 49, Copyright 1997 American Chemical Society.

Au(I) to Au(0) by a second reductant (e.g. $NaBH_4$). Recently, the reduction kinetics[51] and identification of the precursors[52] have been studied in detail, and an improved synthetic scheme has been developed.[16,53]

Synthesis of Au_{55} clusters is also categorized in this method. The Au_{55} clusters gain special interest because of their ultimate size of 1.4 nm and ideal cubo-octahedral structure. $Au_{55}(PPh_3)_{12}Cl_6$ was successfully synthesized through the reduction of Ph_3PAuCl in benzene or toluene by gaseous borane (B_2H_6) as a reducing agent instead of $NaBH_4$,[54] because B_2H_6 not only directly reduced Au(I) to Au(0), but also bound excess PPh_3 to form Ph_3P-BH_3, and removed them from the clusters. The resulting Au_{55} clusters can replace their surface ligands and generate either hydrophilic (with borate and monosulfonated phosphine) or hydrophobic (with silsesquioxane and aryl phosphine) characters.

5.3.2.3 *Synthesis in micelles*

The syntheses involve a two-phase system with a surfactant that causes the micelle formation maintaining a favorable nano- and micro-environment, together with the extraction of metal ions from the aqueous phase to the organic phase.[50] A micelle is an aggregate of surfactant molecules dispersed in a liquid, which have hydrophobic and hydrophilic domains. Although this method uses a similar two phase system, it is distinct from the Brust–Schiffrin method because the nanoparticle formation actually occurs in the micellar structure. Synthetic copolymers are reported to form effective micelle structures for gold nanoparticle synthesis, due to their stabilization of the gold surface and surface functionality control.[55–57] The size and shape of the micelles could be readily tuned by variation of composition, molecular architecture, and constituent block length of the copolymers. Various copolymers including two different polymer units, such as diblock,[58,59] graft,[60] and star-shaped copolymers,[61,62] can form nanometer-sized micelles consisting of a soluble corona and an insoluble core in a solvent selective for one of the blocks, and these micelles were successfully employed as synthetic templates for gold nanoparticles with the size range of 0.5–50 nm. The micelle method is useful for the formation of particles in planar surfaces (Chapter 6).

5.3.2.4 *Polyol process*

The polyol process originally entailed the reduction of metal salts to prepare metal and metal oxide particles.[63] It is a convenient, versatile, and low-cost method for the synthesis of metal nanostructures in a large scale. A metal compound or salt is dispersed in a long-chain diol such as ethylene glycol that acts both as a solvent and as a reductant. After the reduction, the resulting particles are monodisperse and non-agglomerated in a relatively large size range, from several nano- to micrometers. The products are not spherical, but rather polyhedrons or plates with stable surface facets. PVP (poly(vinyl pyrrolidone)) is a representative surface regulating reagent that kinetically controls the growth rates of various facets for the formation of anisotropic gold nanostructures, on the basis of its excellent adsorption ability. Yang *et al.* synthesized so-called "platonic nanocrystals", isotropic gold polyhedrons with sizes of 100–300 nm, by a polyol process.[64] Importantly, these highly symmetric structures can provide fundamental insight into the origin of symmetry and the formation mechanism of nanoscale materials. Song *et al.* employed the polyol process to synthesize gold polyhedrons, nanorods, and nanowires under various reaction conditions (see Section 5.4.3).[37,65–67]

5.3.3 *Nonchemical reduction for preparation of gold nanoparticles*

Although chemical reduction methods with various reductants are simple and convenient, the reaction conditions must be carefully adjusted for particular applications. For example, an excess amount of reducing agents, their oxidation products, and the surfactants used for chemical reduction may contaminate the final nanoparticles. So, in this section, we introduce alternative synthetic processes, including electrochemical, photochemical, sonochemical, and microwave-assisted methods, that do not require the use of chemical reductants and thermal heating. As in Table 5.4, reduction of the gold precursors is assisted by various energy sources, with or without surface stabilizing reagents.

5.3.3.1 *Photochemical and radiolytic methods*

Reduction of gold ions by photochemical and radiolytic methods has the advantage of not using excess reducing agents, which may contaminate the

Table 5.4. Various reduction methods for gold nanoparticle synthesis.

Preparation method	Energy source	Reduction mechanism
Chemical	reductant (sometimes heating)	electrochemical potential difference
Photochemical	Ultraviolet	reductive elimination of ligands from the metal precursors
Electrochemical	Electron	direct reduction by applying a negative potential
Sonochemical	Ultrasonic wave	radicals generated from solvents or other reagents
Microwave-assisted	Microwave	organic reducing agents heated by microwave irradiation

nanoparticles. For this reason, these methods are widely applied to catalytic applications due to their simple and one-step procedures as well as cleanness of nanoparticle surface. Ultraviolet (UV) is a commonly used light source to excite the gold precursor in the presence of surfactants. Irradiation of UV light led to the effective photoreduction of $AuCl_4^-$ in formamide through the dissociation and direct reduction by the photogenerated free radicals from the formamide molecules, and yielded uniform gold nanoparticles of 12 nm at room temperature in the presence of polymer surfactants.[68] The use of near infrared laser irradiation is also reported to prepare thiol-stabilized gold nanoparticles.[69] In the radiolytic method, γ-irradiation generates hydrated electrons or active radicals, which reduce the gold precursors. In order to prevent particle aggregation, surface protecting reagents or surfactants such as polymers and dendrimers are generally added to the particle dispersion (see Section 5.5.2). In the presence of poly(vinyl alcohol) (PVA) or PVP, γ-irradiation of aqueous solutions of $KAuCl_4$ yielded tiny gold nanoparticles with the size between 1.5 to 2.5 nm.[70] The hydrated electrons generated in the aqueous solvent reduced Au(III) to form Au(I), and after a certain period, Au(I) ions accumulated in solution were further reduced to generate Au(0). In the mixture with methanol, the hydroxyl radicals from the radiolysis of water rapidly reacted with methanol, and the resulting hydroxymethyl radicals reduced additional Au(III).[71] Large gold nanoparticles with more than 20 nm were formed by the radiolysis of more concentrated PVA-AuCl$_4^-$ solutions.[72] Controlled enlargement of particle size is carried out by γ-irradiation of the aqueous solutions containing KAu(CN)$_2$ and methanol with gold colloids, and the stepwise radiation growth can grow gold nanoparticles to any desired size up to 120 nm.[73]

5.3.3.2 *Electrochemical methods*

The gold precursors are directly reduced via an electrochemical method without the use of any reducing agents. In the electrochemical method, current flow between two active electrodes by applying a potential can drive oxidation and reduction processes of the materials suspended in solution, generally at room temperature. The particle size is controlled by adjustment of current density. Wang *et al.* demonstrated size-selective synthesis of gold nanorods stabilized by tetraalkylammonium halides such as CTAB by the electrochemical method.[74,75] In this reaction, a bulk gold plate is oxidized from the anode to form gold nanorods most probably either at the interfacial region of platinum cathodic surface or in electrolytic solution.

Anisotropic gold nanorods with a controllable aspect ratio are also yielded via electrodeposition in hard templates such as nanoporous aluminium oxide (see Section 5.4.5). Zande *et al.* prepared monodisperse dispersion of gold nanorods in water by electrodeposition. The length of the gold rods could be easily varied by changing deposition time, while the thickness was determined by the pore diameter of nanoporous template.[76,77] In order to disperse the gold nanorods, the templates and substrates were selectively dissolved, and PVP was added as a stabilizer.

5.3.3.3 *Sonochemical method*

Sonolysis entails the application of ultrasound to the precursor solution. The gold ions are reduced to form colloidal nanoparticles in a standing wave system generated by an ultrasonic generator. In the presence of organic compounds such as alcohol, ultrasound irradiation generates organic radical species, which reduce the gold precursors. The size of the resulting nanoparticles depends on the frequency used and the Au(III) reduction rate.[78–80] Grieser *et al.* controlled gold nanoparticle sizes from 10 to 30 nm via sonochemistry in the presence of 1-propanol.[81] It was found that the rate of $AuCl_4^-$ reduction induced by cavitation was the highest at 213 kHz in the range of 20 to 1062 kHz. The size of the gold nanoparticles produced was closely related to the sonochemical reduction rate of $AuCl_4^-$. Su *et al.* prepared gold nanoparticles by ultrasound irradiation without alcohols and stabilizers.[82] Instead, they found that the citrate dosages gave a drastic effect on the formation of the well-separated, monodisperse gold particles with the diameters of around 20 nm.

5.3.3.4 *Microwave assisted methods*

Microwave dielectric heating is useful for effective and uniform heating, and thus is applied for rapid synthesis of gold nanostructures.[83] Microwaves are in the electromagnetic spectrum with a frequency range of 300 MHz to 300 GHz. In these wavelengths, polar solvents such as water, alcohols, dimethylformamide (DMF), and ionic liquids[84] continuously orientate under an alternating electric field, and lose energy in the form of heat by molecular friction. The microwave heating is rapid and homogeneous, and can even provide superheating beyond the boiling point of solvent. Therefore, it provides uniform nucleation and growth conditions, shorter reaction times, and reduced energy consumption for gold nanoparticle synthesis. Many studies have explored the synthesis of gold nanoparticles by microwave assisted methods, and it is known that the reaction time, temperature, and ramping rate are important factors in enhancement of particle monodispersity. For example, monodisperse gold nanoparticles with 20 nm in diameter could be synthesized within 15 min at 125°C, with a faster ramping rate (>20°C/min) in the presence of citrate.[85] Tsuji *et al.* employed microwave heating to the polyol process conditions for the synthesis of gold nanopolygons, where $AuCl_4^-$ was reduced in ethylene glycol in the presence of PVP.[86]

5.4 Shape Control of Gold Nanoparticles

Spherical gold nanoparticles are commonly employed as electronic reservoirs or supporters for chemical and bio-functionalities. Control over the morphology of gold nanoparticles in terms of size, shape, and surface structures, however, can enhance catalytic, electronic, and optical properties of the nanoparticles such that can be used in a broader range of applications. For example, the presence of sharp edges and tips significantly promotes electric field enhancement, which is useful for optical sensing of chemical and biomolecules.[87–90] Nanoparticle morphology also determines three-dimensionally assembled structures,[91,92] which are very important for real-device applications. Many researchers have successfully developed rational synthetic methods to control gold nanostructures, yielding, for example, polyhedrons, rods, plates, and branched structures. Among a variety of nanostructures, one-dimensional nanostructures such as rods and wires have been intensively studied owing to their unique applications in mesoscopic physics and for the fabrication of nanoscale devices.[93] In particular, by adjusting the aspect ratio of gold nanorods, their optical properties can be tuned in a wide range from visible to NIR. In order to realize anisotropic structures such as nanorods, a rational strategy of symmetry breaking must be developed, because gold has an isotropic crystallographic face-centered cubic (fcc) structure.

5.4.1 *Shaping strategies with seed-mediated growth*

In general, the growth mechanism of metal nanoparticles involves nucleation and growth steps.[94] Seed-mediated growth, wherein tiny seeds or clusters serve as nucleation centers for surface growth, can effectively separate both steps and successfully control the dispersity of the nanoparticles (see Section 5.3.1). Another advantage of seed-mediated growth is that the original seeds can influence the growth direction. For instance, Murphy *et al.* synthesized nanorods by reduction of the gold precursors with NaBH$_4$ in the presence of 4 nm-sized gold seeds. In this reaction, a decahedral shape of gold seeds led to anisotropic growth of gold nanorods by controlling the growth conditions (see Section 5.4.2).[95,96] Huang *et al.* applied this seed-mediated growth to obtain diverse morphology of gold nanorods by changing growth solutions and pH.[97]

5.4.2 *Selective binding of capping reagents*

As noted earlier, Murphy *et al.* synthesized nanorods with a tunable aspect ratio using gold seeds through optimization of the CTAB and ascorbic acid concentrations and by applying a step seeding process.[95,96] Mann *et al.* confirmed that the electron diffraction patterns of the resulting nanorods were consistent with a pentagonally twinned prism with five {100} side faces capped with five {111} faces at both ends, with growth again being observed to be along the [110] direction (Fig. 5.6b).[98] The formation mechanism of one-dimensional rods was proposed to entail preferential binding of the CTAB head group to {110} and {100} faces of gold existing along the sides of twinned rods, as compared to {111} faces at the tips. This is based on the finding that CTAB absorbs onto gold nanorods in a bilayer fashion with the trimethylammonium headgroups of the first monolayer facing the gold surface (Fig. 5.6).

Recently, numerous gold nanostructures, such as rods,[97,99] platonic crystals,[97] triangles,[100] branched structures[101–103] and high index surface structures,[104] have been synthesized using CTAB-based synthetic methods,

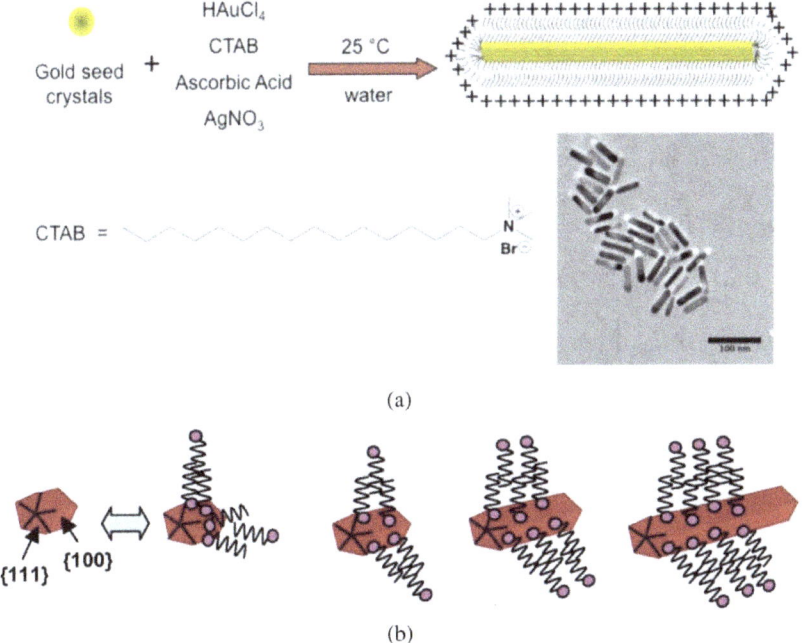

(a)

(b)

Fig. 5.6. (a) Summary of silver-assisted gold nanorod synthesis, showing the CTAB bilayer. A representative TEM image of aspect ratio 4.6 gold nanorods is shown. (b) Proposed mechanism of surfactant-directed metal nanorod growth. (a) Modified with permission from Ref. 95, Copyright 2010 American Chemical Society, (b) Ref. 96, Copyright 2005 American Chemical Society.

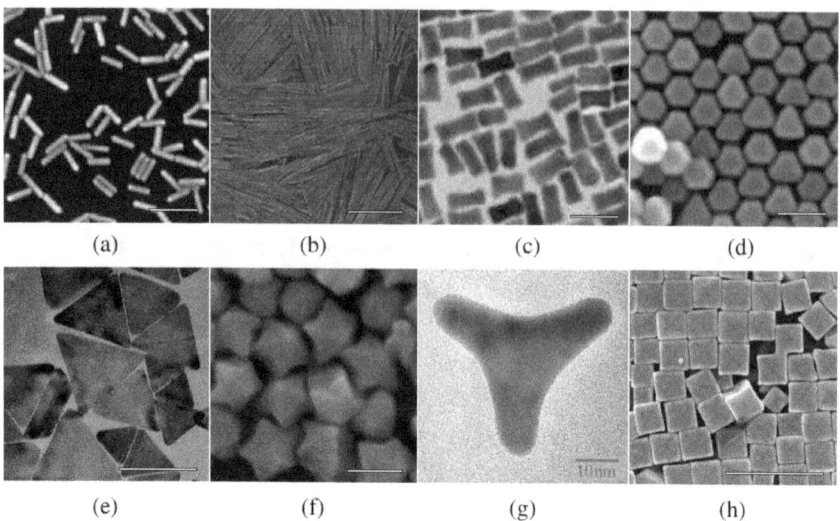

Fig. 5.7. (a, b, d, f, h) SEM, and (c, e, g) TEM images of gold nanostructures of various kinds of morphology synthesized with CTAB as a surfactant in aqueous solutions. The bars represent: (a, c, d, f) 100 nm; (b, h) 500 nm; and (e) 166 nm. (a, d) Modified with permission from Ref. 97, Copyright 2009 American Chemical Society; (b) Ref. 99, Copyright 2007 American Chemical Society; (c) Ref. 98, Copyright 2004 American Chemical Society; (e) Ref. 100, Copyright 2005 American Chemical Society; (f) Ref. 101, Copyright 2009 American Chemical Society; (g) Ref. 102, Copyright 2003 American Chemical Society; (h) Ref. 104, Copyright 2010 American Chemical Society.

where the morphology and dimensions of the Au nanocrystals can be controllably varied by manipulation of the synthetic parameters (Fig. 5.7).

5.4.3 *Underpotential deposition of heterometallic additives*

The presence of silver ions allows better control of the shape of gold nanorods synthesized through an electrochemical method. Murphy *et al.*[105] also reported on the effects of silver on the preparation of gold nanorods and spheroids. It was found that the addition of silver nitrate influences not only the yield and aspect ratio control of the gold nanorods, but also the mechanism of gold nanorod formation and the crystal structure. It has been hypothesized that silver adsorbs at the surface of gold, and the silver deposition takes place in an epitaxial fashion due to close matching of the lattice parameters between gold and silver.[106–109]

Yang *et al.* and Song *et al.* have prepared isotropic cubic, octahedral, truncated tetrahedral, and icosahedral nanostructures by polyol processes in the presence of PVP and silver nitrate.[37,64−66] These highly symmetric structures are very important, because they can provide fundamental insight into the origin of symmetry and the formation mechanism in nanoscale materials. In these reactions, a small amount of silver species ($\sim$1/100 molar ratio than the gold precursor) is introduced as an additive with gold precursor and PVP. During the reaction, the silver species are deposited on a specific surface of the gold nanoparticles through underpotential deposition (UPD).[67] UPD is a well-known phenomenon in electrochemistry, where metal overlayers are electrodeposited onto a foreign metal substrate at a potential that is less negative than the Nernst equilibrium potential in atomic layer scale deposition.[107,110] The underpotential shift of the Au/Ag$^+$ system is particularly facilitated due to a large work function gap (>0.5 eV) between silver and gold.[111] The underpotential shift is very sensitive to the distinct surface structures, and theoretical calculations on the Au/Ag$^+$ system demonstrate selective island growth of Ag on Au(100) but not on Au(111) at intermediate deposition rates.[112] This selective deposition of the silver species on the gold {100} surface appears to suppress {100} surface growth and/or enhance {111} growth. Exploiting this phenomenon, subsequent growth is precisely controlled to prepare polyhedral structures[65] and pentagonal nanorods[67] (Fig. 5.8).

5.4.4 *Galvanic replacement*

The Galvanic replacement reaction is a general and effective method for preparing metallic nanostructures by consuming the more reactive component without additional reducing agents. Since the standard reduction potential of the gold precursor is higher than those of other metals such as silver and cobalt (Table 5.2), the gold precursors are reduced to Au(0) while the other metals are oxidized. When the Au(III) precursors galvanically substitute Ag(0) nanostructures, vacancies are generated inside the structures to form either hollow or shell morphologies. Using this strategy, Xia *et al.* synthesized gold nanoboxes and nanotubes via Galvanic replacement from silver nanocubes and nanowires in an aqueous phase.[113,114] The exterior shapes of the final structures tend to reproduce those of the sacrificial silver counterparts without any change to their overall morphology. Alivisatos *et al.* also demonstrated the same chemical transformation to

Fig. 5.8. Underpotential deposition of heterometallic additives: (a) Shape evolution of cubo-octahedral seeds by changing the growth solutions. SEM images of large octahedrons grown by {100} growth (left), cubo-octahedral seeds (middle), and cubes grown by {111} selective growth. (b) Anisotropic growth of gold nanocrystal from gold decahedron. SEM images of pentagonal nanorods from decahedral seeds by the {111} selective growth method. The bar represents 500 nm for all images. (a) Modified with permission from Ref. 65, Copyright 2008 Wiley-VCH, (b) Ref. 67, Copyright 2009 American Chemical Society.

124

prepare hollow structures of gold from silver particles of $\sim$10 nm size in an organic phase.[115] Partial Galvanic replacement reactions were induced by careful control of the reaction kinetics, and provided anisotropic single hollows as well as symmetric double hollows from silver-gold-silver heterometallic nanorods.[116]

5.4.5 *Template-directed synthesis*

Template-directed synthesis represents a straightforward route to generate 1-D structures. In this strategy, templates present on the surface of a solid substrate and channels in a porous membrane serve as a scaffold within which different materials are generated in situ. The resulting nanostructure has morphology complementary to that of the original template.[117] Many kinds of methods use template-directed preparation of the gold nanostructures. One example in the area of liquid phase preparation is electrodeposition onto step edges on the surface of a solid substrate such as highly oriented pyrolytic graphite.[118] The nanowires were found to preferentially nucleate and grow along the step edges present on a graphite surface into a two-dimensional parallel array that could be transferred onto another surface.

Channels in a porous membrane also provide one-dimensional nanostructures, an approach pioneered by Martin and other researchers.[119] Two types of porous membranes are commonly used, i.e. ion-track-etched membranes and anodic aluminium oxide (AAO) templates, because both are commercially available. Using these channels, gold ions are reduced from electrolytic solutions through the application of a negative potential, typically in a three-electrode electrochemical cell.

5.5 Surface Functionality of Gold Nanoparticles

Surface functionality plays essential and multiple roles in many aspects of nanostructures. Surface organic layers of gold nanoparticles largely influence the properties of nanostructures including solubility, stability, melting point, and electronic structure. During the synthesis of gold nanoparticles, stabilizing agents are employed to prevent nanoparticle aggregation and precipitation. Surface molecules can also behave as linkers, which attach other functional materials. The conjugation of gold nanoparticles with other molecules can be achieved by direct covalent linkage or non-covalent interactions. The most direct covalent approach is using strong Au-S bonds with organothiols, disulfides, and cysteine groups. Non-covalent interaction approaches use physisorption and electrostatic interactions of surface-ionized ligands.

5.5.1 *Monomeric thiol and amine molecules*

Although various organic materials have been utilized as capping materials on the surface of nanoparticles through covalent bonding, ionic interaction, and simple adsorption, surface modification using thiol molecules on the gold surface is a well-established route. In the stabilization of gold nanoparticles by alkanethiols, the thiol molecules strongly bind to gold through a covalent bond due to the soft character of gold and sulfur, respectively.[44] Furthermore, they do not undergo any unusual reactions, and they can readily displace adventitious materials from the surface.[13,45] To incorporate functionalities on the surface, end-group functionalized (ω-functionalized) organothiols are utilized or mixed with pre-synthesized gold nanoparticles and other surfactants.[120,121] Figure 5.9 shows the incorporation of bio-functionalities such as oligonucleotides and antibodies on the surface of gold nanostructures through thiol-gold linkages. Another method to introduce surface functionality using thiol molecules is the variation of end groups through well-established substitutions by ligand exchange (see Section 5.5.3).

Other functional ligands such as amines and phosphines also form stable gold colloids, although their binding forces with gold surface are rather weak and the ligands are readily exchanged by thiols. Dodecylamine and oleylamine can behave as surface passivants, and the resulting amine-passivated gold nanoparticles are stable in organic solvents including toluene, hexane, and THF.[124] Reduction of the gold precursors by $NaBH_4$ in a mixture of tri-n-octylphosphine oxide (TOPO) and octadecylamine at $190^\circ C$ yielded uniform gold nanoparticles with the diameter of 8.6 nm, which was stable in toluene for month.[125] The Au_{55} cluster was stabilized by triphenyl phosphine, which could be substituted by various ligands (see Section 5.3.1.2).

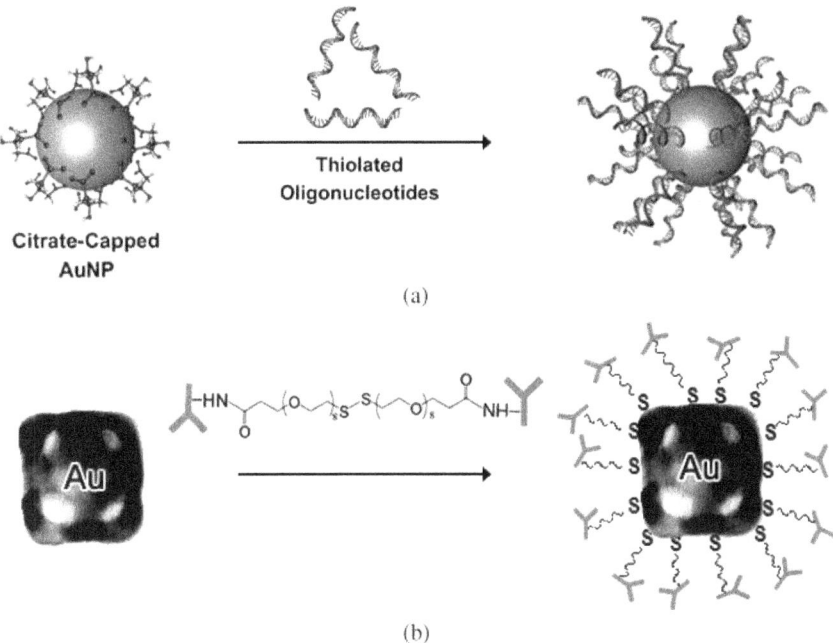

Fig. 5.9. Surface functionalization using thiol linkage, (a) oligonucleotide gold nanoconjugate using alkanethiol-terminated oligonucleotides from citrate-stabilized gold nanoparticles, (b) antibodies conjugate using disulfide molecules (succinimidyl propionyl poly(ethylene glycol) disulfide) to gold nanocages. (a) Modified with permission from Ref. 122, Copyright 2010 Wiley-VCH, (b) Ref. 123, Copyright 2005 American Chemical Society.

5.5.2 *Surface regulating polymers*

For biological applications (see also Chapters 10 and 11), gold nanostructures must be stable and thoroughly dispersed in aqueous media. Gold conjugates functionalized with CTAB, quaternary amine or citrate, have been shown to be toxic.[126-131] Therefore, thiol groups containing carboxylate[132,133] or poly(ethylene oxide) units[126-131] are commonly incorporated in water-soluble, biocompatible gold nanoparticles. Furthermore, these gold-thiol linked conjugates can be used for molecular delivery, owing to cleavable gold-sulfur bonds by light or reductive cyotoplasm.[134,135] However, relatively poor long-term stability of gold-sulfur bonds limits their biomedical and catalytic applications. Alternative ligands of choice are polymers with polar groups such as poly(vinyl pyrrolidone) (PVP),[37,64-67,136] poly(vinyl alcohol) (PVA),[137–140] poly(acrylic acid) (PAA),[141–144] and their copolymers.

In the polyol process, a random copolymer containing a vinylpyrrolidone unit can be utilized for the synthesis of surface functionalized nanomaterials. Song *et al.* have used a copolymer, poly(vinyl pyrrolidone-*ran*-vinyl acetate) (PVP-PVAc), which comprises both vinyl pyrrolidone for gold nanoparticle formation and vinyl acetate for functionalization, and have successfully synthesized gold nanoparticles through a one-step reaction of the gold precursors. The resulting nanoparticles have multiple hydroxyl groups on their surface, derived by acid hydrolysis from the acetate group of PVP-PVAc during the reaction.[145]

Block copolymers have ordered structures in solution with periodic thicknesses between 10 and 100 nm. These structures can provide nanosized domains as nanoreactors, which can be used for gold nanoparticle synthesis. Diblock copolymers such as poly(styrene)-*b*-poly(vinylpyridine) (PS-*b*-P4VP)[146–148] and poly(styrene)-*b*-poly(acrylic acid) (PS-*b*-PAA)[149,150] have been used to synthesize gold nanoparticles and assemble them into polar domains.[151]

Dendrimers, well-defined polymer molecules that are known to act as hosts for guest molecules, are another potential template for the formation of gold nanoparticles.[152,153] Basically, one dendrimer molecule can entrap one or more gold nanoparticles. Poly(amidoamine) (PAMAM) dendrimers are commonly used polymers for incorporating gold nanoparticles, because of their monomodal and well-defined size and shape.[154,155] Esumi *et al.* reported the reduction of gold ions by a chemical or photochemical reduction method in the presence of generation 4 (G4) PAMAM dendrimers (Fig. 5.10). PAMAM-conjugated gold nanoparticles are functionalized with

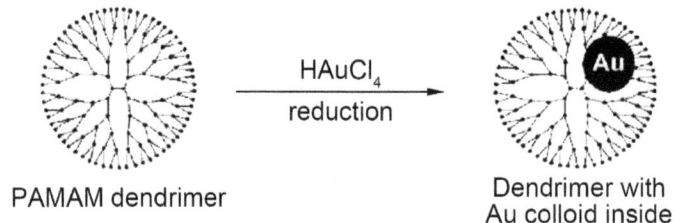

Fig. 5.10.   Schematic representation of dendrimer nanotemplating in an aqueous solution. Modified with permission from Ref. 153, Copyright 2000 American Chemical Society.

hydrophobic groups for solubilization in organic media, and have been employed as pH sensors and in cell imaging.[156–158]

5.5.3 *Secondary modification*

The "place exchange" method in the monomeric thiol system is a well-established route to incorporate various surface functionalities on nanoparticles. Murray *et al.* demonstrated the place exchange of a controlled proportion of thiol ligands by various functional thiols at the end.[159–161] Recently, other thiol-ended ligands containing organic/inorganic dyes,[118] smart polymers,[119] bio-molecules,[134,122,123] drug molecules,[162,163] and other nanoparticles have been used to prepare hybrid gold nanoparticles.[164]

Secondary modification is also commonly employed when the surface functionality on gold nanoparticles has activated groups such as carboxylates and hydroxyls (Fig. 5.11). The secondary reaction on the surface is

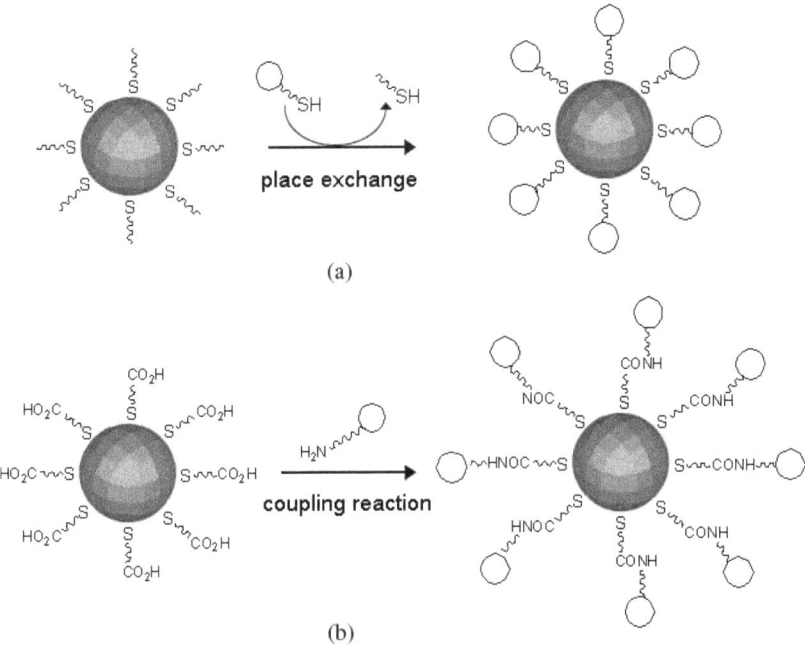

Fig. 5.11. Schematic representation of surface modification method through (a) place exchange in organothiol system, and (b) secondary modification of ligand end groups.

accomplished using chemical coupling,[161,165,166] polymerization,[167] electrostatic interaction,[168-171] and selective interaction between biomolecules.[122,124,134,172] The most well-established methods are coupling and esterification. The gold nanoparticles functionalized with carboxylic acid-terminated thiol ligands readily form amide linkages with other molecules through EDC (1-ethyl-3-(3-dimethylaminopropyl) carbodiimide) coupling. The hydroxyl group of the surface functionality can directly react with acyl chloride to generate various functionalities through esterification.[145]

5.5.4 *Biological materials*

Conjugates of gold nanoparticles with biological materials such as oligonucleotides are currently generating interest due to the potential use of the programmability of DNA base pairing to organize nanocrystals in space and the multiple ways of providing a signature for the detection of precise DNA sequences. Possible application is found in the field of biosensors, disease diagnosis, and gene expression[122,173] (see also Chapters 10 and 11). Although there are a variety of ways to achieve nanoparticle-biological materials, two different approaches dominate. The first involves the surface modification of prepared nanoparticles using ligand exchange and secondary modification. Murray *et al.* reported on the use of ethidium thiolate modified Au nanoparticles as a probe for the mechanism of DNA intercalation.[174] Similarly, Mirkin *et al.* have developed a method for the sequence specific detection of polynucleotides based on DNA functionalized gold nanoparticles (Fig. 5.9a).[122] The second entails the use of biological materials as synthetic agents including reductants and surfactants. Biological materials such as enzymes and peptides, plant extracts, and fungi have been used for the reduction of metal ions and the formation of nanoparticles.[175-181] The biosynthetic method has the potential to be clean, non-toxic, and environmentally friendly. Although the exact mechanisms for the nanoparticle synthesis are not clear, cystein rich, metal binding polypeptides, phytochelatins and metallothioneins have been relatively well characterized. Recently, in vivo synthesis of gold nanoparticles by a recombinant E. coli system expressing phytochelatin synthase and/or metallothionein was demonstrated.[182]

5.5.5 *SiO$_2$ coating and their derivatives*

One of the important surface treatments is coating with other components. Silica is the mostly widely employed coating material among other materials, such as polymers, carbon, and other metal and metal oxides. Silica coating can enhance stability against aggregation and provide tuneable solubility in various solvents. Generally, gold nanoparticles are coated with a silica shell via the Stöber process using tetraethylorthosilicate (TEOS) as a silica source.[183] The Stöber method was originally developed from sol-gel chemistry of silicon alkoxides to synthesize silica nanostructures in aqueous solutions with alcohols. The porous silica shell can be coated by hydrolysis and subsequent condensation of TEOS as a silica source and octadecyltrimethoxysilane (C$_{18}$TMS) as a pore generator.[184,185] Because of excellent biocompatibility, silica coating on the gold nanoparticle surface is an excellent tool for diverse applications, such as colorimetric diagnostics, photothermal therapy, and surface enhanced Raman scattering (SERS) detection.[186] The silica-coated surface is readily functionalized with amino-, mercapto-, and carboxy-terminated silanes, where secondary modification can be applied for conjugation with other materials, as introduced in gold-thiol systems. Silica-coated gold particles are also a potential framework for heterogeneous catalysis, and have many advantages in comparison to conventional catalysts embedded in bulk supports. Each nanoparticle isolated by a silica shell provides a relatively homogeneous environment around the particle surface.[187,188] The outer shell structure also hinders the aggregation of neighboring particles, even under harsh reaction conditions.

5.6 Conclusion

In this chapter, we briefly reviewed the synthesis, morphology control, and surface modification of gold nanoparticles in ligand phase. Since chemical and physical properties of gold nanoparticles critically depend on not only their size but also their shape and surface structure, synthesis of monodisperse nanoparticles is highly demanding for the improvement of their performances in versatile applications. Although precise control of morphology, surface facets and surrounding molecules have been a major topic with respect to the preparation of gold nanoparticles in liquid phase over the past decade, careful design of nanostructures and their conjugates with consideration of a variety of factors is still necessary to develop specific applications. Because of their outstanding properties, gold nanoparticles can find application in fields in a wide range of areas, including chemical or biological sensors, imaging, medical materials, catalysis for organic chemical synthesis, and energy generation, and due to technological development, the scope will continue to broaden. Consequently, easy, rapid, and large-scale synthesis, and diverse and stable surface treatments of gold nanoparticles are still required. Hybridization with other metals, semiconductors, polymers, and biomolecules with well-designed morphology and properties is another emerging topic being explored to realize the full use of gold nanoparticles.

References

1. Y. Xia and N.J. Halas, *Mat. Res. Soc. Bull.* **30** (2005) 338.
2. C.J. Murphy, *Science* **298** (2002) 2139.
3. P.V. Kamat, *J. Phys. Chem. B* **106** (2002) 7729.
4. M.A. El-Sayed, *Acc. Chem. Res.* **34** (2001) 257.
5. J.J. Mock, M. Barbic, D.R. Smith. D.A. Schultz and S. Schultz, *J. Chem. Phys.* **116** (2002) 6755.
6. P.D. Jadzinsky, G. Calero, C.J. Ackerson, D.A. Bushnell and R. D. Kornberg, *Science* **318** (2007) 430.
7. J. Zhao, J. Yang and J.G. Hou, *Phys. Rev. B* **67** (2003) 85404.
8. J. Li, X. Li, H.-J. Zhai and L.-S. Wang, *Science* **299** (2003) 964.
9. P. Pyykko, *Angew. Chem. Int. Ed.* **43** (2004) 4412.
10. H.-G. Boyen, G. Kastle, F. Eeigl, B. Koslowski, C. Dietrich, P. Ziemann, J.P. Spatz, S. Riethmuller, C. Hartmann, M. Moller, G. Schmid, M.G. Garnier and P. Oelhafen, *Science* **297** (2002) 1533.
11. P. Buffat and J.-P. Borel, *Phys. Rev. A* **13** (1976) 2287.
12. K. Dick, T. Dhanasekaran, Z. Zhang and D. Meisel, *J. Am. Chem. Soc.* **124** (2002) 2312.
13. M. Brust, J. Fink, D.Bethell, D.J. Schiffrin and C. Kiely, *J. Chem. Soc., Chem. Commun.* (1995) 1655.
14. N.R. Jana, L. Gearheart and C.J. Murphy, *Langmuir* **17** (2001) 6782.
15. D.G. Duff, A. Baiker and P. Edwards, *J. Chem. Soc., Chem. Commun.* (1993) 96.
16. F. Fievet, J.P. Lagier, B. Blin, B. Beaudoin and M. Figlarz, *Solid State Ionics* **32** (1989) 198.
17. J. Turkevich, P.C. Stevenson and J. Hillier, *Discuss. Faraday. Soc.* **11** (1951) 55.
18. P. Vanysek, in *CRC Handbook of Chemistry and Physics* 86th edn., (ed D.R. Lide), CRC Press, Tayer & Francis, Boca Raton, 2005, vol. 8, pp. 8–20.

19. M.J. Yacamán, J.A. Ascencio, H.B. Liu and J. Gardea-Torresdey, *J. Vac. Sci. Technol. B* **19** (2001) 1091.
20. C.L. Cleveland and U. Landman, *J. Chem. Phys.* **94** (1991) 7376.
21. Z.L. Wang, *J. Phys. Chem. B* **104** (2000) 1153.
22. A.S. Barnard and L.A. Curtiss, *ChemPhysChem* **7** (2006) 1544.
23. K.L. Kelly, E. Coronado, L.L. Zhao and G.C. Schatz, *J. Phys. Chem. B* **107** (2003) 668.
24. W.-H. Yang, G.C. Schatz and R.P. Van Duyne, *J. Chem. Phys.* **103** (1995) 869.
25. B.T. Draine and P.J. Flatau, *J. Opt. Soc. Am. A* **11** (1994) 1491.
26. L. Novotny, R.X. Bian and X.S. Xie, *Phys. Rev. Lett.* **79** (1997) 645.
27. R.X. Bian, R.C. Dunn, X.S. Xie and P.T. Leung, *Phys. Rev. Lett.* **75** (1995) 4772.
28. J.T.G. Overbeek, *Adv. Colloid Interface Sci.* **15** (1982) 251.
29. V.K. LaMer and R.H. Dinegar, *J. Am. Chem. Soc.* **72** (1950) 4847.
30. J. Park, J. Joo, S.G. Kwon, Y. Jang and T. Hyeon, *Angew. Chem. Int. Ed.* **46** (2007) 4630.
31. H. Yu, P.C. Gibbons, K.F. Kelton and W.E. Buhro, *J. Am. Chem. Soc.* **123** (2001) 9198.
32. C.B. Murray, D.J. Norris and M.G. Bawendi, *J. Am. Chem. Soc.* **115** (1993) 8706.
33. N.R. Jana and X. Peng, *J. Am. Chem. Soc.* **125** (2003) 14280.
34. H. Reiss, *J. Chem. Phys.* **19** (1951) 482.
35. D.V. Talapin, A.L. Rogach, M. Haase and H. Weller, *J. Phys. Chem. B* **105** (2001) 12278.
36. A.L. Smith, *Particle Growth in Suspensions*, Academic Press, London, 1983, pp. 3–15.
37. D. Seo, J.C. Park and H. Song, *J. Am. Chem. Soc.* **128** (2006) 14863.
38. G. Frens, *Colloid & Polymer Science* **250** (1972) 736.
39. G. Frens, *Nature (London), Phys. Sci.* **241** (1973) 20.
40. M.K. Chow and C.F. Zukoski, *J. Coll. Interface Sci.* **165** (1994) 97.
41. H. Xia, S. Bai, J. Hartmann and D. Wang, *Langmuir* **26** (2001) 3585.
42. B.-K. Pong, H.I. Elim, J.-H. Chong, W. Ji, B.L. Trout and J.-Y. Lee, *J. Phys. Chem. C* **111** (2007) 6281.
43. S.D. Perrault and W.C.W. Chan, *J. Am. Chem. Soc.* **131** (2009) 17042.
44. M. Giersig and P. Mulvaney, *Langmuir* **9** (1993) 3408.
45. M. Brust, M. Walker, D. Bethell, D.J. Schiffrin and R. Whyman, *J. Chem. Soc. Chem. Commun.* (1994) 801.
46. M.J. Hostetler, S.J. Green, J.J. Stokes and R.W. Murray, *J. Am. Chem. Soc.* **119** (1996) 4212.
47. R.S. Ingram, M.J. Hostetler and R.W. Murray, *J. Am. Chem. Soc.* **118** (1996) 4212.
48. A.C. Templeton, W.P. Wuelfing and R.W. Murray, *Acc. Chem. Res.* **33** (2000) 27.
49. K.V. Sarathy, G. Raina, R.T. Yadav, G.U. Kulkarni and C.N.R. Rao, *J. Phys. Chem. B* **101** (1997) 9876.
50. M.C. Daniel and D. Astruc, *Chem. Rev.* **104** (2004) 293.
51. M. Zhu, E. Lanni, N. Garg, M.E. Bier and R. Jin, *J. Am. Chem. Soc.* **130** (2008) 1138.
52. P.J.G. Goulet and R.B. Lennox, *J. Am. Chem. Soc.* **132** (2010) 9582.
53. N.R. Jana and X. Peng, *J. Am. Chem. Soc.* **125** (2003) 14280.
54. G. Schmid, *Chem. Soc. Rev.* **37** (2008) 1909.
55. G. Riess, *Prog. Polym. Sci.* **28** (2003) 1107.

56. R.K. Oreilly, M.J. Joralemon, C.J. Hawker and K.L. Wooley, *J. Polym. Sci. Part A: Polym. Chem.* **44** (2006) 5203.
57. S. Abraham, C.S. Ha and I. Kim, *J. Polm. Sci. Part A: Polym. Chem.* **43** (2005) 6367.
58. Y. Kang and T.A. Taton, *Angew. Chem. Int. Ed.* **44** (2005) 409.
59. Y. Kang, K.J. Erickson and T.A. Taton, *J. Am. Chem. Soc.* **127** (2005) 13800.
60. G. Carrot, J.C. Valmalette, C.J.G. Plummer, S.M. Scholz, J. Dutta, H. Hofmann and J.G. Hilborn, *Colloid Polym. Sci.* **276** (1998) 853.
61. A.B. Lowe, B.S. Sumerlin, M.S. Donovan and C.L. McCormick, *J. Am. Chem. Soc.* **124** (2002) 11562.
62. M. Filali, M.A.R. Meier, U.S. Schubert and J.F. Gohy, *Langmuir* **21** (2005) 7995.
63. N. Zheng, J. Fan and G.D. Stucky, *J. Am. Chem. Soc.* **128** (2006) 6550.
64. F. Kim, S. Connor, H. Song, T. Kuykendall and P. Yang, *Angew. Chem. Int. Ed.* **43** (2004) 3673.
65. D. Seo, C.I. Yoo, J.C. Park, S.M. Park, S. Ryu and H. Song, *Angew. Chem. Int. Ed.* **47** (2008) 763.
66. D. Seo, C.I. Yoo, I.S. Chung, S.M. Park, S. Ryu and H. Song, *J. Phys. Chem. C* **112** (2008) 2469.
67. D. Seo, J.H. Park, J. Jung, S.M. Park, S. Ryu, J. Kwak and H. Song, *J. Phys. Chem. C* **113** (2009) 3449.
68. M.Y. Han and C.H. Quek, *Langmuir* **16** (2000) 362.
69. K. Mallick, Z.L. Wang and T. Pal, *J. Photochem. Photobiol.* **140** (2001) 75.
70. J. Westerhausen, A. Henglein and J. Lilie, *Ber. Bunsen-Ges. Phys. Chem.* **85** (1981) 182.
71. A. Henglein, *Langmuir* **15** (1999) 6738.
72. E. Gachard, H. Remita, J. Khatouri, B. Keita, L. Nadjo and J. Belloni, *New J. Chem.* (1998) 1257.
73. A. Henglein and D. Miesel, *Langmuir* **14** (1998) 7392.
74. Y.-Y. Yu, S.-S. Chang, C.-L. Lee and C.R.C. Wang, *J. Phys. Chem. B* **101** (1997) 6661.
75. G.-T. Wei, F.-K. Liu and C.R.C. Wang, *Anal. Chem.* **71** (1999) 2085.
76. B.M.I. van der Zande, M.R. Bohmer, L.G.J. Fokkink and C. Schonenberger, *J. Phys. Chem. B* **101** (1997) 852.
77. B.M.I. van der Zande, M.R. Bohmer, L.G.J. Fokkink and C. Schonenberger, *Langmuir* **16** (2000) 451.
78. Y. Nagata, Y. Mizukoshi, K. Okitsu and Y. Maeda, *Radiat. Res.* **146** (1996) 333.
79. K. Okitsu, A. Yue, S. Tanaba, H. Matsumoto and Y. Yobiko, *Langmuir* **17** (2001) 7717.
80. R.A. Caruso, M. Ashokkumar and F. Grieser, *Langmuir* **18** (2002) 7381.
81. K. Okitsu, M. Ashokkumar, F. Grieser *J. Phys. Chem. B* **109** (1005) 20673.
82. C.-H. Su, P.-L. Wu and C.-S. Yeh, *J. Phys. Chem. B* **107** (2003) 14240.
83. M. Tsuji, M. Hashimoto, Y. Nishizawa, M. Kubokawa and T. Tsuji, *Chem. Eur. J.* **11** (2005) 440.
84. Y. Jiang and Y.-J. Zhu, *Chem. Lett.* **33** (2004) 1390.
85. F.-K. Liu, C.-J. Ker, Y.-C. Chang, F.-H. Ko, T.-C. Chu and B.-T. Dai, *Jpn. J. Appl. Phys., Part 1* **42** (2003) 4152.
86. M. Tsuji, N. Miyamae, M. Hashimoto, M. Nishio, S. Hikino, N. Ishigami and I. Tanaka, *Coll. Surf. A* **302** (2007) 587.

87. K.L. Kelly, E. Coronado, L.L. Zhao and G.C. Schatz, *J. Phys. Chem. B* **107** (2003) 668.
88. G.C. Schatz and R.P. Van Duyne, in *Handbook of Vibrational Spectroscopy* (edited by J.M. Chalmers and P.R. Griffiths), John Wiley, Chichester, 2002, p. 1.
89. Y. Sun and Y. Xia, *Anal. Chem.* **74** (2002) 5297.
90. T. Rindzevicius, Y. Alaverdyan, A. Dahlin, F. Hook, D.S. Sutherland and M. Kall, *Nano Lett.* **5** (2005) 2335.
91. O.C. Compton and F.E. Osterloh, *J. Am. Chem. Soc.* **129** (2007) 7793.
92. A. Tao, P. Sinsermsuksakul and P. Yang, *Nat. Nanotech.* **2** (2007) 435.
93. Y. Xia, P. Yang, Y. Sun, Y. Wu, B. Mayers, B. Gates, Y. Yin, F. Kim and H. Yan, *Adv. Mat.* **15** (2003) 353.
94. X. Peng, J. Wickham and A. P. Alivisatos *J. Am. Chem. Soc.* **120** (1998) 5343.
95. C.J. Murphy, L.B. Thompson, A.M. Alkilany, P.N. Sisco, S.P. Boulos, S.T. Sivapalan, J.A. Yang, D.J. Chernak and J. Huang, *J. Phys. Chem. Lett.* **1** (2010) 2867.
96. C.J. Murphy, T.K. Sau, A.M. Gole, C.J. Orendorff, J. Gao, L. Gou, S.E. Hunyadi and T. Li, *J. Phys. Chem. B* **109** (2005) 13857.
97. K. Sohn, F. Kim, K.C. Pradel, J. Wu, Y. Peng, F. Zhou and J. Huang, *ACS Nano* **3** (2009) 2191.
98. C.J. Johnson, E. Dujardin, S.A. Davis, C.J. Murphy and S. Mann, *J. Mat. Chem.* **12** (2002) 1765.
99. S.-Y. Wu, W.-L. Huang and M.H. Huang, *Crystal Growth & Design* **7** (2007) 831.
100. J.E. Millstone, S. Park, K.L. Shuford, L. Qin, G.C. Schatz and C.A. Mirkin, *J. Am. Chem. Soc.* **127** (2005) 5312.
101. H.-L. Wu, C.-H. Chen and H. Huang, *Chem. Mater.* **21** (2009) 110.
102. S. Chen, Z.L. Wang, J. Ballato, S.H. Foulger and D.L. Carroll, *J. Am. Chem. Soc.* **125** (2003) 16186.
103. T.K. Sau and C.J. Murphy, *J. Am. Chem. Soc.* **126** (2004) 8648.
104. J. Zhang, M.R. Langille, M.L. Personick, K. Zhang, S. Li and C.A. Mirkin, *J. Am. Chem. Soc.* **132** (2010) 14012.
105. C.J. Orendorff and C.J. Murphy, *J. Phys. Chem. B* **110** (2006) 3990.
106. D. Seo, C.I. Yoo, J. Jung and H. Song, *J. Am. Chem. Soc.* **130** (2008) 2940.
107. M.C. Gimenez, M.G. Del Popolo, E.P.M. Leiva, S.G. Garcıa, D.R. Salinas, C.E. Mayer and W.J. Lorenz, *J. Electochem. Soc.* **149** (2002) E109.
108. S. Garcia, D. Salinas, C. Mayer, E. Schmidt, G. Staikov and W.J. Lorenz, *Electrochim. Acta.* **43** (1998) 3007.
109. A. Kuzume, E. Herrero, J.M. Feliu, R.J. Nichols and D.J. Schiffrin, *J. Electroanal. Chem.* **570** (2004) 157.
110. M.C. Gimenez, M.G. Del Popolo and E.P.M. Leiva, *Electrochim Acta* **45** (1999) 699.
111. L.B. Rogers, D.P. Krause, J.C. Griess, Jr and D.B. Ehrlinger, *J. Electrochem. Soc.* **95** (1949) 33.
112. M.C. Giménez, M.G. Del Pópolo, E.P.M. Leiva, S.G. García, D.R. Salinas, C.E. Mayer and W.J. Lorenz, *J. Electrochem. Soc.* **149** (2002) E109.
113. Y. Sun, B. Wiley, Z.-Y. Li and Y. Xia, *J. Am. Chem. Soc.* **126** (2004) 9399.
114. S.E. Skrabalak, L. Au, X. Li and Y. Xia, *Nature Protocols* **9** (2007) 2182.

115. Y. Yin, C. Erdonmez, S. Aloni and A.P. Alivisatos, *J. Am. Chem. Soc.* **128** (2006) 12671.
116. D. Seo and H. Song, *J. Am. Chem. Soc.* **131** (2009) 18210.
117. S.J. Hurst, E.K. Payne, L. Qin and C.A. Mirkin, *Angew. Chem. Int. Ed.* **45** (2006) 2672.
118. E.C. Water, B.J. Murray, F. Favier, G. Kaltenpoth, M. Grunze and R.M. Penner, *J. Phys. Chem. B* **106** (2002) 11407.
119. C.R. Martin, *Chem. Mater.* **8** (1996) 1739.
120. M.J. Hostetler, S.J. Green, J.J. Stokes and R.W. Murray, *J. Am. Chem. Soc.* **118** (1996) 4212.
121. P.V. Kamat, S. Barazzouk and S. Hotchandani, *Angew. Chem. Int. Ed.* **41** (2002) 2764.
122. D.A. Giljohann, D.S. Seferos, W.L. Daniel, M.D. Massich, P.C. Patel and C.A. Mirkin, *Angew. Chem. Int. Ed.* **49** (2010) 3280.
123. J. Chen, F. Saeki, B.J. Wiley, H. Cang, M.J. Cobb, Z.-Y. Li, L. Au, H. Zhang, M.B. Kimmey, X. Li and Y. Xia, *Nano. Lett.* **5** (2005) 473.
124. J.R. Heath, L. Brandt and D.V. Leff, *Langmuir* **12** (1996) 4723.
125. M. Green and P. O'Brien, *Chem. Commun.* (2000) 183.
126. W.P. Wuelfing, S.M. Gross, D.T. Miles and R.W. Murray, *J. Am. Chem. Soc.* **120** (1998) 12696.
127. M. Zheng. Z. Li and X. Huang, *Langmuir* **20** (2004) 4226.
128. M.K. Corbierre, N.S. Cameron and R.B. Lennox, *Langmuir* **20** (2004) 2867.
129. E. Glogowski, R. Tangirala, J. He, T.P. Russell and T. Emrick, *Nano. Lett.* **7** (2007) 389.
130. B.C. Mei, K. Susumu, I.L. Medintz and H. Mattoussi, *Nature Protocols* **4** (2009) 412.
131. N.F. Steinmetz and M. Manchester, *Biomacromolecules* **10** (2009) 784.
132. H. Yoo, O. Momozawa, T. Hamatani and K. Kimura, *Chem. Mater.* **13** (2001) 4692.
133. S. Chen and K. Kimura, *Langmuir* **15** (1999) 1075.
134. G.B. Braun, A. Pallaoro, G. Wu, D. Missirlis, J.A. Zasadzinski, M. Tirrell and N.O. Reich, *ACS Nano* **3** (2009) 2007.
135. M. Oishi, J. Nakaogami, T. Ishii and Y. Nagasaki, *Chem. Lett.* **35** (2006) 1046.
136. A. Sánchez-Iglesias, I. Pastoriza-Santos, J. Péresz-Juste, B. Rodríguez-González, F.J. García de Abajo and L.M. Liz-Marzán, *Adv. Mater.* **18** (2006) 2529.
137. L. Prati and G. Martra, *Gold Bull.* **32** (1999) 96.
138. M. Comotti, W.-C. Li, B. Spliethoff and F. Schüth, *J. Am. Chem. Soc.* **128** (2006) 917.
139. Y. Zhou, C.Y. Wang, Y.R. Zhu and Z.Y. Chen, *Chem. Mater.* **11** (1999) 2310.
140. M. Comotti, C.D. Pina, R. Matarrese, M. Rossi and A. Siani, *Appl. Catal. A* **291** (2005) 204.
141. T.S. Ahmadi, Z.L. Wang, T.C. Green, A. Henglein and M.A. El-Sayed, *Science* **272** (1996) 1924.
142. J.M. Petroski, Z.L. Wang, T.C. Green and M.A. El-Sayed, *J. Phys. Chem. B* **102** (1998) 3316.
143. E. Gacgard, H. Remita, J. Khatouri, B. Keita, L. Nadjo and J. Belloni, *New J. Chem.* (1998) 1257.
144. H. Jans, K. Jans, L. Lagae, G. Borghs, G. Maes and Q. Huo, *Nanotechnology*, 21 (2010) 455702.

145. C.I. Yoo, D. Seo, B.H. Chung, I.S. Chung and H. Song, *Chem. Mater.* **21** (2009) 939.
146. J.P. Spatz, S. Mossmer, C. Hartmann, M. Moller, T. Herzog, M. Krieger, H.-G. Boyen and P. Ziemann, *Langmuir* **16** (2000) 407.
147. J.H. Youk, M.-K. Park, J. Kocklin, R. Advincula, J. Yang and J. Mays, *Langmuir* **18** (2002) 2455.
148. J.-F. Gohy, N. Willet, S. Varshney, J.-X. Zhang and R. Jerome, *Angew. Chem. Int. Ed.* **40** (2001) 3214.
149. Y. Kang and T.A. Taton, *Angew. Chem. Int. Ed.* **44** (2005) 409.
150. C.-M. Huang, K.-H. Wei, U-S. Jeng and K.S. Liang, *Macromolecules* **40** (2007) 5067.
151. A. Aqil, C. Detrembleur, B. Gilbert, R. Jerome and C. Jerome, *Chem. Mat.* **19** (2007) 2150.
152. M. Stemmler, F.D. Stefani, S. Bernhardt, R.E. Bauer, M. Kreiter, K. Mullen and W. Knoll, *Langmuir* **25** (2009) 12425.
153. F. Grohn, A.J. Bauer, Y.A. Akpalu, C.L. Jackson and E.J. Amis, *Macromolecules* **33** (2000) 6042.
154. J.-A. He, R. Valluzzi, K. Yang, T. Dolukhanyan, C. Sung, J. Kumar and S.K. Tripathy, *Chem. Mat.* **11** (1999) 3268.
155. X. Shi, K. Sun and J.R. Baker Jr., *J. Phys. Chem. C* **112** (2008) 8251.
156. K. Esumi, K. Satoh and K. Torigoe, *Langmuir* **17** (2001) 6860.
157. K. Hayakawa, T. Yoshimura and K. Esumi, *Langmuir* **19** (2003) 5517.
158. K. Esumi, H. Houdatsu and T. Yoshimura, *Langmuir* **20** (2004) 2536.
159. A.C. Templeton, M.J. Hostetler, C.T. Kraft and R.W. Murray, *J. Am. Chem. Soc.* **120** (1998) 1906.
160. M.J. Hostetler, A.C. Templeton and R.W. Murray, *Langmuir* **15** (1999) 3782.
161. A.C. Templeton, W.P. Wuelfing and R.W. Murray, *Acc. Chem. Res.* **33** (2000) 27.
162. J.D. Gibson, B.P. Khanal and E.R. Zubarev, *J. Am. Chem. Soc.* **129** (2007) 11653.
163. P. Ghosh, G. Han, M. De, C.K. Kim and V.M. Rotello, *Adv. Drug Delivery Rev.* **60** (2008) 1307.
164. K.G. Thomas and P.V. Kamat, *Acc. Chem. Res.* **36** (2003) 888.
165. S. Banerjee and S.S. Wong, *Nano. Lett.* **2** (2002) 195.
166. J. Liu, A.G. Rinzler, H. Dai, J.H. Hafner, R.K. Bradley, P.J. Boul, A. Lu, T. Iverson, K. Shelimov, C.B. Huffmann, F. Rodriguez-Macias, Y.-S. Shon, T.R. Lee, R.E. Colbert and R.E. Smalley, *Science* **280** (1998) 1253.
167. T.K. Mandal, M.S. Fleming and D.R. Walt, *Nano. Lett.* **2** (2002) 3.
168. J. Kolny, A. Kornowski and H. Weller, *Nano. Lett.* **2** (2002) 361.
169. K. Jiang, A. Eitan, L.S. Schadler, P.M. Ajayan and R.W. Siegel, *Nano. Lett.* **3** (2003) 275.
170. W. Cheng and E. Wang, *J. Phys. Chem. B* **108** (2004) 24.
171. Y.-M. Chen, C.-J. Yu, T.-L. Cheng and W.-L. Tseng, *Langmuir* **24** (2008) 3654.
172. B.L. Frankamp, O. Uzun, F. Ilhan, A.K. Boal and V.M. Rotello, *J. Am. Chem. Soc.* **124** (2002) 892.
173. Z. Li, R. Jin, C.A. Mirkin and R.L. Letsinger, *Nucleic Acid Research* **30** (2002) 1588.
174. G. Wang, J. Zhang and R.W. Murray, *Anal. Chem.* **74** (2002) 4320.
175. J.L. Gardea-Torresdey, J.G. Parsons, E. Gomez, J. Peralta-Videa, H.E. Troiani, P. Santiago and M.J. Yacaman, *Nano Lett.* **2** (2002) 397.

176. S.S. Shankar, A. Rai, B. Ankamwar, A. Singh, A. Ahmad and M. Sastry, *Nat. Mater.* **3** (2004) 482.

177. S.P. Chandran, M. Chaudhary, R. Pasricha, A. Ahmad and M. Sastry, *Biotechnol. Prog.* **22** (2006) 577.

178. P. Mukherjee, A. Ahmad, D. Mandal, S. Senapati, S.R. Sainkar, M.I. Khan, R. Ramani, R. Parischa, P.V. Ajayakumar, M. Alam, M. Sastry and R. Kumar, *Angew. Chem. Int. Ed.* **40** (2001) 3585.

179. T.L. Riddin, M. Gericke and C.G. Whiteley, *Nanotechnology* **17** (2006) 3482.

180. R.R. Naik, S.J. Stringer, G. Agarwal, S.E. Jones and M.O. Stone, *Nat. Mater.* **1** (2002) 169.

181. S.A. Kumar, Y.-A. Peter and J.L. Nadeau, *Nanotechnology* **19** (2008) 495101.

182. T.J. Park, S.Y. Lee, N.S. Heo and T.S. Seo, *Angew. Chem. Int. Ed.* **49** (2010) 7019.

183. L.M. Liz-Marzan, M. Giersig and P. Mulvaney, *Langmuir* **12** (1996) 4329.

184. J.Y. Kim, S.B. Yoon and J.-S. Yu, *Chem. Commun.* (2003) 790.

185. K. Yano and Y. Fukushima, *J. Mater. Chem.* **14** (2004) 1579.

186. S. Liu and M.-Y. Han, *Chem. Asian. J.* **5** (2010) 36.

187. J. Lee, J. C. Park and H. Song, *Adv. Mater.* **20** (2008) 1523.

188. J. Lee, J.C. Park, J.U. Bang and H. Song, *Chem. Mat.* **20** (2008) 5839.

Chapter 6
Chemical Preparation of Gold Nanoparticles on Surfaces

Catherine Louis

Laboratoire de Réactivité de Surface, UPMC-CNRS, 4 Place Jussieu, 75005 Paris, France. Email: catherine.louis@upmc.fr.

6.1 Introduction

The size of the gold particles to be prepared on supports depends on the further use of the materials. For catalysis, very small particles are needed, in general smaller than 5 nm (see Chapter 7) while for physics (see Chapters 3 and 4) or biology (see Chapters 10 and 11), larger particles can be required. For academic work in catalysis, supports are in a powder form, essentially an oxide with porosity or without porosity, and chemical methods are involved in preparing catalysts (see below). Gold nanoparticles can also be deposited on organic supports or embedded in organic materials leading to hybrid inorganic-organic materials, for instance for catalytic applications or applications in life sciences. For specific physics studies and as "model" catalysts (Chapter 8), gold particles must be deposited on planar surfaces (thin films or single crystals). Physical methods can be used (see Chapter 8), as well as chemical methods (Section 6.5); a control of the distances between the particles, i.e. ordering, may also be needed.

The chemical methods of preparation of gold nanoparticles onto supports are mainly based upon two principles, each being performed in two steps:

- The deposition of a gold precursor, most often in the oxidation state III (chloro-auric acid or gold chloride) onto the support, and then its reduction into metal nanoparticles thanks to a thermal treatment. This is called

139

"*deposition-reduction*" in the following. It is noteworthy that because of the instability of Au^{III} compounds, a thermal treatment even under an oxidising atmosphere, oxygen or air, usually leads to the formation of metallic gold nanoparticles.

• The pre-formation of gold nanoparticles in the liquid phase, i.e. of gold colloids, by reduction of a gold precursor in the presence of stabilising agents (Chapter 5) then the deposition of the gold colloids onto the support. This is called "*reduction-deposition*" in the following chapter.

The *deposition-reduction* method is essentially used for catalyst preparation when very small particles are needed. The *reduction-deposition* method can be used for any application since the gold particle size can be adjusted within a large range depending on the method of colloid preparation (Chapter 5). An additional step may be required for the second method, which is the decomposition of the stabilising agents.

The characterisation of the gold particles mostly relies on the measurement of the particle sizes to get a distribution of size and an average particle size. For divided supports, transmission electron microscopy (TEM) is used, and for planar supports, atomic force microscopy (AFM) or scanning tunneling microscopy (STM) are used; these techniques also provide information on the degree of ordering of the gold nanoparticles.

6.2 Gold Nanoparticles on Powder Inorganic Supports

As mentioned above, the powder supports are often oxide supports, such as alumina, titania and ceria. The choice of the supports depends on the application, and more specific supports can be used, such as structured mesoporous supports, zeolites and carbon. For more details, the reader can refer to Ref. 1

Only a small number of gold precursors are commercially available. Gold trichloride, tetrachloroauric acid or salt (Na, K), gold acetate, gold nitrate, which are both poorly soluble in water, and gold acetyl acetonate, which must be handled in air-free and water-free conditions.

6.2.1 *Deposition of a gold precursor (deposition-reduction)*

6.2.1.1 *Impregnation with HAuCl₄ and related methods*

a. Impregnation

Impregnation is the simplest method of preparation of metal-supported catalysts, and it was the first used for gold catalyst preparation. It consists of wetting the support with an aqueous solution containing the metal precursor, usually tetrachloroauric acid ($HAuCl_4$) or $AuCl_3$. Afterwards, the sample is dried then a thermal treatment is performed to reduce the precursor into metallic particles (Fig. 6.1). The thermal treatment can be performed under a reducing gas such as hydrogen or an oxidising gas such as oxygen or air; in both cases, gold is reduced because of the instability of Au^{III} compounds. This method can be used with any type of support. The drawback of this method is that the chlorides of the gold precursor are also present on the support, and induce gold particle sintering during calcination and the formation of large gold particles (>10 nm) even for low gold loading ($1-2$ wt %). For catalysis applications, this leads to poorly active catalysts.[2-4] Smaller gold particles can be obtained when reduction under hydrogen is applied, but the residual chlorides may be detrimental to catalytic performances[1] (see Chapter 7).

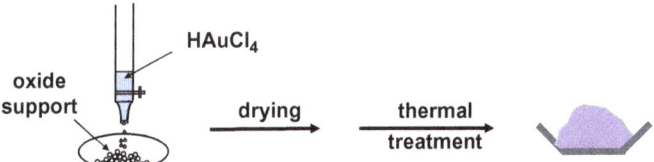

Fig. 6.1. Scheme showing the various steps of the preparation of supported gold samples by impregnation.

b. Anion adsorption

Another way of depositing gold is to induce specific interaction between the gold precursor and the oxide support, so as to be able to wash the sample after deposition and remove the chlorides without leaching gold. The preparation is based on the principle that the hydroxyl groups present on the oxide surface may be protonated (OH_2^+) or deprotonated (O^-) depending on the solution pH in which the oxide is immersed. The pH value at which the total electric charge of the surface is zero is the point of zero charge (PZC) (Fig. 6.2).

Fig. 6.2. Charge on an oxide surface (S) in aqueous solution as a function of pH.

Adsorption of chloroauric anions on oxide supports is in principle possible, provided that the pH of the solution is lower than the PZC of the oxide support, making the support surface positively charged; for example, this excludes supports like silica, the PZC of which is very low (1–2). Practically (Fig. 6.3), the powder is immersed into a solution containing the gold precursor at a pH lower than the support PZC, stirred for a given time (usually around 1 h), most often at room temperature, to allow the interactions to establish. The sample is recovered after filtering or centrifugation, and washed several times with water to get rid of the chlorides and any gold not interacting with the support. In this preparation, the amount of gold deposited is limited by the capacity of adsorption of the oxide supports which mainly depends on the PZC, the surface area of the support and the pH of the solution. For usual supports, the maximum gold loading is around 1–2 wt %.[5,6] The gold particles are reasonably small after calcination ~4 nm, and smaller after reduction under hydrogen, but the chlorides present in the adsorbed gold complex are not totally eliminated.[6]

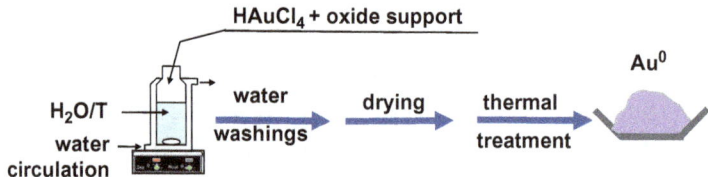

Fig. 6.3. General principle of the preparations in excess of solution of gold supported on oxides.

c. Washing with ammonia

A way to eliminate chlorides is to wash the samples with ammonia solution (0.1–1 M) after deposition.[6–7] This can be used for samples prepared by incipient wetness impregnation or anion adsorption. Most of the gold is retained while chlorides are eliminated, and small gold particles are obtained (Table 6.1). It was proposed that during ammonia washing, gold hydroxychloride anion transformed into amino-hydroxo-aquo gold cation, $[Au(NH_3)_2(H_2O)_{2-x}(OH)_x]^{(3-x)+}$, that could interact with the support.[7]

Table 6.1. Titania-supported gold sample (4 wt % of nominal gold loading) prepared by impregnation: gold and chlorine loadings after ammonia washing and average particle size after reduction in H_2 at 300°C.[7]

	After washing		After thermal treatment	
Washing solution	Au (wt %)	Cl	Average gold particle size (nm)	Standard deviation (nm)
—	3.6	2.7 wt %	7.9	2.4
NH_3 (1M)	3.6	<200 ppm	3.7	1.0
NH_3 (0.1M)	3.0	<200 ppm	2.8	0.8
H_2O^a	0.9	0.2 wt %	3.3	0.9

6.2.1.2 *Deposition-precipitation and related methods*

a. Deposition-precipitation at fixed pH

The method of deposition-precipitation is certainly the method most used for the preparation of gold catalysts since it readily leads to the formation of small gold particles (2–3 nm) with a very low amount of chloride. This method was first proposed by Haruta *et al.*[8–9] It is noteworthy that thanks to gold catalysts prepared by this method, these authors discovered the extraordinary catalytic performance of gold in the reaction of CO oxidation at room temperature (see Chapter 7). The main difference with the procedure used for anion adsorption (see Section 6.2.1.1) is that the pH of the solution containing both $HAuCl_4$ and the oxide support is adjusted to 7–8 by addition of NaOH (or Na_2CO_3 or NH_3), and that the mixture is usually stirred at

70–80°C for around 1 h. As for anion adsorption, the catalyst is washed with water, dried between RT and 100°C, and usually calcined in air. This method is suitable for oxide supports whose PZC is higher than 5, such as MgO, TiO$_2$, Al$_2$O$_3$, ZrO$_2$, CeO$_2^{10-12}$ and not for SiO$_2$ (PZC ~2), SiO$_2$-Al$_2$O$_3$ (PZC ~1)[11] or for activated carbon.[13]

The choice of pH of 7–8 is the result of a compromise between the yield of gold deposition (which usually does not reach 100 % except when targeted gold loading is around 1 wt %) and the gold particle size obtained after thermal treatment. It is noteworthy that the conditions of preparation are such that both the support surface and the gold complex are negatively charged, which raises the question of the mechanism of interaction that cannot be based on electrostatic interaction. The gold uptake on titania supports versus pH looks like a kind of volcano curve with a maximum at pH 6 that is the PZC of TiO$_2$.[14] Quite the same type of trend was observed in the case of gold on alumina.[15,16] It is worth noting that as pH increases, the gold speciation changes as chlorides are hydrolysed, and form, for instance, [AuCl(OH)$_3$]$^-$ at pH 7–8. An explanation for the interaction with the oxide support would be a reaction between anion gold complexes and residual non-deprotonated OHs of the support, leading to the formation of a surface gold complex:

$$TiOH + AuCl(OH)_3^- \rightarrow [(TiO)Au(OH)_2Cl]^- + H_2O$$

the amount of which is maximum when pH is close to the PZC because of the higher number of OH groups.

b. Deposition-precipitation at increasing pH with urea

The gradual rise in pH required in the procedure of deposition-precipitation with urea developed first by Geus *et al.*[17,18] for the preparation of other metallic systems, is obtained by the addition of urea (CO(NH$_2$)$_2$). Applied to the preparation of gold catalysts, the procedure consists of stirring an aqueous solution containing HAuCl$_4$ and urea with the oxide support in suspension at ~80°C (the same procedure as in Fig. 6.3 apart from the addition of urea); this temperature leads to the decomposition of urea and gradual rise in pH:

$$CO(NH_2)_2 + 3H_2O \rightarrow CO_2 + 2NH_4^+ + 2OH^-.$$

An extensive study performed by Louis *et al.*[5,12] showed that small gold particles (~2 nm) could be obtained on various supports (TiO$_2$, Al$_2$O$_3$ and

CeO_2). In contrast with deposition-precipitation at fixed pH, all the gold of the solution (at least up to 8 wt %, the highest loading studied) can be deposited onto the supports within the first hour of preparation, when the suspension pH is still acidic ($\sim$3). The sample then "matures" as the pH of the suspension increases and the gold speciation changes concomitantly. "Maturation" leads to a decrease in the average gold particle size after the sample is washed, dried and calcined (Table 6.2). When lower gold loading is required ($\sim$1 wt % Au, i.e. lower than the adsorption capacity of the support), it is possible to drastically reduce the preparation time to $\sim$1 h and to get small gold particles (Table 6.2).[6,12]

Table 6.2. Gold loading and average particle size on titania-supported gold sample prepared by deposition-precipitation with urea at 80°C for various durations.[5,6]

Au loading (wt %)	DP time (h)	Final pH	Au loading (wt %)	d Au^0 (nm)
8	1	3.0	7.8	5.6
	2	6.3	6.5	5.2
	4	7.0	7.7	2.7
	16	7.3	6.8	2.5
1	1	3.9	1.1	3.0
	16	7.4	1.1	2.2

6.2.1.3 *Less common preparation methods*

a. Cation adsorption

Gold(III) ethylenediamine cation, $[Au(en)_2]^{3+}$ (*en* = $NH_2CH_2CH_2NH_2$), was used for cation exchange in zeolites[19,20] and for cation adsorption on oxide supports, TiO_2,[5,12] amorphous[21,22] and ordered mesoporous silicas[23,24] and activated carbon.[25] $[Au(en)_2]Cl_3$ is easy to synthesise.[26] Cation adsorption requires that the oxide surface is negatively charged with O^- groups, which means that the pH of the aqueous solution must be higher than the PZC of the support (see Fig. 6.2).

145

$[Au(en)_2]^{3+}$ adsorption must be performed at room temperature since the complex decomposes around 60°C. As in the case of anion adsorption, the gold loading is limited by the adsorption capacity of the support, for instance around 1 wt % on TiO_2 with gold particles of 2–3 nm after calcination at 300°C[12] and up to 9 wt % on SBA-15 of higher surface area with gold particles of 4–5 nm.[22] Again, the particles are smaller after reduction under H_2 than after calcination. However, the gold samples retain residual organics after reduction, and it is necessary to proceed to further calcination at 400°C[22] or to a treatment in $KMnO_4$ solution followed by calcination[27] to eliminate them completely.

Note that for the preparation methods cited above, instead of thermal treatments, chemical agents can be used to reduce gold after deposition on oxide supports, for instance $NaBH_4^{28}$ or dimethylamine borane;[29] in both cases, the reduced gold particles were smaller than 5 nm.

b. Preparations involving organogold precursors

There are only a few examples of preparations involving organogold precursors probably because they are less easy to handle or must be synthesised.

The phosphine gold complex, $[Au^I(PPh_3)]NO_3$, and cluster, $[Au_9(PPh_3)_8](NO_3)_3$ (Au^I and Au^0), in CH_2Cl_2 or acetone were used to prepare supported gold catalysts by impregnation. Impregnation of freshly precipitated metal hydroxides led to smaller gold particles ($\sim$3 nm) than on oxides such as TiO_2, SiO_2 and α-Fe_2O_3 ($\sim$30 nm).[30,31] $[Au_6(PPh_3)_6](BF_4)_2$ was also used for the impregnation of TiO_2 in CH_2Cl_2.[32] Other organogold compounds were used for the impregnation of TiO_2 (1 wt % Au): $[Au_2^I(dppm)_2](PF_6)_2$, $Au_3^I(Ph_2pz)_3$, $[Au_4^I(dppm)_2(3,5-Ph_2pz)_2](NO_2)_2$, and $Au_4^I(form)_4$, (dppm = bis(diphenylphosphino)methane, form = formamidinates, pz = pyrazolate).[33] In the latter case, after reduction in hydrogen at 500°C then calcination in oxygen at 400°C, the gold particles were 7.7, 4.8, 3.1 and 3.0 nm, respectively. The advantage of using phosphine organogold complexes or clusters is that they do not contain chlorides, but they must be synthesised, and handled with care because of their toxicity and instability. Moreover, traces of phosphorus may remain on the support.

Another type of organogold compound is the di-methyl gold(III) acetylacetonate ($Au(CH_3)_2(acac)$); it is commercially available, but expensive,

146

and it must also be handled in the absence of moisture and air. Impregnation in dried pentane or hexane was performed with various supports, MgO,[34] γ-alumina,[35] TiO_2,[36] La_2O_3[37] and Na-Y zeolites.[38] Chemical vapour deposition (CVD) is another manner to deposit this compound on various oxides, TiO_2, Al_2O_3, SiO_2, silica MCM-41, and activated carbon.[39] In both cases, impregnation and CVD, gold particles around 3 nm were obtained after decomposition of the grafted complex. The mechanism of $Au(CH_3)_2(acac)$ adsorption and decomposition into gold particles was studied in Refs. 35, 40.

c. Solid grinding

This method has been recently applied by Haruta and colleagues to the preparation of supported gold catalysts.[41] It is very simple since it consists of grinding a mixture of support and dimethyl gold acetylacetonate in the absence of solvent. Grinding can be performed manually in a mortar or by ball-milling in air at room temperature.

This method has been developed first for the deposition of gold nanoparticles on porous coordination polymers (PCPs) such as Metal-Organic Frameworks (MOFs).[41] Gold loading of 1 wt % could be deposited with an average size of 1.5 nm. Several other supports have also been used, such as carbon, oxide, cellulose, and porous coordination polymers.[42–46] The amount of gold that can be deposited depends on the support and on the establishment of bonding between the gold precursor and the OH of the support. The preparation must be performed under low humidity levels ($<50\%$) (mesoporous titano-silicate) or under high humidity levels (cellulose) or in acetone (alumina or zirconia). Au^{III} is then thermally reduced under hydrogen (around $120°C$) or in air depending on the support. For all supports, particles around 2 nm can be obtained except in the case of semiconductive supports such as ceria or titania because of uncontrolled reduction of gold during grinding.

6.2.2 *Deposition of preformed gold nanoparticles (reduction-deposition)*

Deposition of preformed gold nanoparticles in solution is another strategy to prepare supported gold samples since these metal particles have in principle controlled size and a narrow size distribution (see Chapter 5).

6.2.2.1 Gold colloids

As mentioned in Chapter 5, gold colloids are obtained by reduction of a gold precursor, $HAuCl_4$ or $AuCl_3$ in solution in the presence of stabilising or capping agents, which can be molecules, such as CO, citrate, thiol, amine and glucose, or polymers, such as poly-vinylpyrrolidone (PVP), poly-vinylalcohol (PVA), or micelles with diblock copolymers or dendrimers (see Section 6.2.2.2). These stabilisers are added to control the growth of the particles during reduction and avoid aggregation and precipitation. Reduction is usually performed by addition of chemical agents such as sodium borohydride, hydrazine or weaker reducing agents such as amine-borane complexes. The stabiliser may also act as a reducer, for instance, sodium citrate and tetrakis(hydroxymethyl)phosphonium chloride (THPC). Reduction can be assisted by heating, sonolysis, radiolysis or microwaving. The formation of the gold sol is attested by the colour change of the solution. The average size and size distribution of the so-prepared gold particles, as well as shape, strongly depend on the conditions of synthesis, i.e. temperature, ageing time, concentration, nature of the different constituents and reduction technique; they may vary between a few to hundreds of nanometers.

Depending on the preparation method, the colloids may be in the aqueous or organic phase. In the first case, deposition on supports is performed by mere immersion of the support into the colloidal solution, followed by washing with water, and drying. Adsorption is attested by the decolouration of the solution and the colouration taken by the support. This method is especially appropriate for supports such as carbon materials, carbon black or activated carbon, because of their acidic carboxylic groups, which can easily interact with colloids.[43] Some colloid surfaces are negatively charged because of the chemisorption of stabilising agents, allowing them to electrostatically interact with an oxide support provided that the pH of the solution is lower than the support PZC (Fig. 6.2). The deposition step is not a trivial issue, as attested by some results in the literature that appear at variance in terms of pH setting and resulting gold particle size. For instance, according to Porta et al.,[47] adsorption of THPC-stabilised gold colloids on titania and zirconia was successful at pH = 2 because this pH is lower than the PZC of the support, but the same colloids could be immobilised on carbon and alumina, independently of the pH of the sol. On the other hand, PVA is an appropriate stabiliser for depositing gold on carbon, but it is not suitable for silica or alumina.[47] A delicate balance must be achieved between several parameters, such as the nature and the concentration of the stabiliser, the stabiliser/Au ratio, the solution pH and the nature of the support. To favour interactions between gold colloids and an oxide support, it is also possible to functionalise the oxide support. For instance, Fe_3O_4 particles (10 nm) were functionalised with APTSE (3-Aminopropyl trimethoxysilane) with $-NH_2$ terminal groups in ethanol. After addition of HNO_3, they interacted with negatively charged THPC-stabilised gold colloids (3.4 nm). It appeared *in fine* that only the smaller gold particles were adsorbed (2.5 nm).[48]

In the case of gold sols in organic medium, the problem is to conciliate colloid hydrophobicity with support hydrophilicity. One solution is to proceed to a mere impregnation of the support with colloids in organic solution, followed by solvent evaporation: for instance, dodecanethiol-stabilised gold colloids (1.9 nm, prepared with the Brust method, see Chapter 5) in ethyl acetate were impregnated on ceria ($\sim$2 wt % Au);[49] other colloids (1.8, 3.9 and 9.9 nm) in hexane were impregnated on silica, titania and carbon ($\sim$2 wt % Au).[50] However, this method does not permit the control of the distribution of the hydrophobic gold particles over the hydrophilic supports because of the absence of specific interaction. One solution, proposed by Zheng and Stucky,[51] is to use an aprotic solvent such as chloroform or methylene chloride: when oxide powders, TiO_2, SiO_2, ZnO, and Al_2O_3, were added to dodecanethiol-stabilised gold colloids in these solvents, the decolouration of the solution and the darkening of the supports were observed, attesting to their interaction. Herranz *et al.*[52] used this method to adsorb hexadecanethiol-stabilised gold colloids (2, 3.7 and 4.8 nm) in chloroform on SiO_2 and TiO_2 (1–2 wt % Au).

Depending on the applications of these supported gold particles, stabilising agents may have to be removed; this is the case for further gas phase catalytic reactions. This is a critical step because gold particles may sinter, and residues of stabiliser may poison the catalyst. Thermal treatment under air or oxygen is the most frequent method used. For reasons that remain unclear, some studies report the growth of the particles during thermal treatment[49,53−55] while others do not, in spite of the same rather high temperature applied ($\sim$300°C).[51,52,56−59] For instance, the size of PVA- and glucose-protected gold particles adsorbed on TiO_2 (3.0

and 3.8 nm, respectively) remained unchanged after calcination in air at 250°C.[57] Decanethiol-stabilised gold particles of 3.0 nm after adsorption on TiO_2 from a solution of methylene chloride were of 3.4 nm after treatment under H_2/N_2 at 290°C.[59] In contrast, 2 nm decanethiol-capped gold nanoparticles adsorbed on carbon drastically increased in size after calcination at 300°C (3–20 nm).[55] Sintering can be prevented using a stabilising agent that decomposes at low temperature. This holds for lysine, an α-amino acid, which does not strongly interact with gold, and can be decomposed at 200°C without growth of the particles (<5 nm) supported on α–Fe_2O_3.[58]

Alternatively, techniques other than thermal treatments can be used to decompose the stabilisers: ozone treatment or plasma activation, both performed at RT. For some reasons, these techniques are most often applied to planar surfaces (see Section 6.5), and only in a few cases were used for powder supports. For instance, the removal of ligands of $Au_{13}[PPh_3]_4[S(CH_2)_{11}CH_3]_4$ adsorbed on TiO_2 from toluene solution was performed by flowing ozone at RT,[60] and much smaller particles (1.2 nm) were obtained than after calcination at 400°C (2.7 nm). Another example is given by dodecanethiol-stabilised gold nanoparticles (1.8 nm of diameter) deposited on the channel walls of cordierite monoliths and on surfaces of cleaved mica (001) and gypsum (010) single crystals.[61] After oxygen plasma treatment at RT, the gold nanoparticles were also much smaller (3.5 nm) than after thermal treatment at 250°C in He (10–20 nm).

6.2.2.2 *Gold in micelles or in dendrimers*

Micelles as stabilisers are most often used for planar surfaces as they allow quasi-hexagonal ordering of the gold nanoparticles (see Section 6.5.2), and only in a few cases were used for powder supports. Diblock copolymers [polystyrene-block-poly(2-vinylpyridine)] dissolved in toluene solution formed micelles with a hydrophilic core into which $HAuCl_4$ can diffuse. These micelles can then be deposited onto oxides by impregnation. After drying, the samples were heated to 300°C in air to reduce gold and to remove the polymers, and gold particles of various sizes were obtained (8.4, 12.4 and 22 nm), depending on the support (TiO_2, ZnO, and ZrO_2, respectively).[62]

The stabilisers of the gold colloids can also be dendrimers, often PAMAM (polyamidoamine, commercially available), which are hyperbranched polymers that ramify from a single core and form a porous sphere[63,64] (Fig. 5.10 in Chapter 5). Dendrimer-encapsulated nanoparticles (DENs) are synthesised by sequestering metal ions within dendrimers, which are afterwards chemically reduced most often by addition of $NaBH_4$. DENs can be synthesised in water or alcohols. The resulting metal nanoparticles are usually small and nearly monodispersed, and their size can be tuned by varying the metal to dendrimer ratio and the dendrimers nature (composition, size ("generation") and terminal function). As in the case of colloids, they can be deposited onto supports or used as a templating agent in the preparation of embedded nanoparticles (Section 6.3).

As for colloids, it may be desirable to remove the dendrimers from the gold particles, but this may lead to an increase of the metal particle size. For instance, the calcination temperature of 500°C required to remove PAMAM dendrimers from gold DENs impregnated on titania led to particle growth from 1.7 to 7.2 nm.[65] As for colloids, lower temperature routes, such as oxygen plasma treatment, must permit the retention of the particle size. Another way to prevent nanoparticle sintering is to extract them from dendrimers using an appropriate thiol, which after deposition onto a support, can be decomposed at lower temperature. For instance, gold nanoparticles in PAMAM dendrimers were extracted with toluene containing decanethiol.[59] After separation and purification, the thiol-protected Au particles of 3.0 nm were adsorbed on TiO_2 in a solution of methylene chloride. After thermal treatment under hydrogen at 120°C only, thiols were eliminated, and the gold nanoparticles were 3.4 nm, i.e. almost the same size as those stabilised with thiols.

6.2.3 *One-pot deposition-reduction of gold*

The advantage of such a type of preparation is that deposition and reduction take place in a single step and that the extent of gold deposition is almost quantitative.

6.2.3.1 *Photo-deposition*

Photo-deposition is based on the principle that metal cations with appropriate redox potentials can be reduced by the photoelectrons created by bandgap illumination of semiconductors used as supports. Titania is the most appropriate support[9,66,67] UV irradiation of de-aerated solutions containing $HAuCl_4$ and titania led to both the deposition of gold and its reduction. Rather high gold loadings could be achieved ($\sim$4 wt %) with a 100% deposition yield. The average gold particle sizes were in the range of 5–10 nm, even when organic capping agents (poly-vinylalcohol (PVA), poly-vinylpyrrolidone (PVP) or poly-ethylene glycol (PEG) were added to the solution.[68] Smaller gold particles were obtained (4 instead of 8 nm) by raising the pH of the suspension to 8–10 to increase the reductive potential of photo-generated electrons.[69] According to other studies, gold particles can be either smaller (1.9 nm)[70] or larger (8.4 nm)[29] than those obtained after deposition-thermal reduction (3 or 5 nm). Mechanisms of photoreduction assisted by the titania were proposed.[66,67] Gold photo-deposition was also efficient on ZnO (d_{Au} = 4–20 nm),[71] but not on tin and iron oxides although they are also semiconductors.[68] Gold photo-deposition was also efficient on composites such as ZnO-CNT (carbon nanotube) (d_{Au} = 5–8 nm)[72] or C-TiO_2 composites (d_{Au} = 9–20 nm).[73]

This method was also applied to gold deposition on other types of supports such as calcium alginate hydrogel beads; the carboxylate and hydroxyl functionalities help them to act as both a reductant and a stabiliser. UV activation helps the reduction to occur within a short time period, leading to gold particles of 5 nm.[74]

6.2.3.2 Sonication

Like UV irradiation, sonication is used to simultaneously reduce and deposit gold on a support. This method is also used to synthesise gold colloids (see Chapter 5). Sonication induces chemical changes due to the cavitation phenomena, which favours the formation, growth and implosive collapse of bubbles in liquid, and leads to the decomposition of water molecules in H^+ and OH^+ radicals. The H^+ radicals can trigger the reduction of Au^{III} species in solution.

The Gedanken's group first developed this method, and applied it to the deposition of gold on silica submicrospheres,[75] and on powders of silica and titania.[76] The supports were immersed in a solution of $HAuCl_4$ in water or ethylene glycol basified with ammonia. After sonication under an Ar or Ar and H_2 atmosphere, washing and drying, gold particles of $\sim$4–5 nm in average were obtained, and all gold was deposited onto the supports (5–7 wt %). Sonication was also applied to the preparation of gold on maghemite (γ-Fe$_2$O$_3$) in the presence of poly-ethylene glycol monostearate (PEG-MS), gold particles of 6.7 nm on average were obtained. PEG-MS was supposed to act both as a surfactant and as a reducing agent in the sonochemical reaction.[77] Gold on carbon (d$_{Au}$ = 4–8 nm)[78] and gold on ZnO (d$_{Au}$ between 5 and 10 nm in average depending on the type of ZnO) were also prepared using this technique.[71]

Note that most often, sonication is simply used in the preparation of supported gold to disperse the powder support in solution[71] or to avoid the aggregation of gold colloids and favour their interaction with a support.[58]

6.2.3.3 *Microwave irradiation*

This method is also used to prepare colloids in solution (see Chapter 5). Again a reducer is added and it can also act as a nanoparticle stabiliser: PVP, PEG, EG. The advantage of microwave dielectric heating over convective heating is that microwaves can heat a substance quickly and uniformly and generate rapidly more homogeneous nucleation sites. When microwave irradiation is applied in the presence of the support in suspension, oxide or carbon, the metal particles formed can interact with the support.

For deposition-reduction on carbon, ethylene glycol was used to assist microwave irradiation: the dielectric constant and the dielectric loss of ethylene glycol are high, so reduction occurs rapidly under microwave irradiation. Small gold particles were obtained, 2 nm with ethylene glycol[79] and 2.6 nm with poly-ethylene glycol.[80]

Additionally, microwaves can assist reduction. This is the case for gold reduction by citrate. After pH adjustment with NaOH, a solution containing HAuCl$_4$ and sodium citrate was irradiated with microwaves for 10 min.[81] After cooling the solution to room temperature, activated carbon was immersed and stirred, then washed and dried (0.5 wt % Au). The smallest gold particles (10 nm) were obtained for a pH of 6.8 or 8.

In some cases, support and colloids can be prepared in one pot. This was the case for instance of Au/CeO$_2$:[82] HAuCl$_4$ was mixed with cerium nitrate in ethanol to obtain the desired gold loading (2%, 5% or 10%) in the presence of PEG or PVP as a protective polymer. After microwaving, the gold particles on ceria, too small to be detected by X-ray diffraction, were probably smaller than 5 nm.

6.3 Gold Embedded into an Inorganic Matrix or in the Porosity of Materials

There are several strategies for the preparation of gold nanoparticles embedded in matrices. Such types of material may be needed for studies of the optical properties (see Chapter 3). When these matrices contain residual porosity or ordered or disordered porous structures, and gold particles are accessible to reactants, these materials can be used for catalysis. This section is divided into two subsections, gold in a matrix and gold in a porous material. Note that the preparation of gold-oxide core-shell nanoparticles is reported in Section 5.5.5 of Chapter 5.

6.3.1 *Gold embedded into a matrix*

Most of the methods described below involve one-pot preparation of both the matrix and the gold particles. Usually, these methods generate porosity in the matrix. These methods can also generate oxide particles of the same size range as the gold particles, allowing the gold particles to be located on the external surfaces of the oxide particles.

One method is *co-precipitation*, which has been used to prepare gold particles on or in oxide supports, such as α-Fe_2O_3, Co_3O_4, MnO_2, CuO, CeO_2, ZnO.[83–89] Co-precipitation is generally performed by addition of sodium carbonate to aqueous solutions containing both $HAuCl_4$ and the nitrate precursor of the oxide support at controlled pH and temperature. After aging, the co-precipitate is washed, dried, and calcined to obtain metallic gold and to transform the oxy-hydroxy-carbonate into oxide. It is not always clear in the papers whether all the gold in the solution is precipitated, which fraction of gold particles is located on the support surface, and which fraction is embedded into the matrix. Co-precipitation can lead to small gold particles (≤ 5 nm), but is limited to the supports that can precipitate at a rate consistent with that of gold precipitation or deposition; the gold particle size depends on the affinity between gold and the support, and on the temperature and pH of precipitation.

Another one-pot preparation is the *sol-gel method*. It involves the hydrolysis of an alkoxide precursor of the oxide support (for instance, tetraethoxysilane (TEOS) or tetrabutoxytitanate), in a water-alcohol solution containing HAuCl$_4$[90-92] or other gold precursors such as gold acetate and hydrogen tetranitratoaurate.[93] Most of the preparations involve a gelation step performed at controlled pH and temperature, sometimes in an autoclave, followed by washing and drying steps (xerogel), and finally by thermal treatments to remove the organics, transform the hydroxide into oxide (aerogel) and reduce gold. Gold particle size varies between 10 and 100 nm depending on the preparation conditions. As in the case of coprecipitation, a fraction of the gold particles may be embedded into the oxide, and another one located on the surface.

Gold can also be introduced as colloids into the solution mixture used for the sol-gel synthesis of the oxide support (*reduction-deposition*). For instance, gold-titania aerogels were synthesised from alkanethiol-stabilised gold particles (2–3 nm, Brust method, see Chapter 5) added to a titania sol.[94] After calcination at 425°C to crystallise the amorphous titania into anatase and to burn off the ligands, gold particles of 5–10 nm were found on the surface of the titania nanoparticles (10–12 nm). Mesoporous Au–TiO$_2$ nanocomposite microspheres were obtained by hydrothermal treatment at 180°C of the precipitate resulting from mixing tetrabutyl titanate (Ti(OC$_4$H$_9$)$_4$) into a solution of water-ethanol containing citrate-stabilised gold colloids (10 nm).[95] Gold particles of 6–7 nm in size were well distributed and in close contact with the elemental anatase particles (7–8 nm) that constitute TiO$_2$ microspheres (1.2–1.5 μm).

Budroni *et al.*[96] used a modified sol–gel approach to incorporate gold nanoparticles into an "open shell" of silica, so as to limit gold particle sintering at high temperature while keeping the particles accessible to reactants for catalysis. The synthesis involved the formation of gold nanoparticles capped with both 1-dodecanethiol and 3-mercaptopropyltrimethoxysilane (MPTS) and dispersed in ethanol containing TEOS. The hydrolysis of TEOS was catalyzed by NH$_4$F, and the alkoxysilane groups of MPTS promoted the hydrolysis and condensation of TEOS around the gold particles and provided a link between the metal particles and the silica. The inert alkane chains of dodecanethiol behaved as physical spacers between the particles

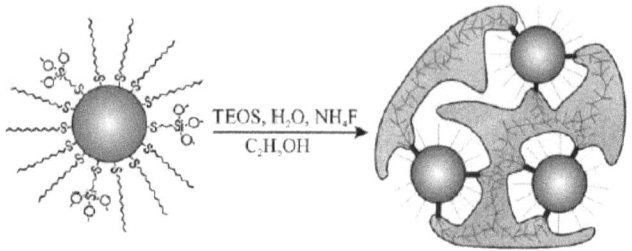

Fig. 6.4. Schematic representation of gold nanoparticles capped with 1-dodecanethiol (DT) and MPTS and the AuNP-organic-SiO$_2$ structure. MPTS (thick lines) on the right-hand side of the figure, provides the organic links between the metal particles and the inorganic mesoporous silica,[96] reproduced by permission of John Wiley and Sons.

and the silica shell. Calcination in air at 450°C resulted in gold particles of around 3.5 nm in pores of 9 nm (Fig. 6.4).

Self-assembly of water-soluble gold nanoparticle micelles with silica permits the formation of periodically ordered 3D gold nanoparticles in silica.[97] Water-soluble gold nanoparticle micelles were obtained by addition of 1-dodecanethiol-stabilised gold nanoparticles to an aqueous solution containing CTAB (cetyltrimethyl bromide) surfactant. Under vigorous stirring, a hybrid bilayer shell with defined primary (1-dodecanethiol) and secondary layers (CTAB) formed, which led to an oil-in-water microemulsion. After organic solvent evaporation, the water-soluble gold nanoparticle micelles self-assembled in aqueous medium, and formed hexagonally ordered gold nanoparticle arrays. The addition of tetraethyl orthosilicate (TEOS) under basic conditions formed an ordered gold nanoparticles/silica mesophase composed of gold nanoparticles organised in a periodic FCC lattice within a dense silica matrix. Under acidic conditions, siloxane condensation was inhibited and ordered thin-film NC/silica mesophases can be obtained by spin coating or casting that can be readily integrated into electronic devices.

6.3.2 *Gold in a porous matrix*

To form gold nanoparticles in a porous matrix, HAuCl$_4$ may also be incorporated during the sol-gel synthesis of structured mesoporous supports (*deposition-reduction*). The main difference with the sol-gel method described in the former section is that a surfactant is added to the solution mixture to structure the oxide, for instance, CTAB (cetyltrimethyl bromide) to TEOS to form MCM-41[98] or triblock copolymer to Ti tetrapropoxide to form mesoporous titanium oxide.[99] To better insure that the gold particles are located in the porosity of the matrix and are of controlled size, bifunctional amino-organosilane then $HAuCl_4$ were added to the MCM-41 synthesis mixture (CTAB and TEOS);[100] the amino group complexes the gold anions while the organosilane binds to the porous silica matrix. Thermal reduction of gold followed by removal of the templating surfactants via ion-exchange reaction led to ordered mesoporous materials containing uniform gold nanoparticles of 2–5 nm in the pores with a gold loading as high as 7 wt %.

Gold can also be introduced as colloids either into the solution mixture of the precursors of mesoporous oxides or during the gelation step (*reduction-deposition*). Gold particles smaller than the pore size were incorporated into mesoporous materials such as MCM-41, MCM-48 or SBA-15.[101–103] After calcination at 550°C, the gold particles were slightly larger (for instance 4 nm instead of 2 nm), but still homogeneously distributed in the mesoporous materials. Phenylethylthiol-coated gold nanoparticles of 2.1–2.4 nm were added to a sol-gel mixture of tetramethoxysilane, phenyltriethoxysilane, water and THF acidified with HCl[104] (Fig. 6.5). The use of phenyltriethoxysilane allows optimising the dispersion of the gold nanoparticles, which can interact via their capping agent with the phenyl groups of the

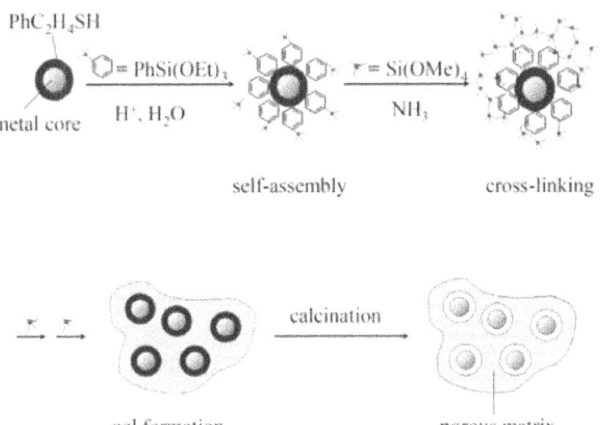

self-assembly   cross-linking

gel formation   porous matrix

Fig. 6.5. Reaction scheme for the encapsulation of metal clusters in a porous silica matrix,[104] reproduced by permission of the Royal Chemical Society.

matrix. After gelation, the organics were removed by calcination at 600°C in air. The encapsulated metal particles, which grew during the sol–gel process, reached an average diameter of 6 nm after calcination.

It is also possible to incorporate dendrimer-encapsulated gold nanoparticles (see Section 6.2.2.2) into amorphous matrices during sol-gel preparation, then to calcine to remove the organics. With this approach, the dendrimer not only determines the size and monodispersity of the gold nanoparticles, but also that of the porosity of the material. For instance, gold particles initially of 2 nm in PAMAM dendrimers barely increased after incorporation in the sol-gel matrix of titania and calcination at 500°C (2.7 nm), and they were found located in the 4.5 nm pores, which are the imprint of the PAMAM dendrimers[105] (Fig. 6.6).

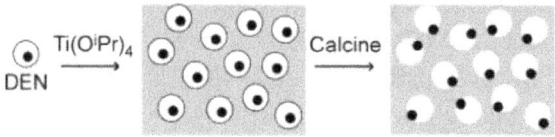

Fig. 6.6. Incorporation of DENs into an amorphous titania network prepared using sol-gel chemistry. Both the inside and the outside of the dendrimer act as templates: the exterior of the dendrimer templates pores within the sol-gel matrix, and the interior templates the metal nanoparticles. The dendrimer can be removed by calcination to leave behind just the metal nanoparticles; Reprinted with permission from Ref. 105. Copyright (2004) American Chemical Society.

Another way to incorporate gold nanoparticles into a porous matrix is to functionalise the mesoporous materials during or after synthesis before gold introduction. Mesoporous silica thin films were also functionalised with a thioether ((1,4)-bis(triethoxysilyl)propyl tetrasulfide (BPTS)) during synthesis.[106] After gelation, $[Au(en)_2]^{3+}$ were adsorbed on the SO_3^- groups, and after reduction at 300°C under hydrogen, the nanoparticles located in the mesopores of the thin film had a narrow size distribution with diameters in the size range of 3–7 nm. SBA-15 was functionalised with PAMAM dendrimers before HAuCl$_4$ was introduced and reduced with NaBH$_4$, leading to gold particles not exceeding 5 nm.[107] Various silicic structured mesoporous materials, HMS, SBA-15, SBA-16, MCM-41 and MCM-48 were functionalised with amine ligands ((3-aminopropyl)triethoxysilane (APTS) or N-[3-(trimethoxysilyl)propyl]ethylenediamine (DAPTS)), in water or ethanol.[108] Depending on the structure and size of the pores and also the calcination temperature (200 or 550°C), different sizes of gold particles could be obtained. Gold nanoparticles (around 6 nm) were also synthesised in preformed mesoporous SBA-15 materials by reduction of the gold ions by imines or hemiaminals grafted inside the channel pores of mesoporous silica.[109]

6.4 Gold Nanoparticles on/in Organic Materials

Gold nanoparticles can also be deposited on organic polymers. In this case, it is not easy to make the difference between deposition on the surface and embedding into the materials. In both cases, hybrid inorganic-organic materials are obtained, for applications in catalysis, life science or analytical science. As in the case of the inorganic supports (Section 6.2), it is possible to proceed via *deposition-reduction* or *reduction-deposition* using most often the same type of preparation method.

6.4.1 *Deposition-reduction*

As in the case of the oxide supports, very simple methods can be used. For instance, the *impregnation* of a cation-exchanged resin pretreated with NaOH, with HAuCl$_4$ followed by drying, led to gold already reduced into nanoparticles of around 10 nm.[110] Gold nanoparticles could also be generated within the interconnected macropores of poly(styrene-co-divinylbenzene) matrices (polyHIPE with pore size around 200–300 nm);[111] HAuCl$_4$ dissolved in water-acetone mixture was forced to fill up the void spaces of the matrix by applying and releasing a vacuum. Spontaneous reduction led to gold particles of around 50–200 nm.

The second method involves anion adsorption thanks to functional sites on the polymer that can interact with the gold precursor. The cationic sites of the solid polystyrene core, onto which long linear cationic polyelectrolyte chains (poly(2-aminoethyl methacrylate hydrochloride, PAEMH)) were grafted, were used to immobilise AuCl$_4^-$ ions in water;[112] reduction of these immobilised metal ions with NaBH$_4$ led to nanoparticles of 1 to 3 nm diameter in average. Miyamura *et al.*[113] developed a way to form gold nanoparticles stabilised by the benzene rings of polystyrene derivatives obtained by reduction of chlorotriphenylphosphine gold (AuClPPh$_3$) with NaBH$_4$ in the presence of the polymer. After thermal treatment at 150°C to cross-link the side chains and after washings, gold nanoparticles of around 1 nm were obtained. The thioether (R-S-R) group in gel-type polyacrylic resin[114] and related cross-linked co-polymers[115] was used as a coordination site for the $AuCl_3$ precursor prior to reduction to Au^0 with $NaBH_4$. According to the authors, the slightly smaller size of the gold nanoparticles (2.2 nm) with respect to the pore size of the polymer gel suggests that the cavities in the polymer gel (2.5 nm) prevent the gold nanoparticles from aggregating. A crosslinked poly(4-vinylpyridine) matrix of polymer beads on which poly(propyleneimine)-G2 dendrimers were grafted, was used to stabilise gold nanoparticles arising from the addition of $HAuCl_4$, which was then reduced with $NaBH_4$.[116] This led to gold particles of 10–20 nm.

The thermal grinding method was also used to deposit gold on polymers (see Section 6.2.1.3).

6.4.2 *Reduction-deposition*

The second case is based on gold colloid deposition. For instance, gold particles of 2.5 nm were adsorbed on a conducting polymer, polyaniline (PANI), by immersion of the latter in a gold sol stabilised with THPC, followed by washings with water.[117] Citrate-capped gold sol of different average sizes from 8 to 55 nm were immobilised on resin beads of polystyrene-based anion exchangers by stepwise introduction of portions of colloidal gold to replace gradually the chlorides on the resin, followed by washing with water to expel the Cl^-.[118] Adsorption of gold colloids of 20 nm size stabilised by citrate on polystyrene beads (107 μm) with surface positively charged by grafting of $-NH_2$ groups, led to a uniform coating.[119]

6.5 Gold Nanoparticles on Planar Surfaces

Materials based on gold nanoparticles supported on planar surfaces can be used as model catalysts (thin films or single crystals) for surface sciences studies (Chapter 8), electrodes for electrocatalysis, materials for physics studies (applications in surface enhanced Raman spectroscopy (SERS) and nanophotonics) and biosensors. As already mentioned, depending on the applications targeted, the gold particle size required may be between a few nm for catalytic applications and tens of nm for electronics and biology. In addition to the control of the particle size and shape, periodic arrangements of gold particles on these planar surfaces may be needed for instance for nanoelectronic applications. This can be done using various nanolithography techniques (physical techniques, top-down approach, see Chapter 8) or using chemical methods involving self-assembled block copolymer nanostructures as templates (bottom-up approach, see below in Section 6.5.2). Moreover, depending on the further use of the samples, the stabiliser molecules may be decomposed, often by ozone treatments or oxygen plasma treatments.

As for powder supports, the chemical preparations are based on *reduction-deposition* or *deposition-reduction*. Other deposition techniques are more specifically adapted to planar surfaces, e.g. *galvanic displacement* and *electrodeposition* (see below). The main characterisation techniques of the gold particle size and distribution on the substrate are atomic force microscopy (AFM) and scanning tunneling microscopy (STM).

This section is divided into two parts, non-ordered and ordered deposition of gold nanoparticles on planar surfaces.

6.5.1 *Non-ordered deposition*

The simplest method to deposit gold colloids on planar surfaces is probably the one which consists of depositing droplets of colloidal solution onto these surfaces, followed by water evaporation or spinning. The excess particles can be removed by mere rinsing with water or assisted by ultrasonication. Such a kind of preparation was performed on substrates like ITO (indium tin oxide), for electrochemical reactions[120] or for the fabrication of polyelectrolyte/gold nanoparticles multilayer films as memory devices with adjustable electronic properties.[121] In both cases, citrate-stabilised gold colloids were used; 40, 60 and 80 nm in the first case and 16 nm in the second. More tiny gold entities were also deposited by this method. Six-atom gold clusters in the form of $[Au_6(PPh_3)_6][BF_4]_2$ (Au_6L_6) were deposited onto a $TiO_2(110)$ single crystal after an acetone pretreatment of the substrate, which favours the dispersion of the clusters.[122] After irradiation of the $TiO_2(110)$ surface with an electron beam to remove the ligands, STM revealed single-unit entities of around 1 nm in height, corresponding to the six-atom gold clusters.

Another method involves the adsorption of gold colloids on function-alised substrates. Gold colloids stabilised by citrate (3, 12 and 50–60 nm) were adsorbed on ITO-coated glass substrate pre-functionalised with amine APTMS ((3-amino-propyl)tri-methoxysilane)[123] and on ITO and Pt electrodes after surface functionalisation with 1,12-dodecanediamine.[124] The same principle was used to bond preformed citrate-capped gold nanoparticles of various sizes (2.6, 12.6, 20, 40 and 60 nm)[125] and of different shapes (nanoprisms, nanoperiwinkles and flower-like)[126,127] on gold electrodes after self-assembling of 3-mercaptopropyltrimethoxysilane (MPTS). MPTS forms a sol-gel network around the electrode in which the gold particles are trapped.

Other procedures consist in *deposition-reduction*. One method is based on the *electrodeposition* of gold particles. A deposition potential (0 V vs SCE) was applied to an ITO substrate in the presence of a solution of $HAuCl_4$ and KCl as the supporting electrolyte.[124] The size of the gold particles depended on the electrodeposition time, 150 nm for 15 s and 525 nm for 30 min. Another method involves the formation of gold nanoparticles on semiconductive substrates by *galvanic displacement*, a type of electroless deposition for which the metal ions in aqueous solution are reduced by the electrons arising from the semiconductor itself. Immersion of an InP(100) wafer in an aqueous solution of $HAuCl_4$ at RT resulted in the rapid deposition of strongly adhering gold nanoparticles of several tens of nm.[128] With both these methods, there is no organics around the gold particles in the sample.

6.5.2 Ordered deposition

As mentioned above, self-organisation of diblock copolymers into micellar structures in an appropriate solvent is a way to end up with well-ordered arrays of metal nanoparticles on planar surfaces with narrow distributions in particle size and interparticle spacing.

This method was developed by Spatz *et al.*[129] Inversed micelles were generated by dissolution of amphiphilic PS-*b*-P2VP diblock copolymers (poly(styrene)-block-poly(2-vinylpyridine)) in toluene. Micelles formed with a P2VP core surrounded by a PS corona. When $HAuCl_4$ was added, the $AuCl_4^-$ anions bound to the polar core of the micelles by protonating the pyridine units. Reduction by addition of anhydrous hydrazine (N_2H_4) resulted in the formation of metal particles. Dip coating of planar substrates (mica, thin glass plates, n^+-doped Si wafers, carbon grids) led to self-assembled monolayers of micelles whereby the particles arranged in a mesoscopic quasi-hexagonal two-dimensional lattice (Fig. 6.7). Oxygen plasma treatment yielded naked gold particles of identical size and in the same quasi-hexagonal order as in the micellar film. The size of the clusters could be varied between 1 and 15 nm depending on the concentration of the metal salt, and the interparticle distance, between 30 and 140 nm, depending on the length of the blocks.[129] This type of preparation involving PS-*b*-P2VP diblock copolymers has been extended to other planar substrates from conducting to insulating substrates: natively oxidised n-doped Si(001), Ti on n-doped Si(001), Ti foil and ITO coated glass,[130] glassy carbon[131] and also to single crystal surfaces such as rutile $TiO_2(110)$.[132,133] The main difference with the Spatz method described above is that oxygen or argon plasma treatments, which were performed to decompose the micelles, were also used to reduce gold. The work performed on single crystals[133] also showed that the degree of order and the interparticle distance depended sensitively on the dip-coating velocity, and less on the deposition temperature and the nature of the substrate. It confirmed that the oxygen plasma treatment does not affect the gold particle size and the distance between the particles.

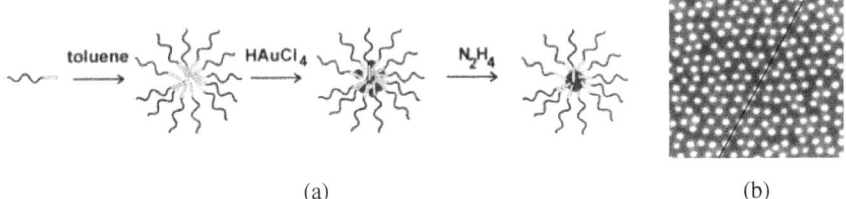

(a)   (b)

Fig. 6.7. (a) Schematic drawing of the micelle formation of poly(styrene)-block-poly(2-vinylpyridine) (PS-*b*-P2VP) block copolymers in toluene; (b) SFM topography image of a monomicellar film cast from a PS(1700)-*b*-P[2VP(HAuCl4)0.3(450)] solution onto a glass substrate. Reprinted with permission from Ref. 129. Copyright (2000) American Chemical Society.

Another strategy is to self-assemble diblock-copolymer micelles on surfaces, before filling them with gold ions and before reduction. Semiconductive surfaces of doped Si(100), Ge(100), GaAs(100) and InP(100) coated with a self-assembled monolayer of PS-*b*-P2VP or PS-*b*-P4VP, were immersed in an aqueous solution of $HAuCl_4$, and HF was added to initiate galvanic displacement, i.e. gold reduction.[134] Elimination of the micelles was performed by dissolution in toluene. According to the authors, this strategy leads to better patterning of the gold particles than the former one. Depending on the molecular weight of the polymer, the gold particle size was varied between 8 and 60 nm and the interparticle distance between 90 and 170 nm. Laskar *et al.*[135] employed the same strategy. On a fluorinated surface of silicium wafer, diblock copolymers PEO_m-b-$PMA(Az)_n$ (amphiphilic block copolymers consisting of a hydrophilic polyethylene oxide block and a hydrophobic polymethacrylate block bearing an azobenzene unit), formed ordered arrays of spherical micelles consisting of PEO cores surrounded by PMA(Az) coronas. After immersion of the plates in an aqueous solution containing gold ions, excimer vacuum UV irradiation to reduce gold and eliminate the micelles, AFM images showed regular hexagonal arrays of gold particles of 3.8 nm in height.

6.6 Conclusion

The actual trend is to make these materials more complex. The supports can themselves be of nanosize; it is possible to prepare nanosized silica-core-gold shell particles; their preparation has already been described in Chapter 5. Gold can be associated with another metal and form bimetallic particles, for instance, for the purpose of catalysis. Still more complex nanomaterials combining several properties can be also synthesised: for instance multifunctional microspheres which consist of a core of nonporous silica-protected magnetite particles, a transition layer of active gold nanoparticles, and an outer shell of mesoporous silica.[136,137]

Acknowledgements

The author thanks Dr Rachel Doherty who read the chapter and revised the English.

References

1. G.C. Bond, C. Louis and D. Thompson, *Catalysis by Gold*, vol. 6, Imperial College Press, London, 2006.
2. S. Lin and M.A. Vannice, *Catal. Lett.* **10** (1991) 47.
3. M. Haruta, S. Tsubota, T. Kobayashi, H. Kageyama, M.J. Genet and B. Delmon, *J. Catal.* **144** (1993) 175.
4. H.S. Oh, J.H. Yang, C.K. Costello, Y.M. Wang, S.R. Bare, H.H. Kung and M.C. Kung, *J. Catal.* **210** (2002) 375.
5. R. Zanella, S. Giorgio, C.R. Henry and C. Louis, *J. Phys. Chem. B* **106** (2002) 7634.
6. A. Hugon, N. El Kolli and C. Louis, *J. Catal.* **274** (2010) 239.
7. L. Delannoy, N. El Hassan, A. Musi, N. Nguyen Le To, J.-M. Krafft and C. Louis, *J. Phys. Chem. B* **110** (2006) 22471.
8. S. Tsubota, D.A.H. Cunningham, Y. Bando and M. Haruta, *Stud. Surf. Sci. Catal.* **91** (1995) 227.
9. G.R. Bamwenda, S. Tsubota, T. Nakamura and M. Haruta, *Catal. Lett.* **44** (1997) 83.
10. A. Wolf and F. Schüth, *Appl. Catal. A* **226** (2002) 1.
11. M. Haruta, *Cattech* **6** (2002) 102.
12. R. Zanella, L. Delannoy and C. Louis, *Appl. Catal. A* **291** (2005) 62.
13. L. Prati and G. Martra, *Gold Bull.* **32** (1999) 96101.
14. F. Moreau and G.C. Bond, *Catal. Today* **122** (2007) 260.
15. C.-K. Chang, Y.-J. Chen and C.-T. Yeh, *Appl. Catal. A* **174** (1998) 13.
16. S.-J. Lee and A. Gavriilidis, *J. Catal.* **206** (2002) 305.
17. L.A. Hermans and J.W. Geus, *Stud. Surf. Sci. Catal.* **4** (1979) 113.
18. J.A. van Dillen, J.W. Geus, L.A. Hermans and J. van der Meijden, in *Proc. 6th Intern. Congr. Catal., London, 1976* (eds P.B.Wells, G.C. Bond and F.C. Tompkins), The Chemical Society, London, 1977, p. 677.
19. D. Guillemot, V.Y. Borovskov, V.B. Kazansky, M. Polisset-Thfoin and J. Fraissard, *J. Chem. Soc., Faraday Trans.* **93** (1997) 3587.
20. D. Guillemot, M. Polisset-Thfoin and J. Fraissard, *Catal. Lett.* **41** (1996) 143.
21. R. Zanella, A. Sandoval, P. Santiago, V.A. Basiuk and J.M. Saniger, *J. Phys. Chem. B*, **110** (2006) 8559.
22. H. Zhu, Z. Ma, J.C. Clark, Z.Pan, S.H. Overbury and S. Dai, *Appl. Catal. A*, **326** (2007) 89.
23. H. Zhu, C. Liang, W. Yan, S.H. Overbury and S. Dai, *J. Phys. Chem. B* **110** (2006) 10842.
24. Y. Guan and E.J.M. Hensen, *Appl. Catal. A* **361** (2009) 49.
25. D.A. Bulushev, I. Yuranov, E.I. Suvorova, P.A. Buffat and L. Kiwi-Minsker, *J. Catal.* **224** (2004) 8.
26. B.P. Block and J. J. C. Bailar, *J. Am. Chem. Soc.* **73** (1951) 4722.
27. H. Yin, Z. Ma, S.H. Overbury and S. Dai, *J. Phys. Chem. C* **112** (2008) 8349.
28. K. Mallick, M.J. Witcomb and M.S. Scurell, *Appl. Catal. A* **259** (2004) 163.

29. R. Isono, T. Yoshimura and K. Esumi, *J. Coll. Inter. Sci.* **288** (2005) 177.
30. Y. Yuan, A.P. Kozlova, K. Asakura, H. Wan, K. Tsai and Y. Iwasawa, *J. Catal.* **170** (1997) 191.
31. A.I. Kozlov, A.P. Kozlova, H. Liu and Y. Iwasawa, *Appl. Catal. A* **182** (1999) 9.
32. T.V. Choudhary, C. Sivadinarayana, C.C. Chusuei, A.K. Datye, J.P.F. Jr and D.W. Goodman, *J. Catal.* **207** (2002) 247.
33. Z. Yan, S. Chinta, A.A. Mohamed, J.J.P. Fackler and D.W. Goodman, *Catal. Lett.* **111** (2006) 15.
34. J. Guzman and B.C. Gates, *Nano Lett.* **1** (2001) 689.
35. J. Guzman and B.C. Gates, *Langmuir* **19** (2003) 3897.
36. J.C. Fierro-Gonzalez and B.C. Gates, *J. Phys. Chem. B* **109** (2005) 7275.
37. J.C. Fierro-Gonzalez, V.A. Bhirud and B.C. Gates, *Chem. Comm.* (2005) 5275.
38. J.C. Fierro-Gonzalez and B.C. Gates, *J. Phys. Chem. B* **108** (2004) 16999.
39. M. Okumura, S. Tsubota and M. Haruta, *J. Mol. Catal. A* **199** (2003) 73.
40. M. Hisamoto, R.C. Nelson, M.-Y. Lee, J. Eckert and S.L. Scott, *J. Phys. Chem. C* **113** (2009) 8794.
41. T. Ishida, M. Nagaoka, T. Akita and M. Haruta, *Chem. Eur. J.* **14** (2008) 8456–8460.
42. T. Ishida, N. Kinoshita, H. Okatsu, T. Akita, T. Takei and M. Haruta, *Angew. Chem. Int. Ed.* **47** (2008) 9265.
43. H. Okatsu, N. Kinoshita, T. Akita, T. Ishida and M. Haruta, *Appl. Catal. A* **369** (2009) 8.
44. J. Huang, T. Takei, T. Akita, H. Ohashi and M. Haruta, *Appl. Catal. B* **95** (2010) 430.
45. T. Ishida, H. Watanabe, T. Bebeko, T. Akita and M. Haruta, *Appl. Catal. A* **377** (2010) 42.
46. T. Ishida, N. Kawakita, T. Akita and M. Haruta, *Stud. Surf. Sci. Catal.* **175** (2010) 839.
47. F. Porta, L. Prati, M. Rossi, S. Coluccia and G. Martra, *Catal. Today* **61** (2000) 165.
48. D. Caruntu, B.L. Cushing, G. Caruntu and C.J. O'Connor, *Chem. Mater.* **17** (2005) 3398.
49. N. Hickey, P. Arneodo Larochette, C. Gentilini, L. Sordelli, L. Olivi, S. Polizzi, T. Montini, P. Fornasiero, L. Pasquato and M. Graziani, *Chem. Mater.* **19** (2007) 650.
50. H. Yin, Z. Ma, M. Chi and S. Dai, *Catal. Lett.* **136** (2010) 209.
51. N. Zheng and G.D. Stucky, *J. Am. Chem. Soc.* **128** (2006) 14278.
52. T. Herranz, X. Deng, A. Cabot, Z. Liu, G. Soler-Illia and M. Salmeron, *Catal. Today* **143** (2009) 158.
53. J.D. Grunwaldt, M. Maciejewski, O.S. Becker, P. Fabrizioli and A. Baiker, *J. Catal.* **186** (1999) 458.
54. J. Chou and E.W. McFarland, *Chem. Comm.* (2004) 1648.
55. M. Tominaga, T. Shimazoe, M. Nagashima and I. Taniguchi, *J. Electroanal. Chem.* **615** (2008) 51.
56. G. Martra, L. Prati, C. Manfredotti, S. Biella, M. Rossi and S. Coluccia, *J. Phys. Chem. B* **107** (2003) 5453.
57. M. Comotti, W.-C. Li, B. Spliethoff and F. Schüth, *J. Am. Chem. Soc.* **126** (2006) 917.
58. Z. Zhong, J. Lin, S.-P. Teh, J. Teo and F.M. Dautzenberg, *Adv. Funct. Mater.* **17** (2007) 1402.

59. C.G. Long, J.D. Gilbertson, G. Vijayaraghavan, K.J. Stevenson, C.J. Pursell and B.D. Chandler, *J. Am. Chem. Soc.* **130** (2008) 10103.
60. L.D. Menard, F. Xu, R.G. Nuzzo and J.C. Yang, *J. Catal.* **243** (2006) 64.
61. J. Llorca, A. Casanovas, M. Domınguez, I. Casanova, I. Angurell, M. Seco and O. Rossell, *J. Nanopart. Res.* **10** (2008) 537.
62. J. Chou, N.R. Franklin, S.-H. Baeck, T.F. Jaramillo and E.W. McFarland, *Catal. Lett.* **95** (2004) 107.
63. R.W.J. Scott, O.M. Wilson and R.M. Crooks, *J. Phys. Chem. B* **109** (2005) 692.
64. B.D. Chandler and J.D. Gilbertson, in *Nanoparticles and Catalysis* (ed D. Astruc), Wiley-VCH, Weinhein, 2007.
65. R.W.J. Scott, O.M. Wilson and R.M. Crooks, *Chem. Mater.* **16** (2004) 5682.
66. A. Fernandez, A. Caballero, A.R. Gonzalez-Elipe, J.-H. Herrmann, H. Dexpert and F. Villain, *J. Phys. Chem.* **99** (1995) 3303.
67. C.-Y. Wang, C.-Y. Liu, X. Zheng, J. Chen and T. Shen, *Coll. Surf. A* **131** (1998) 271.
68. D. Li, J.T. McCann, M. Gratt and Y. Xia, *Chem. Phys. Lett.* **394** (2004) 387.
69. R. Kydd, K. Chiang, J. Scott and R. Amal, *Photochem. Photobiol. Sci.* **6** (2007) 829.
70. T. Soejima, H. Tada, T. Kawahara and S. Ito, *Langmuir* **18** (2002) 4191.
71. S.A.C. Carabineiro, B.F. Machado, R.R. Bacsa, P. Serp, G. Drazic, J.L. Faria and J.L. Figueiredo, *J. Catal.* **273** (2010) 191.
72. J. Khanderi, R.C. Hoffmann, J. Engstler, J.J. Schneider, J. Arras, P. Claus and G. Cherkashinin, *Chem. Eur. J.* **16** (2010) 2300.
73. N.R. de Tacconi, K. Rajeshwar, W. Chanmanee, V. Valluri, W.A. Wampler, W.-Y. Lin and L. Nikiel, *J. Electrochem. Soc.* **157** (2010) B147.
74. S. Saha, A. Pal, S. Kundu, S. Basu and T. Pal, *Langmuir* **26** (2009) 2885.
75. V.G. Pol, A. Gedanken and J. Calderon-Moreno, *Chem. Mater.* **15** (2003) 1111.
76. N. Perkas, V.G. Pol, S.V. Pol and A. Gedanken, *Cryst. Growth Design* **6** (2006) 293.
77. Y. Mizukoshi, Y. Tsuru, A. Tominaga, S. Seino, N. Masahashi, S. Tanabe and A. Yamamoto, *Ultrason. Sonochem.* **15** (2008) 875.
78. H. Bunazawa and Y. Yamazaki, *J. Power Sources* **190** (2009) 210.
79. L. Ma, H. Zhang, Y. Liang, D. Xu, W. Ye, J. Zhang and B. Yi, *Catal. Comm.* **8** (2007) 921.
80. H. Zhu, Y. Liu, L. Shen, Y. Wei, Z. Guo, H. Wang, K. Han and Z. Chang, *Int. J. Hydr. Energ.* **35** (2010) 3125.
81. P. Zhang, B. Zhang and R. Shi, *Front. Environ. Sci. Engin. China* **3** (2009) 281.
82. G. Glaspell, L. Fuoco and M.S. El-Shall, *J. Phys Chem. B* **109** (2005) 17350.
83. M. Haruta, H. Kageyama, N. Kamijo, T. Kobayashi and F. Delannay, *Stud. Surf. Sci. Catal.* **44** (1988) 33.
84. G.J. Hutchings, M.R.H. Siddiqui, A. Burrows, C.J. Kielyb and R. Whymana, *J. Chem. Soc., Faraday Trans.* **93** (1997) 187.
85. S.-J. Lee, A. Gavriilidis, Q.A. Pankhurst, A. Kyek, F.E. Wagner, P.C.L. Wong and K.L. Yeung, *J. Catal.* **200** (2001) 298.
86. M. Khoudiakov, M.-C. Gupta and S. Deevi, *Appl. Catal. A* **291** (2005) 151.
87. B.E. Solsona, T. Garcia, C. Jones, S.H. Taylor, A.F. Carley and G.J. Hutchings, *Appl. Catal. A* **312** (2006) 67.
88. B. Qiao, J. Zhang, L. Liu and Y. Deng, *Appl. Catal. A* **340** (2008) 220.

89. S. Kudo, T. Maki, M. Yamada and K. Mae, *Chem. Eng. Sci.* **65** (2010) 214.
90. F.B. Li and X.Z. Li, *Appl. Catal. A* **228** (2002) 15.
91. H. Kozuka and S. Sakka, *Chem. Mater.* **5** (1993) 222.
92. P. Innocenzi, G. Brusatin, A. MArtucci and K. Urabe, *Thin Solid Films* **279** (1996) 23.
93. E. Seker and E. Gulari, *Appl. Catal. A* **232** (2002) 203.
94. J.J. Pietron, R.M. Stroud and D.R. Rolison, *Nano Lett.* **2** (2002) 545.
95. J. Yu, L. Yue, S. Liu, B. Huang and X. Zhang, *J. Coll. Interf. Sci.* **334** (2009) 58.
96. G. Budroni and A. Corma, *Ang. Chem., Int. Ed.* **45** (2006) 3328.
97. H. Fan, K. Yang, D.M. Boye, T. Sigmon, K.J. Malloy, H. Xu, G.P. Lopez and C.J. Brinker, *Science* **304** (2004) 567.
98. G. Lu, R. Zhao, G. Qian, Y. Qi, X. Wang and J. Suo, *Catal. Lett.* **97** (2004) 115.
99. A.A. Ismail, D.W. Bahnemann, I. Bannat and M. Wark, *J. Phys. Chem. C* **113** (2009) 7429–7435.
100. H. Zhu, B. Lee, S. Dai and S.H. Overbury, *Langmuir* **19** (2003) 3974.
101. Y.-S. Chi, H.-P. Lin, C.-N. Lin, C.-Y. Mou and B.-Z. Wan, *Stud. Surf. Sci. Catal.* **141** (2002) 329.
102. S. Cheng, Y. Wei, Q. Feng, K.-Y. Qiu, J.-B. Pang, S.A. Jansen, R. Yin and K. Ong, *Chem. Mater.* **15** (2003) 1560.
103. Z. Konya, V.F. Funtes, I. Kiricsi, J. Zhu, J.W. Ager, M.K. Ko, H. Frei, A.P. Alivisatos and G.A. Somorjai, *Chem. Mater.* **15** (2003) 1242.
104. N.M. Wichner, J. Beckers, G. Rothenberg and H. Koller, *J. Mater. Chem.* **20** (2010) 3840–3847.
105. R.W.J. Scott, O.M. Wilson and R.M. Crooks, *Chem. Mater.* **16** (2004) 5682.
106. J. Gu, L. Xiong, J. Shi, Z. Hua, L. Zhang and L. Li, *J. Solid State Chem.* **179** (2006) 1060–1066.
107. H. Li, Z. Zheng, M. Cao and R. Cao, *Microp. Mesop. Mater.* **136** (2010) 42.
108. B. Lee, Z. Maa, Z. Zhang, C. Park and S. Dai, *Microp. Mesop. Mater.* **122** (2009) 160.
109. Y.W. Xie, S. Quinlivan and T. Asefa, *J. Phys. Chem. C* **112** (2008) 9996–10003.
110. F. Shi and Y. Deng, *J. Catal.* **211** (2002) 548.
111. C. Feral-Martin, M. Birot, H. Deleuze, A. Desforges and R. Backov, *React. Funct. Polym.* **67** (2007) 1072–1082.
112. M. Schrinner, F. Polzer, Y. Mei, Y. Lu, B. Haupt, M. Ballauff, A. Goldel, M. Drechsler, J. Preussner and U. Glatzel, *Macromol. Chem. Phys.* **208** (2007) 1542–1547.
113. H. Miyamura, R. Matsubara, Y. Miyazaki and S. Kobayashi, *Angew. Chem. Int. Ed.* **46** (2007) 4151–4154.
114. B. Corain, C. Burato, P. Centomo, S. Lorac, W. Meyer-Zaikad and G. Schmidd, *J. Mol. Catal. A* **225** (2005) 189–195.
115. C. Burato, P. Centomo, G. Pace, M. Favaro, L. Prati and B. Corain, *J. Mol. Catal. A* **238** (2005) 26.
116. E. Murugan and R. Rangasamy, *J. Polym. Sci. A* **48** (2010) 2525–2532.
117. F. Klasovsky, M. Steffan, J. Arras, J. Radnik and P. Claus, *Open Phys. Chem. J.* **1** (2007) 1.
118. S. Panigrahi, S. Basu, S. Praharaj, S. Pande, S. Jana, A. Pal, S.K. Ghosh and T. Pal, *J. Phys. Chem. C* **111** (2007) 4596.

119. Y.-C. Cao, Z. Wang, X. Jin, X.-F. Hua, M.-X. Liu and Y.-D. Zhao, *Coll. Surf. A* **334** (2009) 53.
120. T. Miyazaki, R. Hasegawa, H. Yamaguchi, H. Oh-Oka, H. Nagato, I. Amemiya and S. Uchikoga, *J. Phys. Chem. C* **113** (2009).
121. J.-S. Lee, J. Cho, C. Lee, I. Kim, J. Park, Y.-M. Kim, H. Shin, J. Lee and F. Caruso, *Nature Nanotech.* **2** (2007) 790.
122. C.C. Chusuei, X. Lai, K.A. Davis, E.K. Bowers, J.P. Fackler and D.W. Goodman, *Langmuir* **17** (2001) 4113.
123. S. Kumar and S. Zou, *J. Phys. Chem. B* **109** (2005) 15707.
124. P. Diao, D.F. Zhang, M. Guo and Q. Zhang, *J. Catal.* **250** (2007) 247.
125. P. Kalimuthu and S.A. John, *J. Electroanal. Chem.* **617** (2008) 164–170.
126. B.K. Jena and C.R. Raj, *J. Phys. Chem. C* **111** (2007) 15146.
127. B.K. Jena and C.R. Raj, *Langmuir* **23** (2007) 4064.
128. M.R.H. Nezhad, M. Aizawa, L.A.P. Jr., A.E. Ribbe and J.M. Buriak, *Small* **1** (2005) 1076.
129. J.P. Spatz, S. Mossmer, C. Hartmann, M. Moller, T. Herzog, M. Krieger, H.-G. Boyen, P. Ziemann and B. Kabius, *Langmuir* **16** (2000) 407.
130. B.R. Cuenya, S.-H. Baeck, T.F. Jaramillo and E.W. McFarland, *J. Am. Chem. Soc.* **125** (2003) 12928.
131. S. Kumar and S. Zou, *Langmuir* **25** (2009) 574.
132. S. Kielbassa, A. Habich, J. Schnaidt, J. Bansmann, F. Weigl, H.-G. Boyen, P. Ziemann and R.J. Behm, *Langmuir* **22** (2006) 7873.
133. J. Bansmann, S. Kielbassa, H. Hoster, F. Weigl, H.G. Boyen, U. Wiedwald, P. Ziemann and R.J. Behm, *Langmuir* **23** (2007) 10150.
134. M. Aizawa and J.M. Buriak, *Chem. Mater.* **19** (2007) 5090.
135. I.R. Laskar, S. Watanabe, M. Hada, H. Yoshida, J. Li and T. Iyoda, *Surf. Sci.* **603** (2009) 625.
136. J. Ge, Q. Zhang, T. Zhang and Y. Yin, *Angew. Chem. Int. Ed.* **47** (2008) 8924 –8928.
137. Y. Deng, Y. Cai, Z. Sun, J. Liu, C. Liu, J. Wei, W. Li, C. Liu, Y. Wang and D. Zhao, *J. Am. Chem. Soc.* **132** (2010) 8466–8473.

Chapter 7

Catalytic Properties of Gold Nanoparticles

Geoffrey C. Bond

Emeritus Professor, Brunel University, Uxbridge UB8 3PH, UK.
Email: geoffrey10bond@aol.com

7.1 Introduction

For many years it was believed that gold had no catalytic properties of any significance; there were only a few claims in the patent literature, and some academic studies using high temperature. The situation changed in the 1970s and 1980s when it was discovered that very small gold particles showed activity for alkene hydrogenation, and, particularly when supported on transition metal oxides (see Chapter 6), for CO oxidation. Since that time there has been a tremendous expansion in the range of reactions that can be effectively catalysed by gold; these include the selective oxidation of functionalised organic molecules, often showing remarkably high selectivity to the desired product. Some of these reactions are already achieving commercial exploitation. CO oxidation has been intensively investigated, and gold catalysts are applied to the production of pure H_2 by removing CO traces. These exciting and unexpected developments have challenged our ability to understand how catalysts really work: gold often succeeds where other metals fail by adsorbing reactants with just the right strength to facilitate desired reactions, and not so strongly as to form unreactive intermediates. This chapter also considers the possible role of gold catalysts in environmentally important reactions such as those occurring in the exhaust of internal combustion engines, and briefly treats their role in electrocatalytic and photocatalytic processes.

7.2 The Origin of Gold's Catalytic Properties

Research performed over the past two decades has shown beyond doubt that significant properties of gold as a catalyst are only shown by very small particles. This requirement is not merely to increase the available gold surface, but additional features are then revealed. As size decreases, the fraction of surface atoms rises. This affects a whole range of physical properties, such as melting temperature, but more atoms of low coordination number at the corners and edges of particles are also created, and these are often claimed to be important for catalysis (see Fig. 7.1). Particles of the required size are made either as colloidal dispersions (Chapter 5) or as supported catalysts, where the particles are supported on a material such as TiO_2, SiO_2 or carbon (Chapter 6) to enable them to be used at moderate temperatures. In this latter case, gold atoms at the periphery of the particle where they are adjacent to the support have been assigned a prominent role. Supported gold catalysts can be prepared either by chemical routes (Chapter 6) or for surface science studies by atom deposition under vacuum conditions (Chapter 8), but the particles are not equivalent because the state of hydroxylation of the support and the metal-support interactions are different.

Many of the physicochemical properties that are size-dependent have also been claimed as essential criteria for catalysis, but since so many of them

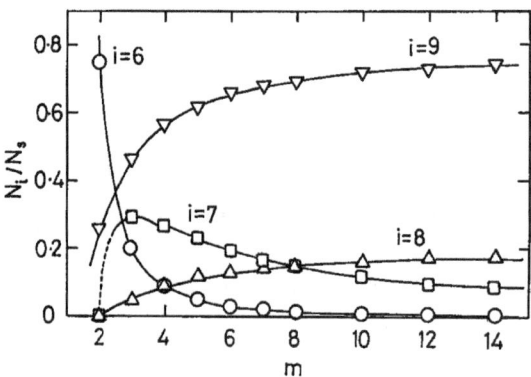

Fig. 7.1. Fraction of surface atoms having the indicated coordination numbers i (i = 6 for corners, i = 7 for edges, i = 8 for (100) plane and i = 9 for (111) plane) as a function of size m (in nm) for cubo-octahedral particles.

are associated it is difficult to identify any one of them as having over-riding importance. All of them are in some way influenced by the relativistic effect (Chapter 2), which determines gold's unique chemical behaviour, and since catalysis is essentially a chemical phenomenon (Section 7.3) we may expect effects that related metals (Cu, Ag) do not show. For reactions of small molecules (CO, H_2 with O_2, H_2O), activity increases markedly when size falls below about 2.5 nm (see Section 7.5.1), but small size is apparently less important for the selective oxidation of larger molecules. Early attempts to observe useful catalysis by gold[1-3] therefore failed because it was used as wire or foil or another form that did not contain particles of the requisite small size. We must now examine in more detail exactly how the catalytic activity of gold is determined.

7.3 Concepts of Catalysis as Applied to Gold

The *activity* of a catalyst for a given reaction is properly expressed as a *rate* of reactant conversion or product formation under specified conditions.[1,4,5] It will depend on the number of active sites present on the catalyst, and its units will then be mole of reactant transformed or product made per mole total metal per second; this is the *specific rate*. Quite often the exposed area of the metal can be estimated, and the rate can then be given as mole per m^2 of metal per s or as mole per mole of surface metal per s. This last measure is sometimes named the *turnover frequency*,[4] but its use in the case of gold is not recommended, as it is not certain, indeed in many cases unlikely, that all surface atoms are equally active.

To assess the true activity of a catalyst, it is essential to use it under conditions such that the surface reaction is rate-limiting, rather than processes of molecular transport in the fluid phase to and from the surface. In a gaseous flow system, the reaction may well become mass-transport limited at conversions above 75%, and it is quite meaningless to evaluate activity by the temperature at which 100% conversion is achieved.[5,6] In the kinetic regime (<75% conversion), conversion should be inversely proportional to flow-rate.[4]

High activity for the desired reaction requires the reactants to be *chemisorbed* on the surface in the form that is relevant to that reaction, and in the highest possible concentration. The structure and composition of the catalyst control the way in which reactants become chemisorbed, and hence determine the activity. Attempts to correlate catalytic activity directly with any parameter of the catalyst without considering how the reactants are chemisorbed are scientifically unsound,[5] because they do not offer an explanation of the effect. The act of chemisorption is a chemical process in which the reactant is prepared for reaction. It may involve changes to its electronic structure as new bonds are created between it and the surface gold atoms; sometimes, as with H_2, it is necessary to dissociate it completely to chemisorb it. The more energy released in the process, the greater will be the fraction of surface covered, but species that are too strongly adsorbed will be unreactive. Gold often succeeds where other metals fail by not holding the reactants too strongly, but well enough to utilise most of the suitable surface sites.

A gold catalyst can act *selectively* in reducing or oxidising a molecule in three ways.[1] (i) The reaction can effectively stop when a useful intermediate product is formed; e.g. when 1,3-butadiene is reduced to a butene and not to *n*-butane.[1] (ii) *Chemoselectivity* is shown when one of two or more functional groups is affected; e.g. crotonaldehyde is reduced to crotyl alcohol and not to *n*-butanol.[7] (iii) When two different reducible or oxidisable molecules are present, one may react preferentially to the other, e.g. the preferential oxidation of CO in the presence of excess H_2. We shall see that gold catalysis offers examples of all three types of selectivity.

7.4 The Chemisorption of Simple Molecules on Gold Nanoparticles

Since the catalytic activity of small gold particles depends critically on the way in which reactant molecules are chemisorbed, it is worth examining the process for simple diatomic molecules (H_2, O_2, CO), the reactions of which are of greatest interest (see Section 7.5). Chemisorption is always exothermic, the heat released being greater the strength of the new bonds formed. On many transition metals active in catalysis, the process occurs without significant activation energy, and can be observed down to very low temperatures, but in the case of small gold particles the dissociative chemisorption of H_2 is activated,[8-11] so that its rate increases with temperature, and H-Au bonds are formed.[11] To be catalytically effective, the O_2 molecule must also separate into atoms, and its chemisorption may also be activated[12,13] but CO adsorbs without dissociation (see below). Provided these processes occur selectively on the metal and not partly on the support, the maximum number of adsorbed molecules measured volumetrically can be used to estimate the metal's surface area, and hence its dispersion (the fraction of atoms on the surface) and then, with some assumptions, its size.[4] Such measurements have to be made with care (especially with CO and O_2) some adsorption may also take place on the support, particularly in the neighbourhood of the metal particles (Fig. 7.2). Moreover it is by no means certain that all surface atoms are capable of chemisorbing, but provided the gold particles are small enough and great care is taken, size estimates in good agreement with those obtained by direct observation with transmission electron microscopy (TEM) can be obtained (Table 1).[1]

The chemisorption of the CO molecule on gold particles has been intensively investigated both experimentally[1,14,15] and theoretically.[16] The vibrational mode of the free molecule at $2143\,cm^{-1}$ is readily observed in the infrared spectrum because of its large extinction coefficient, and when it is adsorbed via the carbon atom the frequency is remarkably sensitive to the chemical state of the atom to which it is attached.[1,15,17] The various possible adsorption sites shown in Fig. 7.2 are to some degree identifiable by the frequency of this mode. Those species formed by the interaction of

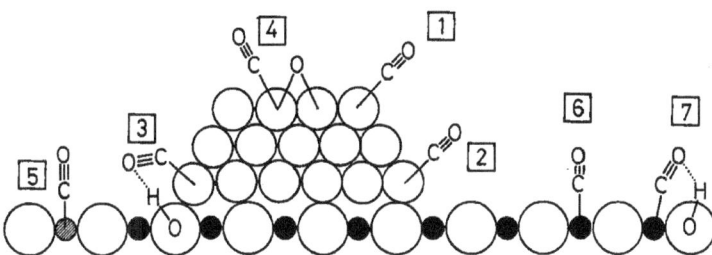

Fig. 7.2. Sites for the adsorption of carbon monoxide on a supported gold catalyst. (1) On a gold atom of low coordination number (corner or step). (2) On a gold atom close to the support. (3) The same, but hydrogen-bonded to a support hydroxyl. (4) On a gold atom adjacent to an oxygen atom. (5) On an oxidised gold species (Au^{3+}, Au^+ etc.). (6) On a support cation. (7) The same, but hydrogen-bonded to a support hydroxyl.

O^{2-} and OH^- ions on the support, such as CO_3^{2-}, HCO_3^- and HCO_2^-, have absorption maxima between 1300 and $1700\,cm^{-1}$; on neutral gold particles (Au^0), the peak occurs at $2100\text{–}2130\,cm^{-1}$, and on negatively charged particles ($Au^{\delta-}$) below $2100\,cm^{-1}$. The main source of confusion lies with positively charged species ($Au^{\delta+}$ and support cations) where absorption maxima at $2130\text{–}2180\,cm^{-1}$ overlap. Such information is of course unavailable for molecules that dissociate on adsorption, so CO uniquely reveals its locations on the surface.

7.5 Reactions Catalysed by Gold Nanoparticles

7.5.1 *The oxidation of carbon monoxide*

The oxidation of CO to CO_2 is a formally simple reaction that is easy to study and has numerous environmental applications, but it is very exothermic $(-\Delta H^0 = 238\,kJ\,mol^{-1})$, so that it is not always easy to keep a constant catalyst bed temperature. It was the discovery by M. Haruta and colleagues in 1987[18] that small gold particles (<3 nm) supported on reducible transition metal oxides (TiO_2, Fe_2O_3 etc) would catalyse the reaction at or below ambient temperature that initiated widespread interest in gold's catalytic properties. Figure 7.3 shows early results for rate dependence on particle size using different oxide supports,[1] and Fig. 7.4 shows the same effect with just TiO_2 as support, with two gold concentrations.[20] Single crystal surfaces (e.g. Au(111)) do not catalyse the reaction unless O atoms are pre-adsorbed,[21] but then the reaction is non-catalytic. It must be stressed that gold atoms on massive metal or larger particles have much lower activity than those on small particles; it is not just a question of the latter having a higher surface area, but some of their atoms, perhaps particularly those at corners and edges having low coordination numbers, are intrinsically more effective. The rise in activity shown in these figures must therefore be due mainly to an increase in the number of surface gold atoms that are able to chemisorb CO sufficiently but not too strongly. Low coordination number atoms have often been identified as the preferred location for CO to adsorb on, due to a favourable local electronic structure.[6]

We may wonder, looking at Figs. 7.3 and 7.4, whether there is any limit to gold's activity for CO oxidation, and whether there is any limit to the

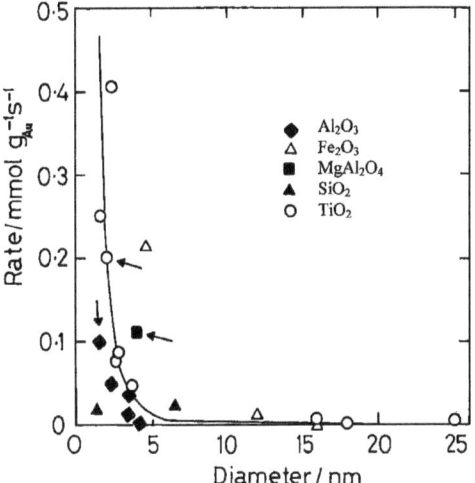

Fig. 7.3. Rates of CO oxidation at 273 K over gold supported on various oxides as a function of particle diameter.[1]

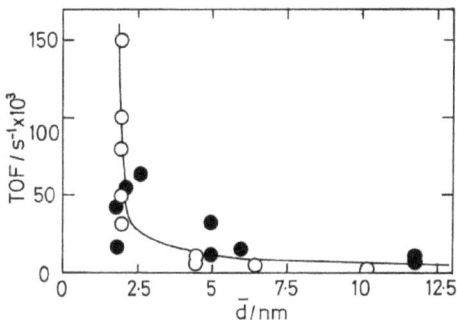

Fig. 7.4. Dependence of turnover frequency (s^{-1}) at 298 K for CO oxidation over Au/TiO$_2$ on mean particle size;[20] open points, 7.2% Au; filled points, 4.5% Au.

beneficial reduction in size. One answer is provided by 'model' catalysts made by depositing gold atoms on a TiO$_x$/Mo(112) surface[22] (Chapter 8); the fastest rate was obtained at about 3.5 nm, but careful electron microscopy revealed that such particles are in fact double layers of atoms ('bilayers') where every atom is on or close to the surface. Similar observations have been made with chemically prepared Au/Fe$_2$O$_3$.[23] Very recently it has been suggested[24,25] that with catalysts of this type, where there is inevitably a range of particle sizes, it is only the very smallest (<2.5 nm) that have significant activity. It is therefore likely that with bilayers and other very small particles all the gold atoms provide sites for acceptable for CO adsorption, and that the transition between low and high activity corresponds to the point where particles cease to behave as metals, but become semiconductors.[24,25] Below the point of transition, particles act as giant molecules, having sets of discrete energy levels rather than bonds of overlapping levels. This probably affects favourably the strength with which reactants can adsorb; similar changes are not shown by metals such as platinum having d-band vacancies. For a tentative molecular orbital interpretation of this effect, see Ref. 25 and Chapter 8.

The early observations that the best activity was found when reducible oxides (TiO_2, CeO_2, Fe_2O_3) were used as supports naturally led to speculation that the lattice oxide ions were somehow involved, or that anion vacancies caused by surface reduction of the support by CO provided possible sites for the adsorption of O_2 molecules adjacent to the gold particles (see Fig. 7.5).[26,27] It now appears the ceramic oxide supports (Al_2O_3, SiO_2, MgO) can also be effective if the gold particles are small enough (see for example Table 7.1).[6] In these cases it is necessary to postulate that O_2 molecules can adsorb dissociatively, for which there is good evidence.[1,12,13] A further significant observation is that rates are accelerated by the presence of water vapour, which presumably ensures that support surfaces are kept fully hydroxylated.[6] Support OH^- ions may in fact initiate reaction by

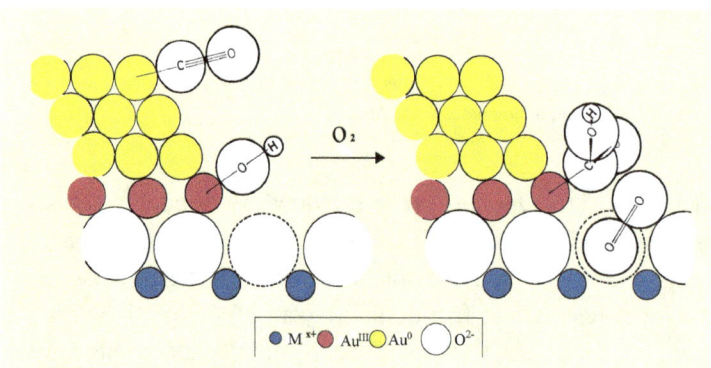

Fig. 7.5. Mechanism of oxidation of CO over gold supported on a reducible oxide.[27]

Table 7.1. Comparison of gold particle size (nm) estimates by various methods.[1]

Support	[Au]/%	O_2 chemisorption	H_2 titration*	TEM
Al_2O_3	0.7	2.0	2.5	1.2
SiO_2	1.24	6.1	—	6.0
MgO	3.46	8.7	8.3	—

*Using the process $O_{ads} + H_2 \rightarrow H_2O$.

forming a carboxyl ion ($COOH^-$) which then reacts with an O atom and restores the OH^- ion[27]:

$$CO + OH^- \rightarrow COOH^- \tag{1}$$

$$COOH^- + O \rightarrow CO_2 + OH^- \tag{2}$$

There have been many attempts to explain reaction mechanisms with the various types of support[1,6,17,27]; an example for a ceramic oxide support[28] is shown in Fig. 7.6. These attempts involve a good deal of speculation, because although they are backed up with spectroscopic and other physical evidence there is a general lack of sound *kinetic* information (e.g. orders of reaction, activation energies etc.), which is essential to the formulation of a plausible mechanism.[5,29]

The one serious difficulty with gold-catalysed CO oxidation is that activity frequently decreases rapidly with time; this is not predominantly due to loss of metal area by sintering, but is more often caused by formation of species such as CO_3^{2-}, HCO_3^- and HCO_2^- which block reaction sites.[6,30] Deactivation can be controlled if not entirely eliminated by inclusion of additives such as Fe_2O_3,[31] or by water vapour,[32] which limit the formation of toxic species.

7.5.2 Selective oxidation of carbon monoxide in hydrogen

There are very large actual and potential industrial requirements for very pure H_2, especially for certain types of fuel cell.[1,19] Steam-reforming of hydrocarbons or alcohols, for example,

$$CH_4 + H_2O \rightarrow 3H_2 + CO \tag{3}$$

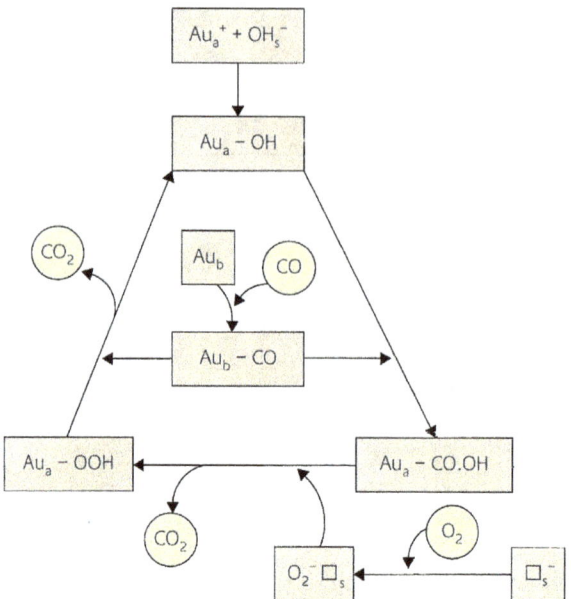

Fig. 7.6. Catalytic cycle for the mechanism of oxidation of CO over gold supported on a ceramic oxide.[6] The subscripts a and b denote different but equivalent atoms.

$$CH_3OH + H_2O \rightarrow 3H_2 + CO_2 \qquad (4)$$

actually leads to mixtures of H_2 plus the carbon oxides[1] because of the sequential processes

$$CO + H_2O \leftrightarrow H_2 + CO_2 \qquad (5)$$

The concentration of the CO can be lowered by the forward reaction, named the *water-gas shift* (see Section 7.5.3), but ultimate purification requires the removal of remaining small amounts of CO by *PReferential OXidation* (PROX) with O_2.

Practical considerations associated with the use of H_2 in fuel cells determine the conditions under which catalysts must be used and the criteria they must meet.[3,33] Tests are usually performed with an *idealised reformate* containing 1% CO, 1% O_2 and 75% H_2 plus N_2; a successful catalyst must lower the CO to below 10 ppm and remove less than 0.5% H_2; this means that the O_2 selectivity for CO must be above 50%, and conversion of CO greater than 99.5%. Many gold catalysts meet these criteria at around room temperature, because H_2 is less strongly adsorbed than CO. Unfortunately

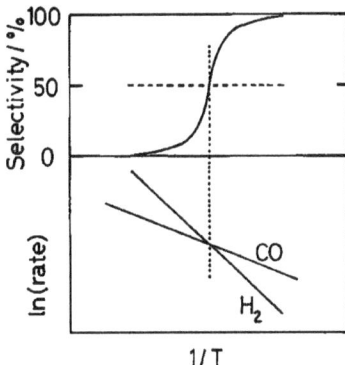

Fig. 7.7. Top part, the dependence of selectivity for the oxidation of CO in H_2 on reciprocal absolute temperature T; bottom part, Arrhenius plots (ln(rate) vs. T^{-1}) for the oxidations of CO and of H_2, showing how the fall in CO oxidation selectivity with increasing temperature is due the lower activation energy for this reaction.[1]

the working of the fuel cell demands a reaction temperature of about 353 K, but here H_2 is able to gain access to the surface, the activation energy for its oxidation being higher than that for CO oxidation (Fig. 7.7).[1] The preference of O_2 for CO therefore falls progressively with rising temperature as the H_2 oxidation starts to compete, and the desired criteria become harder to meet. However an Au/Fe_2O_3 catalyst calcined at 673 K and then at 823 K, having gold particles of average size 6.7 nm, just meets them.[34] Success depends on the catalyst *not* catalysing the reverse water-gas shift (reaction 5; see also Section 7.5.3), i.e. by *not* reforming CO by reaction of CO_2 with H_2.

In PROX systems a little water is inevitably formed by H_2 oxidation, and this tends to prevent formation of species such a CO_3^{2-} and HCO_3^- that cause deactivation (see Section 7.5.1)[30]; H_2 actually increases the rate of CO oxidation,[35] the first 10% by volume of H_2 being most effective. Its introduction therefore opens up a new and easier route for oxidation of CO[35] by creating a way of forming the carboxyl ion (reaction 1) that is not available when H_2 and water are absent.

7.5.3 *Catalysis of the water-gas shift*

The requirements of industry for large quantities of pure H_2 are met by reacting steam with hydrocarbons at high temperature (see for example reaction 3), followed by the water-gas shift (reaction 5, see Section 7.5.2); the

CO$_2$ formed is easily removable. The forward reaction is exothermic ($-\Delta H° = 41.2\,\text{kJ mol}^{-1}$), so the equilibrium conversion rises as temperature falls, but at 423 K the equilibrium concentration is still about 1% for a stoichiometric reactant mixture. Very active catalysts are therefore needed to lower the CO content to an acceptable level. In large-scale operation, an iron-chromium oxide is first used, then a copper-zinc-alumina, but gold particles supported on a number of reducible oxides (TiO$_2$, Fe$_2$O$_3$, CeO$_2$, ZrO$_2$, etc.) have activities at least as good and sometimes significantly better.[36,37] Continued redox changes tend to weaken ceria particles, which ultimately disintegrate, but in the mixed oxide CeO$_2$-ZrO$_2$ the zirconia component provides a stabilising framework within which cerium ions can oscillate between the 3+ and 4+ states. This is now one of the most preferred supports.[37]

There has been much discussion concerning the mechanism and particularly the oxidation state of the gold in the working catalyst.[38,39] The reaction is less easy than the oxidation of CO, due to the greater difficulty of activating the water molecule and persuading it to release its H atoms. Rates can be measured down to 373 K, but good conversions are generally only obtainable above 423 K. At one time it was thought that cationic gold (Au^{3+}) might be the active form because treatment of catalysts with solutions containing the cyanide ion dissolved the metallic particles, but hardly affected the activity.[40] However it is now established that very small zero-valent particles are the active component, with some cooperation from the support.[39] The probable first step is reaction 1, and the rate-controlling step may be its decomposition as

$$\text{Au-COOH} \rightarrow \text{Au-H} + \text{CO}_2 \qquad (6)$$

followed by recombination of the H atoms to give H$_2$ (see Fig. 7.8). Parasitic reactions leading to strongly adsorbed species can lead to deactivation, as with CO oxidation, but although deactivation rates have been reduced to a low level they are not yet of the standard required for industrial use.

7.5.4 *Synthesis of hydrogen peroxide*

Hydrogen peroxide (H$_2$O$_2$) in aqueous solution is an important and effective oxidising agent, its major uses being for bleaching and disinfecting; the

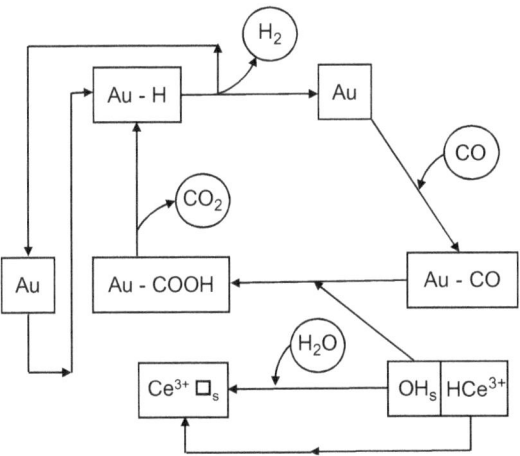

Fig. 7.8. Catalytic cycle for the mechanism of the water-gas shift over Au/CeO$_2$.[39] The reactants and products are encircled, and the surface species are in boxes; starting with CO and H$_2$O, follow the arrows.

current usage is about two million tonnes per year, and it is still made by an old process based on anthroquinone that is only economical on a very large scale. There is however a vast unrealised potential for its production on a smaller scale by the direct reaction of H$_2$ with atmospheric O$_2$, especially if this could be achieved in high concentration at the point of final use, thus avoiding its costly and hazardous transportation. It is only quite recently that the ability of gold, and particularly its alloy with palladium, to catalyse the synthesis of H$_2$O$_2$ in this way has been recognised.[1,41,42]

H$_2$O$_2$ is however an unstable molecule and changes to water either by further hydrogenation or by decomposition with the release of O$_2$; the hydrogenation of O$_2$ to water is the more exenthalpic reaction[41] (see Table 7.2), and for this reason the reaction has to be attempted at low temperature (273–283 K) with the catalyst suspended in water or methanol. Reactions are usually conducted with low concentrations of H$_2$, i.e. below the lower explosion limit. Many supported gold catalysts have given disappointingly low selectivities to H$_2$O$_2$, but the important advance was the observation[43] that 1:1 gold-palladium alloys on these supports gave much higher activities, typically twice that shown by palladium alone, and H$_2$ selectivities to H$_2$O$_2$ of 70–80% (see Table 7.3). Much effort has been devoted to optimisation of possible catalysts by Professor Graham Hutchings and his collaborators at

Table 7.2. Standard reaction enthalpies for processes met in the synthesis of hydrogen peroxide.[36]

Reaction	$-\Delta H^0/kJ\,mol^{-1}$
$H_2 + O_2 \rightarrow H_2O_2$	135.9
$H_2O_2 + H_2 \rightarrow 2H_2O$	211.5
$H_2 + \frac{1}{2}O_2 \rightarrow H_2O$	241.6
$H_2O_2 \rightarrow H_2O + \frac{1}{2}O_2$	105.8

Table 7.3. Rates and selectivities for hydrogen peroxide synthesis using 1:1 Pd/Au catalysts on various supports.[43]

Support	Rate/mol H_2O_2 $(kg_{cat}h)^{-1}$	H_2 selectivity/%
SiO_2	108	80
C	110	80
Al_2O_3	15	14
TiO_2	64	70

*2.5% of each metal in all cases; T = 275 K.

Cardiff University.[41-44] Catalysts made by simple impregnation work better than those made by deposition-precipitation (Chapter 6), probably because they have larger alloy particles. Uncalcined catalysts are very active (e.g. with 1:1 Pd:Au/TiO_2, the synthesis rate was 202 mol $(kg_{cat}h)^{-1}$, and H_2 selectivity 89%), but they are unstable as the metals dissolve in the liquid medium.[41] This is avoided by calcination at 673 K, which increases particle size, but decreases activity by some 70%. With PdAu/C catalysts, performance is improved by acid-washing the carbon before use. The best catalysts for H_2O_2 synthesis are therefore *large* PdAu particles, whereas for CO oxidation very small gold particles have excelled (Section 7.5.1, Figs. 7.3 and 7.4).

The mechanism of this reaction has not yet been investigated in detail. It seems likely that the O-O bond in O_2 is not broken and re-formed, because breaking it would probably lead to water formation. The hydroperoxy radical (HOO^-) is a probable intermediate. Electron microscopy studies[42–44] of catalyst structure reach perplexing conclusions: on oxidic supports, the surface of the alloy particles become enriched in palladium, but with carbon as support the surface composition is the same as the bulk.

7.5.5 *Selective oxidation of organic molecules*

The principles of *selective* catalytic reactions were outlined in Section 7.3. The organic molecules to be considered are either hydrocarbons (mainly alkenes) or oxygenated molecules having one or two hydroxyl or aldehyde groups; attempts to oxidise alkanes selectively with gold catalysts have not so far been successful. With oxygenated molecules, the desired reactions include

$$R-CH_2OH \rightarrow R-CHO \rightarrow R-COOH \tag{7}$$

and it is usually hoped to operate on only one of the molecule's functional groups if more than one is present.[45,46] There are strong economic motivations for these processes; alkenes are easily and quite cheaply available, but oxidised products derived from them are much more valuable and useful. Photosynthesis in plants leads first to formaldehyde (methanal, HCHO) and thence to *D*-glucose, which polymerises to either starch (needing to store energy) or cellulose (which provides structural strength); these and other molecules formed naturally constitute biomass and the routes by which it can be transformed into products capable of further valorification by catalytic processes have recently been comprehensively summarised.[47]

Most metals apart from gold are too vigorous in catalysing oxidations of organic molecules, and tend to give deep oxidation to CO_2 and water, but there have however been many attempts to develop gold catalysts that could be used for the selective oxidation of propene to propene oxide (methyloxirane).[3] Early work showed that Au/TiO_2 at 323 K gave selectivity of 99%, but only at 1% conversion and provided H_2 was also present ($H_2 : O_2 = 1 : 1$).[48] Subsequent studies have explored the importance of particle size and choice of support. Particles smaller than 2 nm tend to form propane, and, since it is particularly important for the product to be able to diffuse away from the surface before reacting further, much attention has been given to the use of crystalline mesoporous supports such as Ti-MCM-41 and TS-1 (a SiO_2-TiO_2 material).[1] The presence of titanium ions in the support is considered important, because such supports alone are effective oxidising catalysts with H_2O_2 as oxidant. Since gold can catalyse its formation from H_2 and O_2 (see Section 7.5.4), it has been supposed that the reaction proceeds as

$$H_2 + O_2 + C_3H_6 \rightarrow H_2O_2 + C_3H_6 \rightarrow C_3H_6O + H_2O \qquad (8)$$

More recent work does however cast some doubt on this idea, as gold on sodium-treated TS-1 affords 85% selectivity at 7.9% conversion and works equally well with either water or H_2+O_2.[49] Commercial application requires a conversion exceeding 10%, which is almost in sight, but problems of catalyst stability have still to be overcome. An interesting recent development is the observation that H_2O_2 is a by-product of the oxidation of CO when the catalyst (Au/TiO_2 or Au/TS-1) is suspended in water; it can be used for the simultaneous oxidation of butene.[50]

The catalysts that are effective for the selective oxidation of alcohols and aldehydes differ from those that perform well in CO oxidation. Rates are often independent of particle sizes and carbon is an effective support; colloidal gold suspensions are also active, and particular benefit is obtained when gold is combined with either palladium or platinum in either of these forms.[1,51] Primary and secondary alcohols are converted into respectively aldehydes and ketones in excellent selectivity either in the liquid or vapour phase. There is a large industrial demand for acetic acid ($\sim 6 \times 10^6$ tonnes yr^{-1}), needed for example for the synthesis of vinyl acetate monomer (VAM) and methods better than the currently used carbonylation of methanol, which uses chemicals derived from fossil fuels, are eagerly sought. Bioethanol obtained by fermentation is therefore the preferred starting point. Aqueous ethanol (5%) is oxidised to acetic acid (ethanoic acid) over Au/TiO_2 and Au/$MgAl_2O_4$ at 360–470 K with 100% conversion and 90% selectivity in 4 h under optimum conditions.[52,53] The reaction proceeds through acetaldehyde (ethanal), which can also be quite selectively oxidised to acetic acid; the small amount of CO_2 arises from the aldehyde and not from the acid. When the ethanol concentration is above 80%, ethyl acetate is a major product ($\sim 50\%$).[53] Above 470 K, the reaction becomes non-selective.[52]

Another important derivative of biomass is glycerol (HOCH$_2$–CHOH–CH$_2$OH), which is a by-product from the manufacture of biodiesel.[47] One of its terminal hydroxide groups is readily oxidised to give glyceraldehyde (HOCH$_2$–CHOH–CHO), which can be obtained in 100% selectivity, but its further oxidation to glyceric acid (HOCH$_2$-CHOH-COOH) is frequently observed.[54] With *D*-glucose the aldehydic group is oxidised to carboxyl to give *D*-gluconic acid, but under certain conditions oxidised products having only two or three carbon atoms are obtained.[55] At the moment it appears to be a matter of sheer luck as to what products of a given reaction will arise from any chosen set of conditions; the products obtained depending on the form of the catalyst and its support, on the solvent, if any, and in particular on temperature, in a manner that has not yet been rationalised. Nevertheless certain reactions, e.g. the *D*-glucose to *D*-gluconic acid conversion, are approaching commercial realisation,[51] as this product or its salts are already manufactured by a somewhat unsatisfactory fermentation on a scale approaching 10^5 tonnes per year; they find application in water-soluble cleansing agents and as an additive to food and drinks.[47] The role of the Group 10 metal when present remains unclear, but the great activity of gold and gold-containing catalysts cannot be over-emphasised; in *D*-glucose oxidation, the activity of the best gold catalysts approach that shown in enzymatic oxidation.[56]

7.5.6 *Hydrogenation*

For many years it was believed that gold catalysts would not be effective for the hydrogenation of unsaturated functions, because of their apparent inability to chemisorb H$_2$ dissociatively, although a recent survey[3] of the early patent literature (before 1978) has revealed extensive efforts, occasionally successful, to use gold for this purpose. However the cause of many failures was undoubtedly due to the use of catalysts containing gold particles too large to be effective.

It is now well established that suitably made gold catalysts are able to hydrogenate multiple carbon-carbon bonds below 473 K.[1,14,58,59] In the first significant academic study it was shown[60] that Au/SiO$_2$ prepared by impregnation and reduction catalysed the hydrogenation of 1-pentene at 373 K, the turnover frequency increasing markedly as the gold loading was decreased below 1%. This higher activity was associated with materials having a purple colour, and showing an Electron Spin Resonance signal (g = 2.07 − 2.21) at 100 K, but the presence of particles smaller than about 4 nm, which these observations suggested, was not observable by X-ray diffraction or TEM due to the low metal loading. These Au/SiO_2 catalysts showed another unusual feature; only the larger particles found at loadings above 1% gave simultaneous double-bond isomerisation during 1-pentene hydrogenation. Au/Al_2O_3 catalysts also showed hydrogenation activity, but TOF did not rise at low loadings as with Au/SiO_2. The reaction of ethene with D_2 was followed on 5% Au/SiO_2; the distribution of D atoms in the ethene and ethanes resembled that given by Pt/SiO_2 catalysts, but the gold source ($HAuCl_4$) was very pure and contained almost no trace of platinum. Another technique used in this early work,[60] but not widely adopted, was to estimate the amount of chemisorbed H atoms by exchange with D_2; this showed clearly that there were more of them on the smaller particles.

Supported gold catalysts of this type also affect the selective hydrogenation of 1,3-butadiene, i.e. the reaction stops at the butene stage and little or no butane is formed; 1-butene is the major product.[1,58,61] Study of this reaction is complicated by oligomerisation of the hydrocarbons leading to loss of activity; in ethene hydrogenation, molecules containing from three to six carbon atoms were observed as by-products.[61] More recently, bimetallic $PdAu/Al_2O_3$ catalysts shown higher activity without loss of selectivity.[62] Alkynes are also hydrogenated to the corresponding alkenes with high selectivity;[58,63] 2-butyne gives mainly Z-2-butene.[1,61] The aromatic nucleus can also be hydrogenated in the presence of supported gold catalysts.[1]

An important industrial use of hydrogenation catalysis is the hardening of polyunsaturated natural oils to produce margarine. Current processes use nickel catalysts, but palladium and gold catalysts have also been tried,[1] and yield satisfactory results, but the industry is conservative and loath to change.

A further long-standing challenge for the application of heterogeneous catalysis is the chemoselective reduction of unsaturated aldehydes to unsaturated alcohols:

$$R - CH{=}CH - CHO + H_2 \rightarrow R - CH{=}CH - CH_2OH. \qquad (9)$$

On most metallic catalysts the C=C bond is much more easily reduced, unless selective poisoning is used, but gold catalysts supported on oxides such as TiO_2 and ZnO have given selectivities to the desired alcohols[1,7,58,64] (R=H, propenol (acrolein); R=CH$_3$, crotonaldehyde (propenal); R=C$_6$H$_5$, cinnamyl alcohol) between 50 and 80%, but complete selectivity remains to be achieved.

A final challenge in hydrogenation catalysis is the chemoselective reduction of aromatic nitro-compounds containing another substituent that is removable by hydrogenolysis, e.g. the reduction of *p*-chloronitrobenzene to *p*-chloroaniline. Most metals bring about an unacceptable degree of hydrogenolysis of the C–Cl bond to give aniline, but Au/TiO$_2$ and Au/Al$_2$O$_3$ give very high selectivity to the desired product.[65] The mechanism of nitrobenzene hydrogenation has been discussed in some depth,[57] and it has been concluded that it proceeds by a direct route through nitrosobenzene (C$_6$H$_5$NO) and the oxime (C$_6$H$_5$NHOH) to give aniline.[66]

7.5.7 *Reactions of environmental importance*

These reactions may be classified as (i) those that purify the air in domestic or other enclosed places (e.g. factories, aircraft, submarines) or air that is immediately required for breathing (e.g. respirators for firefighters), and (ii) those that prevent emission into the atmosphere of molecules that contribute to localised pollution (e.g. photochemical smog) or indirectly to global warming (e.g. by destruction of ozone). Supported gold catalysts are already used in the first class; their high activity for CO oxidation (Section 7.5.1) makes them attractive for use in respirators,[1] as they are not much affected by water vapour, but lengthy shelf life is essential, and stability in use has to be maximised. Their use in cigarette filters is being investigated, but the heavy organic tar-forming molecules that accompany CO affect the catalyst's activity.[67] Gold catalysts are already commercialised in Japan for elimination of domestic odours, e.g. in kitchens and toilets, through their ability to oxidise completely the organic molecules that are responsible.

Historically speaking, the first significant use of catalysts for environmental protection involved the control of CO and the oxides of nitrogen emitted from the exhaust of internal combustion engines. Elimination of CO proved to be quite easy, but removal of the nitrogen oxides (NO, NO_2, N_2O, collectively known as NO_x) required their reduction to N_2 either by the CO or by other reductants present in the exhaust, such as H_2 or unburnt hydrocarbons. To achieve this it was necessary to ensure that the O_2 supplied to the engine was only just sufficient to oxidise all the fuel; in a final stage, catalysts were devised to assist this control. Thus the important reactions needed to be catalysed were

$$CO + NO \rightarrow CO_2 + \tfrac{1}{2}N_2 \tag{10}$$

$$H_2 + NO \rightarrow H_2O + \tfrac{1}{2}N_2 \tag{11}$$

$$\text{Hydrocarbon} + NO \rightarrow CO_2 + H_2O + \tfrac{1}{2}N_2 \tag{12}$$

One of the limitations suffered by catalysts containing a metal of Groups 8 to 10 was the somewhat high minimum temperature that they needed to start working; significant emissions occur before the engine has warmed enough to heat the catalyst. The activity of gold for CO oxidation and the exothermicity of this reaction represent a possible means of tackling this problem (Section 7.5.1).

Gold catalysts (e.g. Au/Al_2O_3) are more active for NO_x reduction by CO than those having metals of Groups 8 to 10;[1] the presence of water vapour in the exhaust is again beneficial, as it was with CO oxidation. They also catalyse NO_x reduction by H_2. There have been numerous studies of NO_x reduction by hydrocarbons, many of them using propene as a prototypical reductant.[1] Au/ZnO, Au/Fe_2O_3 and Au/ZrO_2 are amongst the most effective gold catalysts, and they have the advantage over other metals in that they show higher selectivity to N_2; other catalysts direct the reaction towards N_2O, which doesn't really solve the problem. The Anglo-American Research Laboratories (South Africa) have developed a complex catalyst composition that appears to perform well under net reducing conditions,[68] and Nanostellar Inc. has recently announced the commercialisation of a gold-containing catalyst for treating the exhaust of diesel engines.[69]

Many industrial processes also release a variety of organic compounds into the atmosphere; these include alcohols, aldehydes, ketones and esters, and are known collectively as 'volatile organic compounds' (VOCs).[1] They

characteristically occur in low concentrations (100–1000 ppm). Gold catalysts are effective for their complete oxidation, especially when supported on
a transition metal oxide that itself has some oxidation activity (e.g. Co_3O_4,
CeO_2, V_2O_5). One particularly vicious carcinogenic class of compounds is
the dioxins; members of this class are found in, for example, 'Agent Orange',
a defoliant that has been used in chemical warfare; they are also formed in
combustion of domestic refuse. Catalysts such as Au/Fe_2O_3-La_2O_3 have
been used for their destructive oxidation.[70] Gold catalysts may also find
use in the purification of waste water containing organic compounds, the
process being named 'Catalytic Wet Air Oxidation' (CWAO).[71]

7.6 Electrocatalysis

A great deal of work has been done using massive gold as an electrocatalyst,
including studies of single crystal surfaces, but activity appears to be associated with defects, i.e. atoms of low coordination number;[72] for this reason,
interesting behaviour is expected to be shown by nanoparticles, provided
they can be incorporated into a conducting electrode. This idea has indeed
been explored, largely in connection with electrodes for fuel cell systems.

The principle of the fuel cell can be simply illustrated as follows. The
chemical energy released by the reaction of H_2 with O_2 is captured as electric
current; at the anode the oxidation of H_2 provides electrons that are used at
the cathode to reduce the oxygen, thus:

$$H_2 \rightarrow 2H^+ + 2e^- \tag{13}$$

$$O_2 + 4e^- + 4H^+ \rightarrow 2H_2O \tag{14}$$

The current so produced gives the electrical energy. In practice however
there are difficulties to overcome. Most sources of H_2 are liable to contain
traces of CO that strongly inhibit the electrode reaction when catalysed
by platinum; it has to be purified by selective oxidation (Section 7.5.5).
Bimetallic electrocatalysts such as platinum-ruthenium have been tried, and
the ability of gold to oxidise CO under mild conditions (Section 7.5.1) has
suggested that it might be a partial or even complete replacement for platinum in both the anode and cathode zones.[1] Methanol is another potential

fuel, and graphene-supported platinum-gold particles have recently shown to be effective both for its oxidation and for O_2 reduction.[73]

Gold is one of the most active catalysts for O_2 reduction in basic media,[74] but in the massive form near the rest potential it gives predominantly two-electron reduction (i.e. to HO_2^- or H_2O_2), so that only half the possible energy release is obtained. In acidic media, reduction occurs only below about 0.4V, both the potential range and the rate being far too low to be useful. There have therefore been a number of studies designed to explore whether nanoparticulate gold on a suitably conducting support has any benefit; marked particle size effects have been revealed. With Au/C in basic media, 3 nm particles were 2.5 times more effective than 7 nm particles, and gave four-electron reduction unlike the larger particles, which gave only two-electron reduction.[74] Similar results have been obtained with Au/C in acidic medium, the extent of four-electron reduction increasing as the potential became more negative, to a maximum value at sizes below 3 nm.[75] This type of particle size effect has been rationalised theoretically.[76] It seems most likely however that gold will find use when alloyed with platinum in bimetallic nanoparticles; their preparation and use in O_2 reduction have been studied.[72,73] The presence of the platinum secures complete four-electron reduction to water. Although the phase diagram of platinum and gold in the massive state shows a very significant miscibility gap, it is absent in nanoparticles, which therefore often form homogeneous alloys. This change has been associated with the move from metallic to non-metallic character, i.e. to the breakdown of the band structure.[77]

The electro-oxidation of CO has also been studied,[78] in the hope that gold's activity for this reaction in the gas phase will also apply here, perhaps occurring competitively with H_2 reduction and thus eliminating its toxicity to platinum in bimetallic electrodes. In perchloric acid, Au/C is active only at the same potential as massive gold, the activity falling at particle sizes below 2.5 nm; with Au/TiO$_2$ reduction occurs at much lower overpotentials, and the activity is a strong function of particle size, being a maximum at 3 nm and falling to zero below 1.5 nm.[78] It is also active at high potentials, where poisoning normally occurs. The behaviour of electrocatalytic oxidation therefore bears a marked similarity to the gas-phase reaction, although the implications of this have not extended to discussions of reaction mechanism.

7.7 Photocatalysis

Gold nanoparticles (called 'nanodots' in some of the materials science literature[79]) exhibit a plasmon resonance absorption band at about 550 nm, the precise frequency depending on particle size and shape, and on the dielectric nature of the surrounding medium. This band is caused by the collective motion of conductive electrons induced by the electric field of the incident light (Chapter 3). A Schottky barrier of about 1 eV at the interface between the gold particles and a semiconducting support is expected to inhibit electron transfer, but it is nevertheless observed.[79–82] Photoactive materials have been made by simple admixture of colloidal gold with TiO_2 (Degussa P-25)[79] or by photodecomposition of $HAuCl_4$ on CeO_2. Attention has focused on the rate and mechanism of transfer of photoelectrons from gold to the supporting oxide.[79,80,82] Excitation of the plasmon band by $\sim$150 fs laser pulses has shown that electron injection is complete within 50 fs, and is detected by infrared absorption caused by the TiO_2 conduction band at 3440 nm.[81] Numerous applications in the fields of sensors and biosensors are foreseen, and the strong electromagnetic field near gold particles is responsible for the great molecular sensitivity in Surface Enhanced Raman Scattering (SERS), in fluorescence, and in non-linear optical spectroscopies (see Chapter 10). Catalytic applications have not yet been fully exploited, but Au/CeO_2 efficiently catalyses formic acid decomposition in aqueous solution,[80] so photocatalytic water purification is an area for possible development. Photocatalysis by Au/TiO_2 has recently been reviewed.[83] Work so far does not seem to have employed nanoparticles of the smallest size ($<$5 nm), and the skills developed for catalyst preparation (Chapter 6) have not yet been widely applied.

7.8 Conclusion

The early industrial work did not generate any useful applications, so that the widespread occurrence of catalytic activity by supported gold *nanoparticles* came as a complete surprise. In the pure form it has remarkable activity for oxidation of CO and for the water-gas shift, and in addition its specificity in selective oxidations makes it an attractive member of the armoury of heterogeneous catalysts. Much yet remains to be done in exploring reaction mechanisms, but there is distinct promise of significant applications in environmental control and fine chemicals processing.[84] In many applications, however, deactivation remains an important problem to be solved; however, the success shown by Nanostellar Inc. in developing a system for diesel exhaust treatment illustrates the opportunities that exist for the application of gold catalysts in environmental control.

Acknowledgements

I am grateful to all those who have sent me copies of their work, and particularly to Dr Richard Holliday (World Gold Council) and Prof. Gary Attard (Cardiff) for their advice and assistance.

References

1. G.C. Bond, C. Louis and D.T. Thompson, *Catalysis by Gold*, Imperial College Press, London, 2006.
2. G.C. Bond and P.A. Sermon, *Gold Bull.* **6** (1973) 102.
3. G.C. Bond, *Gold Bull.* **41** (2008) 235.
4. G.C. Bond, *Metal-Catalysed Reactions of Hydrocarbons*, Springer, New York, 2005, Chapter 2.
5. G.C. Bond, *Catal. Rev. Sci. Eng.* **50** (2008) 532.
6. G.C. Bond and D.T. Thompson, *Gold Bull.* **42** (2009) 47.
7. R. Zanella, C. Louis, S. Giorgio and R. Touroude, *J. Catal.* **223** (2004) 328.
8. E. Bus, J.T. Miller and J.A. van Bokhoven, *J. Phys. Chem. B* **109** (2005) 14581.
9. P. Serna, P. Concepción and A. Corma, *J. Catal.* **265** (2009) 19.
10. C. Kartusch and J.A. van Bokhoven, *Gold Bull.* **43** (2009) 343.
11. M. Boronat, F. Illas and A. Corma, *J. Phys. Chem. A* **113** (2009) 3750.
12. H. Bernt, I. Pitsch, S. Evert, K. Struve, M.M. Pohl, J. Radnik and A. Martin, *Appl. Catal A: Gen.* **244** (2003) 169.
13. S.S. Pansare, A. Sirijaruphan and J.G. Goodwin Jr., *J. Catal.* **234** (2005) 151.
14. G.C. Bond and D.T. Thompson, *Catal. Rev. Sci. Eng.* **41** (1999) 319.
15. M. Boronat, P. Concepción and A. Corma, *J. Phys. Chem. C* **113** (2009) 16772.
16. R. Coquet, K.L. Howard and D.J. Willock, *Chem. Soc. Rev.* **37** (2008) 2046.
17. C. Louis, in *Nanoparticles and Catalysis* (ed D. Astruc), Wiley-VCH, Weinheim, 2007, p. 475.
18. M. Haruta, T. Kobayashi, H. Sano and N. Yamada, *Chem. Lett.* **2** (1987) 405.
19. S.A.C. Carabineiro and D.T. Thompson, in *Gold: Science and Applications* (eds C. Corti and R. Holliday), CRC Press, Baton Rouge, FL., 2010, p. 89.

20. S.H. Overbury, V. Schwartz, D.R. Mullins, W.F. Yan and S. Dai, *J. Catal.* **241** (2006) 56.
21. B.K. Min, A.R. Alemozofar, D. Pinnaduwage, X. Deng and C.M. Friend, *J. Phys. Chem. B* **110** (2006) 19833.
22. M. Chen and D.W. Goodman, *Chem. Soc. Rev.* **37** (2008) 1860.
23. A.A. Herzing, C.J. Kiely, A.F. Carley, P. Landon and G.J. Hutchings, *J. Mater. Chem.* **19** (2009) 1.
24. G.C. Bond, *Gold Bull.* **43** (2010) 88.
25. G.C. Bond, *Faraday Discuss.* **152** (2011) 277.
26. H. Liu, A.I. Kozlov, A.P. Kozlova, T. Shido, K. Akasura and Y. Iwasawa, *J. Catal.* **185** (1999) 252.
27. G.C. Bond and D.T. Thompson, *Gold Bull.* **33** (2000) 41.
28. C.K. Costello, J.H. Yang, H.Y. Law, Y. Wang, J.-N. Liu, L.D. Marks, M.C. Kung and H.H. Kung, *Appl. Catal. A Gen.* **243** (2003) 15.
29. M.A. Bollinger and M.A. Vannice, *Appl. Catal. B: Env.* **8** (1996) 417.
30. Y. Denkwitz, B. Schumacher, G. Kučerová and J.R. Behm. *J. Catal.* **267** (2009) 78.
31. G.C. Bond and F. Moreau, *Catalysis Today* **114** (2006) 362.
32. M. Date, M. Okumura, S. Tsubota and M. Haruta, *Angew. Chem. Int. Ed.* **43** (2004) 2129.
33. C.W. Corti, R.J. Holliday and D.T. Thompson, *Top. Catal.* **47** (2007) 331.
34. P. Landon, J. Ferguson, B.E. Solsona, T. Garcia, A.F. Carley, A.A. Herzing, C.J. Kiely, S.E. Golunski and G.J. Hutchings, *Chem. Commun.* (2005) 3385.
35. E. Quinet, L. Piccolo, H. Daly, F.C. Meunier, F. Morfin, A. Vacarel, F. Diehl, P. Avenier, V. Caps and J.-L. Rousset, *Catal. Today* **138** (2008) 43.
36. D. Mendes, H. Garcia, V.B. Silva, A. Mendes and L.M. Madeira, *Ind. Eng. Chem. Res.* **48** (2009) 430.
37. D. Tibiletti, A. Ameiro-Fonseca, R. Burch, Y. Chen, J.M. Fisher, A. Gouget, C. Hardacre, P. Hu and D. Thomsett, *J. Phys. Chem. B* **109** (2005) 22553.
38. R. Burch, *Phys. Chem. Chem. Phys.* **8** (2006) 5483.
39. G.C. Bond, *Gold Bull.* **42** (2009) 337.
40. Q. Fu, H. Saltsburg and M. Flytzani-Stephanopoulos, *Science* **301** (2003) 935.
41. G.J. Hutchings, *Chem. Commun.* (2008) 1148.
42. J.K. Edwards, B. Solsona, E. Ntainju N., A.F. Carley, A.A. Herzing, C.J. Kiely and G.J. Hutchings, *Science* **321** (2008) 1331.
43. J.K. Edwards, B. Solsona, P. Landon, A.F. Carley, A. Herzing, M. Watanabe, C.J. Kiely and G.J. Hutchings, *J. Mater. Chem.* **15** (2005) 4595.
44. A.A. Herzing, M. Watanabe, J.K. Edwards, M. Conte, Z.-R. Tang, G.J. Hutchings and C.J. Kiely, *Faraday Discuss.* 138 (2008) 337.
45. C.D. Pina, E. Falletta, L. Prati and M. Rossi, *Chem. Soc. Rev.* **37** (2008) 2077.
46. A. Corma and H. Garcia, *Chem. Soc. Rev.* **37** (2008) 2096.
47. A. Corma and S. Iborra, *Chem. Rev.* **107** (2007) 2411.
48. M. Haruta, *Chem. Record* **3** (2003) 75.
49. J.-H. Huang, E. Lima, T. Akita, A. Guzman, C.-X. Qi , T. Takei, H. Ohashi and M. Haruta, *J. Catal.* **278** (92011) 8.
50. J.-A. Jiang, H.H. Kung, M.C. Kung and J.-T. Ma, *Gold Bull.* **42** (2009) 280.
51. L. Prati and M. Rossi, *J. Catal.* **176** (1998) 552.

52. S.M. Thembe, G. Patrick and M.S. Scurrell, *Gold Bull.* **42** (2009) 321.
53. B. Jørgensen, S.E. Christiansen, M.C.D. Thomas and C.H. Christensen, *J. Catal.* **25** (2007) 332.
54. G.J. Hutchings, *J. Chem. Soc. Dalton Trans.* **41** (2008) 5513.
55. S. Meenakshisundaram, E. Nowicka, P.J. Miedziak, G.L. Brett, R.L. Jenkins, N. Dimitratos, S.H. Taylor, D.W. Knight, D. Bethell and G.J. Hutchings, *Faraday Disc.* **145** (2010) 341.
56. C.D. Pina, E. Falletta, M. Rossi and A. Sacco, *J. Catal.* **263** (2009) 92.
57. A. Girrane, A. Corma and H. García, *Science* **322** (2008) 1661.
58. L. McEwan, M. Julius, S. Roberts and J.C.Q. Fletcher, *Gold Bull.* **43** (2010) 298.
59. A. Hugon, L. Delannoy and C. Louis, *Gold Bull.* **41** (2008) 127.
60. P.A. Sermon, G.C. Bond and P.B. Wells, *J. Chem. Soc. Faraday Trans.* **75** (1979) 385.
61. D.A. Buchanan and G. Webb, *J. Chem. Soc. Faraday Trans. I* **70** (1978) 134.
62. A. Hugon, L. Delannoy J.-M. Krafft and C. Louis, *J. Phys. Chem. C* **114** (2010) 10823.
63. Y. Azizi, V. Pitchon and C. Petit, *Appl. Catal. A: Gen.* **385** (2010) 170.
64. C. Milone, R. Ingoglia, L. Schipilliti, G. Neri and S. Galvagno, *J. Catal.* **236** (2003) 80.
65. F. Cárdenas-Lizana, S. Gómez-Quero and M.A. Keane, *Chem. Sus. Chem.* **1** (2008) 215.
66. A. Corma, P. Concepcion, and P. Serna, *Angew. Chem. Int. Ed.* **46** (2007) 7266.
67. J. McPherson, P. Branton, S. Roberts, D. Barkhuizen, D. Ramdayal, M. Raphulu and E. van der Lingen, *GOLD* 2009, Heidelberg, abstract, p. 120.
68. J.R. Mellor, A.N. Palazov, B.S. Grigorova, J.F. Greyling, K. Reddy, M.P. Letsoala and J.H. Marsh, *Catal. Today* **72** (2002) 145.
69. http://www.gold.org/media/press_releases/archive/2011/06/first_gold-based_emissions_control_technology_in_commercial_production/ (Accessed 1 February 2012.)
70. O. Kajikawa, X.-S. Wang, T. Tabata and O. Okado, *Organohalogen Compounds* **40** (1999) 581.
71. M. Besson, A. Kallel, P. Gallezot, R. Zanella and C. Louis, *Catal. Commun.* **4** (2003) 471
72. L.D. Burke and A.M. O'Connell, in *Gold: Science and Applications* (eds C. Corti and R. Holliday), CRC Press, Baton Rouge, FL, 2010, p. 51.
73. Y.-J. Hua, Hua Zhang, P. Wu, Hui Zhang, B. Zhou and C.-X. Cai, *Phys. Chem. Chem. Phys.* **13** (2011) 4083.
74. W. Tang, *J. Phys. Chem. C* **112** (2008) 10515.
75. T. Masaki and S. Kobayashi, *Electochim. Acta* **54** (2009) 4893.
76. J. Greeley, J. Rossmeisl, A. Helman and J.K. Nørskov, *Z. Phys. Chem. (Munich)* **221** (2007) 1209.
77. G.C. Bond, *Platinum Metals Review* **51** (2007) 63.
78. B.E. Hayden, D. Pletcher, M.E. Rendall and J.-P. Suchsland, *J. Phys. Chem. C* **111** (2007) 17044.
79. A. Furube, L.-C. Du, K. Hara, R. Katoh and M. Tachiya, *J. Am. Chem. Soc.* **129** (2007) 14852.
80. H. Kominami, A. Tanaka and K. Hashimoto, *Chem. Commun.* **46** (2010) 1287.

81. L.-C. Du, A. Furube, K. Yamamoto, K. Hara, R. Katoh and M. Tachiya, *J. Phys. Chem. C* **113** (2009) 6454.
82. Y. Tiana and T. Tatsuma, *J. Am. Chem. Soc.* **127** (2005) 7632.
83. A. Primo, A. Corma and H. Garcia, *Phys. Chem. Chem. Phys.* **13** (2011) 886.
84. C.W. Corti, R.J. Holliday and D.T. Thompson, *Top. Catal.* **44** (2007) 331.

Chapter 8
Surface Structures of Gold Nanoparticles

Shamil Shaikhutdinov

Department of Chemical Physics, Fritz Haber Institute of the
Max Planck Society, Faradayweg 4–6, Berlin 14195, Germany.
Email: shaikhutdinov@fhi_berlin.mpg.de

8.1 Introduction

Since Roman times it has been recognized that gold dispersed in glasses
gives rise to fascinating optical phenomena, which were (much later, of
course) rationalized on the basis of *physical* properties of gold in a highly
dispersed state (e.g. see Chapter 3). Regarding its *chemical* properties, gold
has long been considered as an inert material since bulk gold does not
react easily with many molecules typically present in the ambient atmosphere, the property that renders gold the most noble metal. Only recently
chemistry of gold has received much attention, owing to unique catalytic
properties observed for highly dispersed gold in a number of industrially
important reactions such as low-temperature CO oxidation, selective oxidation of propene, water-gas shift reaction and selective hydrogenation of
acetylene, to name a few[1-5] (see Chapter 7). It is evident that the performance of gold in heterogeneous catalysts is primarily determined by
geometrical and electronic structures of gold species which are *surface* in
nature. The surface structures become very complex for Au clusters and
nano-sized particles exposing facets of different orientations and substantial amounts of undercoordinated atoms, e.g. on edges and corners. The issue
becomes even more complicated in the presence of a support (typically, an
oxide with the high surface area), which often behaves not as a spectator,

but as a player that may even control reactivity of gold, for example, by stabilizing a particular structure of gold species or gold/support interface.

The atomic-level information on structure, composition and electronic state of surfaces can only be obtained by surface-sensitive techniques, which are primarily (and somewhat routinely) used in surface science. The recent surge in study of gold catalysts[1–5] has been accompanied by a corresponding swell of interest in the surface science of gold. In spite of the enhanced activities worldwide, the surface chemistry of the small gold particle is still far from being well understood at the atomic scale. Partially, this is due to the fact that the "real" systems, typically consisting of Au nanoparticles deposited on oxide supports, having the high surface area (see Chapter 7), but ill-defined surface structures, are too complex for making precise and unambiguous structural characterization of such systems which may then be linked to their properties. This renders careful model studies a necessity. Within the so-called "surface science" approach,[6–15] the structural complexity of a real catalyst is reduced to a well-defined model. The latter, however, should retain the main features of the real system such as nature of a support, high dispersion of a supported metal, etc.

Numerous surface science studies are currently being performed in order to understand the fundamental aspects of catalysis by gold, in particular addressing the key question: why does gold behave as a catalyst when its dimensions are reduced to the nanometer scale? (see, for instance, recent reviews[16–23]).

This chapter will focus on surface structures of gold nanoparticles supported on model oxide supports. A number of model systems will be discussed here in order to illustrate the surface science approach to surface chemistry of gold and also to demonstrate the complexity of the gold/support interaction. After a description of model systems in Section 8.2, we will briefly discuss surface structures of gold single crystals (Section 8.3). Basic structural motifs of Au nanoparticles are discussed in Section 8.4. The preparation of thin film supports will be addressed in Section 8.5. Then in Section 8.6, we show basic principles of gold deposition onto planar supports. Section 8.7 presents case studies of size, support end environmental effects observed on Au model systems.

The chapter will not attempt to be completely comprehensive but rather give the reader a view of the current state of research in the attempts to understand the surface structures of gold nanoparticles which may aid our understanding of the unique catalytic chemistry of gold.

8.2 Background

In planar model systems metal particles are deposited onto an electrically conducting oxide single crystal or a thin metal-oxide film grown on a metal single crystal substrate as schematically shown in Fig. 8.1. The system allows the facile application of a large variety of surface sensitive techniques for precise system characterization such as:

- Electron spectroscopy, e.g. Ultraviolet and X-ray Photoelectron Spectroscopy (UPS/XPS), in particular using the synchrotron light source; Auger Electron Spectroscopy (AES). The methods allow the determining of the elemental composition and surface stoichiometry, the oxidation states of the constituting elements, and the valence band structure.
- Ion spectroscopy, e.g. Low Energy Ion Scattering (LEIS), Secondary Ion Mass-Spectrometry (SIMS), in particular for determining the composition of the topmost surface layers in combination with a depth-profile analysis.
- Vibrational spectroscopy, e.g. High Resolution Electron Energy Loss Spectroscopy (HREELS) and Infrared Reflection-Absorption Spectroscopy (IRAS). The methods monitor the lattice vibrations (phonons) and also the vibrations of adsorbed species which are often sensitive to the surface structure.

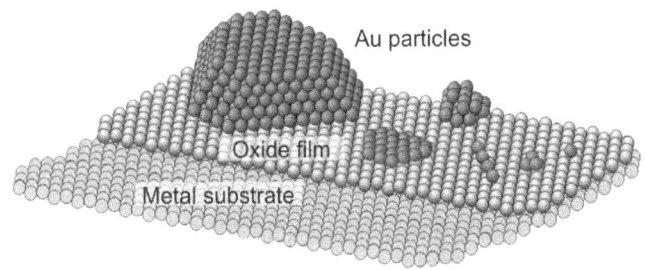

Fig. 8.1. Model planar systems for studying surface structures and reactivity of gold nanoparticles supported on well-defined, thin oxide films.

- Low Energy Electron Diffraction (LEED), Surface X-ray Diffraction (SXRD) and Grazing Incident Small Angle X-ray Scattering (GISAXS). The methods are typically used for ordered structures and are particularly powerful for the epitaxial and self-assembled systems.
- Scanning probe microscopy, e.g. Scanning Tunneling Microscopy (STM) and Atomic Force Microscopy (AFM). The only methods which allow the study of morphology of the surface nanostructures on the atomic scale.

Adsorption and reaction of molecules on these surfaces can readily be studied by Temperature Programmed Desorption/Reaction (TPD/TPR) as the planar systems do not suffer from diffusion limitations of powdered and porous materials. Kinetics and elementary steps of chemical reactions at surfaces can be monitored by Molecular Beam (MB) technique, particularly in combination with IRAS (see, for example, the review in Ref. 24). Although most of the surface sensitive techniques can only be applied in high and ultra-high vacuum (UHV), while catalytic processes occur at ambient pressures, *in situ* methods have recently been developed to bridge the so-called "pressure gap". Sum Frequency Generation (SFG), Polarization-Modulation IRAS (PM-IRAS), High-Pressure XPS (HP-XPS) and High-Pressure STM techniques allow one to carry out studies at more realistic pressure conditions.[25–30] Furthermore, *in situ* structural characterization can be performed simultaneously with reactivity measurements employing gas chromatography (GC) essentially in the same way as it is used in catalytic studies.

Basically, a research strategy of model studies includes the following "elementary" steps:

- Preparation and characterization of a planar support (i.e. morphology, surface composition, defect structure, etc);
- Deposition of metal particles onto the support (nucleation and growth);
- Structural characterization of the metal deposits (size, shape, electronic state, thermal stability, etc);
- Adsorption of molecules of interest (and probe molecules);
- Reactivity studies (kinetics, reaction orders);
- Finding structure-reactivity relationships (size effects, support effects, environmental effects).

202

8.3 Surface Structures of Gold Single Crystals

Although the catalytic nature of gold is most likely linked to size and structure issues, it seems appropriate to begin any discussion with structures
of gold single crystal surfaces. In addition, the extended surfaces are well
suited for studying surface reconstructions, if any, which may occur under
reaction conditions.

Gold shares a face centred cubic (fcc) lattice. The low Miller indexed
surfaces of gold, such as (111), (110) and (100), shown in Fig. 8.2, are all
known to undergo surface reconstruction in high vacuum. Perhaps the most
intriguing is the reconstruction of the Au(111) surface, unique for pure fcc
metals, which is often referred to as a "herringbone" reconstruction after
STM images[31] (see Fig. 8.3a). The reconstruction can be described by a
complex stacking-fault-domain model of the topmost layer having higher
surface density of atoms than in the bulk layers[31-33] and rationalized in
terms of unique surface states that arise due to the interaction of sp and d
states, a consequence of its relativistic nature.[34]

The Au(110) surface reconstructs into the (1×2) surface which is formed
by "missing rows" along $\{1{-}10\}$ direction as shown in inset in Fig. 8.3b.

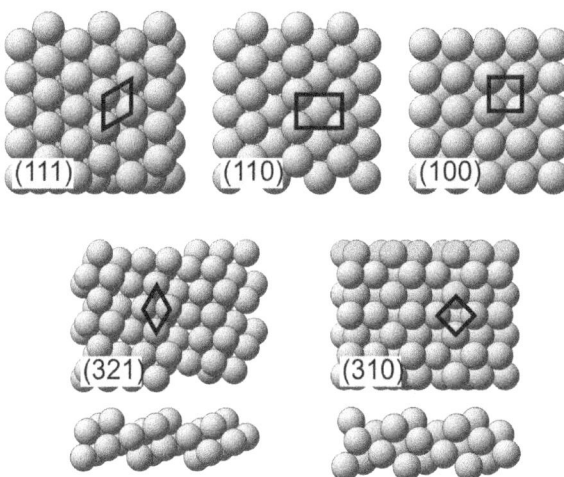

Fig. 8.2. (Top panel) Top views of unreconstructed (111), (110) and (100) surfaces of fcc metals.
The unit cells are indicated. (Bottom panel) Top and cross views of the stepped Au(321) and Au(310)
surfaces. The (111) and (100) unit cells of the microfacets are indicated.

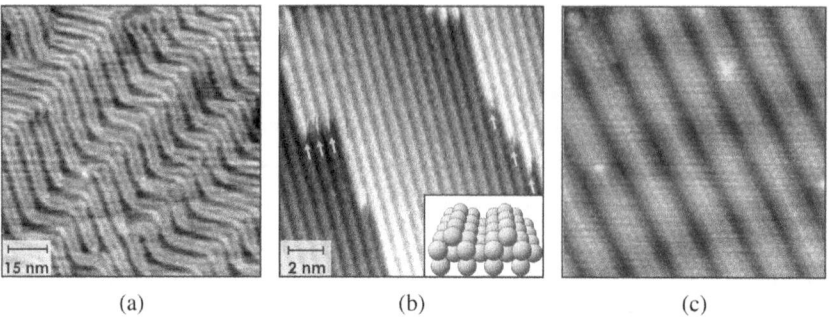

(a)   (b)   (c)

Fig. 8.3. STM images of (a) the "herringbone" reconstruction of Au(111); (b) the (1 × 2) "missing row" reconstruction of Au(110) (adapted from Gimzewski *et al.*[35] Copyright (1992) by the American Physical Society); (c) the "hexagonal" reconstruction of Au(100) (reproduced from Ref. 36).

This reconstruction gives rise to three different types of surface atoms: on top of row, side of row, and trench atoms. The sides of row atoms are arranged in the same manner as the (111) surface and often referred to as (111) microfacets.

The reconstruction of the Au(100) surface apparently depends on preparation and remains controversial. It has originally been assigned to (1 × 5) "adding row" reconstruction.[37] Later, more complex surface structures have been observed (see Fig. 8.3c), such as (5 × 20),[38] a (5 × 20) with rotation,[39] a hexagonal (28 × 5) R0.6°,[40] to name a few.

It has been proposed that the reconstructed surfaces may experience different combinations of these structures depending upon step density and surface temperature.[41] The reconstruction is generally believed to be limited to the first layer, thus indicating that the more compact surface arrangement is favoured to a degree where the energy cost due to a lack of commensuration with the layer underneath can be overcome.[42]

These reconstructions of the gold surfaces are quite stable under vacuum conditions. For example, the herringbone reconstruction of Au(111) was seen at ∼850 K,[43] but can be lifted upon adsorption of certain gases such as CO. Surface XRD studies[44] revealed lattice expansion of the Au(111) surface exposed to CO at 300 K at pressures between 0.1 and 530 mbar, although the herringbone reconstruction was preserved. Even more extensive surface transformations of Au(111) were observed upon exposure to 110 mbar CO at 600 K. Similar behaviour was found for Au(110) at CO pressures above 0.1 mbar.[45] STM results showed significant surface roughening and a lifting of terrace anisotropy. Since the morphological changes observed for both, Au(111) and Au(110) surfaces were remained after evacuation, CO dissociation might have occurred at high pressures accompanied by carbon deposition.

To date, a limited number of studies on gold single crystal surfaces with higher Miller indexes were reported, e.g. Au(991), Au(430), Au(221), Au(332), Au(310) and Au(321).[46-48] The latter two surfaces are shown in Fig. 8.2. The (321) surface forms (111) microfacets, whereas the (310) surface exhibits (100)-like facets, with the step atoms on both surfaces having a coordination number 6. Comparison of adsorption properties of CO (which is very often used for the characterization of the gold-based catalysts) on the two surfaces revealed that CO binding to gold is not only dependent on the coordination number but the exact geometrical surface structure.[48] The studies on the stepped Au surfaces aid our understanding of the reactivity of the low-coordinated gold atoms, which come to dominate the surface of gold nanostructures.

8.4 Morphology of Gold Nanoparticles

In principle, small metal particles exposing different facets resulted from a truncation of single crystals. The surface energies of (hkl) planes, described as the Gibbs free energy per unit area, determine the morphology of particles. Having obtained the free energies of the single crystal surfaces (γ_{hkl}), one can predict an equilibrium crystal shape of a free-standing (unsupported) particle for a given volume using the so-called Wulff construction.[49-51] Representing the surface free energy as a function of direction by vectors of length γ_{hkl} the minimum energy shape at a constant volume is the inner envelope of the normals to this surface (see Fig. 8.4).

Beyond the truncated octahedron, shown in Fig. 8.4, and their twinned variants, the possible structures also include decahedral and icosahedral motifs,[51-53] as schematically shown in Fig. 8.5. The latter structures have a relatively high strain energy.[54]

Obviously, the above described thermodynamic approach is valid only for a particle that reached an equilibrium shape. Note also that this construction neglects "edge" and "corner" energies, and can therefore be applied only to large particles, where the facet dimensions are much larger than the

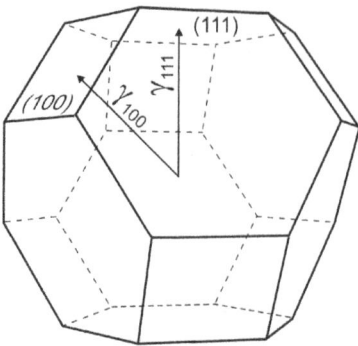

Fig. 8.4. The Wulff construction of (fcc) metal polyhedral particles comprising (111) and (100) faces.

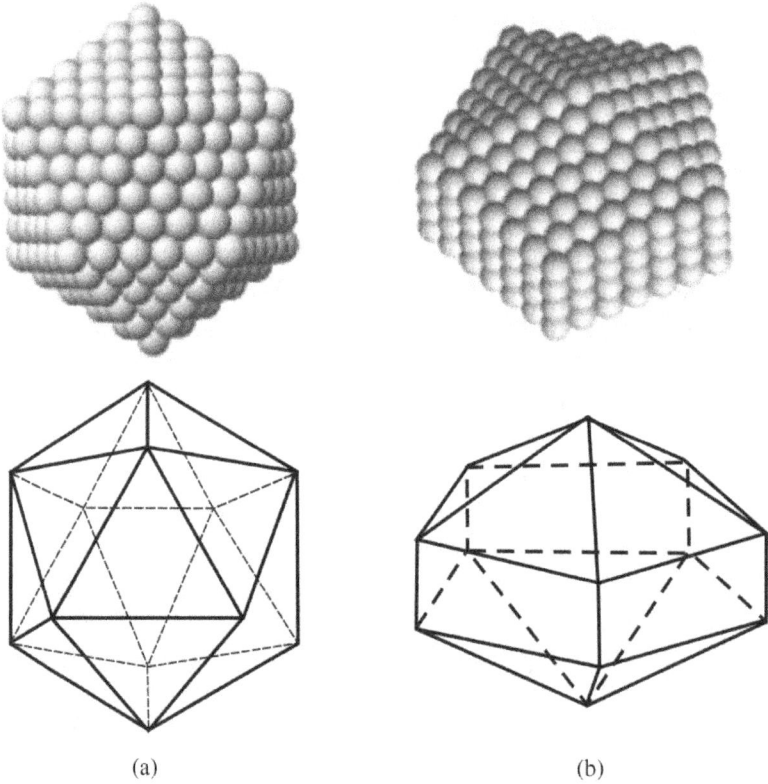

(a)   (b)

Fig. 8.5. Icosahedral (a) and decahedral (b) structures of gold nanoparticles.

surface unit cells. Basically, metal particles at moderate sizes (several tens of nanometers) are generally consistent with the Wulff construction.[51] The situation becomes more complicated upon reducing the size down to several nanometres. Surface stress, size dependent lattice parameter changes, twin boundaries, and ambient conditions are all factors which may influence the morphology of metal particles.[51]

For supported particles, the role of substrate may be included in the construction by replacing the free surface energy of the contact plane with an effective surface energy (γ^*) which is the difference between the interface energy and the surface energy of the substrate:[50]

$$\gamma^* = \gamma_{\text{interface}} - \gamma_{\text{support}} \qquad (1)$$

This gives the equilibrium shape truncated at the interface (see Fig. 8.6):[11]

$$\Delta h/h_i = W_{\text{adh}}/\gamma_{\text{metal}(i)} \qquad (2)$$

where W_{adh} is the adhesion energy, which is defined as the energy per unit area to pull the system apart to its constituents, and relates to the surface energies through the equation:

$$W_{\text{adh}} = \gamma_{\text{metal}(i)} + \gamma_{\text{support}} - \gamma_{\text{interface}} = \gamma_{\text{metal}(i)} - \gamma^* \qquad (3)$$

It is evident from (2) and (3) that, for negative values of γ^*, the particle height becomes smaller than the corresponding radii of the unsupported particle. Clearly, the higher the adhesion energy, the more flattened the supported particle is. Therefore, flattening of metal particles may be considered as the physical manifestation of the strong metal-support interaction.

In addition, the kinetics of particle growth may also influence the morphology of Au nanoparticles. High resolution transmission electron microscopy (HRTEM) studies revealed that the structure of very small Au

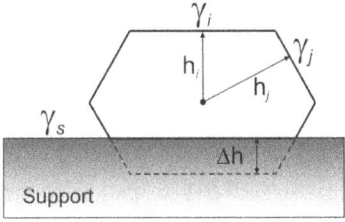

Fig. 8.6. Schematic representation of the Wulff construction for supported particles (see text).

nanoparticles ($d \sim 1$ nm) fluctuated significantly even at room temperature (the phenomenon was later referred to as "quasimelting"[51,55,56]), as the energy difference between conformations calculated for gas phase clusters is comparable with kT.[55] Although several effects may contribute to this phenomenon, including an electron beam induced overheating, it appears that the substrate plays a critical role in the stabilization of the particle in particular states. Indeed, the lack of such structural transitions on Au deposited onto vacuum-cleaved MgO microcubes was explained by the high interfacial energy which must be overcome when the particles are epitaxially oriented to the support.[57]

Monitoring a gold single particle growth on a MgO(100) step by HRTEM in situ revealed that the particle rearranged continuously to maintain the lowest energy structure.[58] Examining the ratio of Au(100) facets to Au(111) facets, repetitions in the truncation pattern were seen to emerge as the cluster grew until reaching a size of over 1,500 atoms at which point rigid epitaxy with the support is lost.

8.5 Planar Supports

Following the research strategy presented in Section 8.2, the first step in the preparation of gold model systems is the preparation of the well-defined supports. Among those, oxide single crystals, such as MgO, TiO$_2$, sapphire (Al$_2$O$_3$), and quartz (SiO$_2$), were of first choice.[11,16] However, the lack of electrical conductivity (most oxides are wide gap semiconductors and insulators) renders their use for surface-science studies very difficult except the partially reduced forms of rutile TiO$_2$(110), which remains the most widely studied support used for gold model catalysts. In contrast, thin oxide films grown on metal single crystals provide good electrical as well as thermal conductivity for a facile application of surface science techniques. In addition, the metal selection rules[59] (i.e. only vibrations with dipole moment changes normal to the surface can be detected by IRAS) often ease assignment of spectral features.

Various methods of preparing well-ordered oxide films were reported in the literature (see reviews in Refs. [8,9,13,60,61]). The first one is the direct oxidation of a metal single crystal as schematically shown in Fig. 8.7. However, this preparation results in amorphous or polycrystalline overlayers

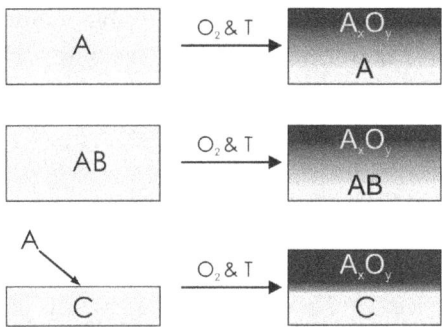

Fig. 8.7. Preparations of well-ordered thin oxide films by oxidation of a metal single crystal (Top), by oxidation of bimetallic surfaces (Middle), and by deposition/oxidation of metal overlayer on the second metal single crystal (Bottom).

(e.g. SiO_2/Si, Al_2O_3/Al)[62,63] or films with a high defect density (e.g. $NiO(100)/Ni(100)$.[64] Formation of a $Cr_2O_3(111)$ film on $Cr(110)$ seems to be the only successful example for this preparation to date.[65]

Another route is the oxidation of bimetallic surfaces where the element having higher affinity for oxygen segregates to the surface and becomes oxidized, ultimately forming well-ordered oxide film at elevated temperatures. Perhaps the best examples of the crystalline films fabricated using this approach are the alumina films grown on $NiAl(110)$, $NiAl(111)$ and $Ni_3Al(111)$ single crystals.[13,66−68]

In the majority of cases, however, oxide films are prepared by vapour deposition of one metal (to be oxidized) onto the second metal (typically noble metal) single crystal either in oxygen ambient or in vacuum followed by post-annealing in oxygen. In addition, a metal substrate can be pre-covered with atomic oxygen to ease oxidation of incoming metal atoms and to prevent their migration into the bulk. Numerous well-defined oxide films can be grown in this way, such as $Fe_3O_4(111)$ and $Fe_2O_3(0001)$ films on $Pt(111)$,[61] $MgO(001)/Ag(001)$,[69] $CeO_2(111)/Ru(0001)$,[70,71] $V_2O_3(0001)/Pd(111)$,[72] $V_2O_5(001)/Au(111)$,[73] $SiO_2/Ru(0001)$,[74] to name a few.

It is noteworthy that also highly oriented pyrolytic graphite (HOPG) has been used as a planar substrate for gold nanoparticles due to its flatness and good electrical conductivity. However, the HOPG substrate exhibits a weak interaction with gold, thus resulting in relatively large Au particles.[75] The particle density and size can be varied by ion sputtering and/or oxidation which result in etched pits, which serve as nucleation centres.[23,75]

8.6 Gold Deposition on Planar Supports

8.6.1 *Physical vapour deposition*

Most commonly, gold particles are deposited onto planar oxide supports in vacuum using *Physical Vapour Deposition* (PVD), which provides the best cleanliness of the systems under study. There are various realizations of PVD depending on how a metal vapour is produced, e.g. by thermal evaporation, by sputtering, by laser ablation, etc.

For the thermal evaporation, gold is placed into a crucible (usually made of high melting point materials such as W and Mo) that is heated resistively or using electron beam (see Fig. 8.8). Since gold has a relatively high vapour pressure, the reasonable flux can be obtained at the temperatures close to the melting point of gold (1337 K). The most simple, "home-built" evaporators can be made of a gold thin wire wrapped around tungsten wire, which is heated by passing the electric current to first form a small droplet of gold. The deposition flux of gold atoms from these sources is typically in the

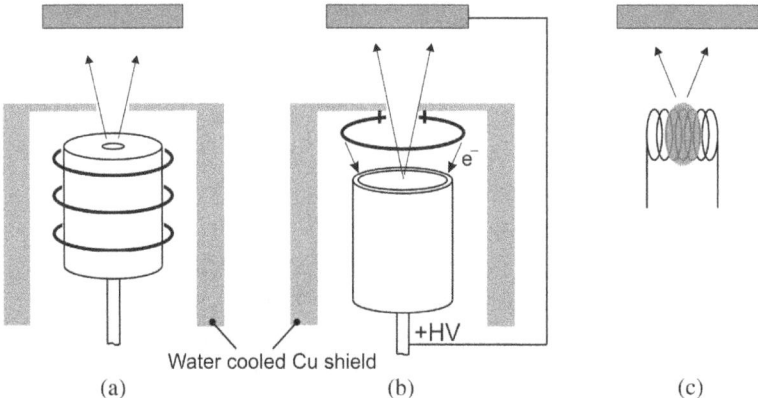

(a) (b) (c)

Fig. 8.8. PVD sources typically used for gold deposition on planar supports. A crucible filled with metallic gold is heated resistively (a) or by electron bombardment (b). In (b) the substrate is biased at the same high voltage potential as the crucible (typically 800–1000 V) to prevent accelerating of charged gold species towards the substrate. Evaporators normally have a water-cooled Cu shield to maintain the UHV conditions during the deposition. (c) The "home-built" evaporator consisting of tungsten filament with a small Au droplet formed upon resistive heating of the Au wire wrapped around the tungsten filament.

range of 10^{14} atoms cm^{-2} sec^{-1}, which can be directly measured by a quartz microbalance or STM.

Note that more recently developed PVD methods include a glow plasma discharge or a cathodic arc to vaporize the target material. In the pulsed laser deposition method, a high power laser ablates material from the target into a vapour. Although the experimental set-up may be fairly simple, the ablation process itself is extremely complex involving the interaction between the laser and a solid target material, plasma formation and the transport of material across the vacuum to the substrate. All these methods are rarely used for PVD of pure metals like gold, but rather for multi-elemental compounds and materials with low vapour pressures.

The incoming Au atoms adsorb onto the surface, diffuse and ultimately form clusters and particles as schematically shown in Fig. 8.9. In principle, two types of nucleation modes may occur at surfaces: homogeneous and heterogeneous. For the *homogeneous nucleation*, an immobile nucleus is formed on regular surface sites by aggregation of several atoms. The critical island size is defined as the size above which the islands are stable. In other words, by addition of further atoms, the nuclei will grow, whereas islands up to this size can dissolve again. For the simplest situation, when a dimer is already stable species (the critical size is one atom), theoretical considerations show[9,76] that the saturation density of islands N depends on the diffusion coefficient D and the deposition flux F as $N \sim (F/D)^{1/3}$. Then no further nuclei form, and all diffusing atoms stick to existing islands. (For a more sophisticated analysis the reader should refer to Ref. 9 and references therein.)

However, if defects are present on the surface, the ad-atoms may be trapped at these sites, thus forming nucleation centres for subsequent growth, which is called *heterogeneous nucleation*. If the interaction of ad-atoms with defects is strong and the defect density is relatively high,

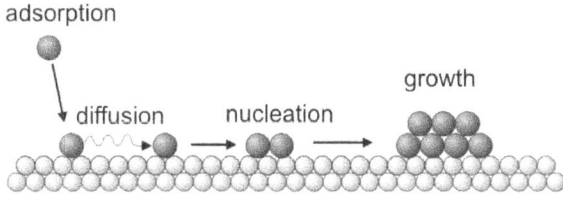

Fig. 8.9.   Elementary steps of the formation of Au particles by physical vapour deposition.

the metal particle density will be independent of the deposition flux. Therefore, it is instructive here to note that, when using electron beam assisted evaporators and other methods resulting in high-energy charged species, one has to take precautions against acceleration of the ionized metal atoms towards a substrate, which may create additional defects upon collision. To prevent this, the sample is biased at the same potential as the metal source (typically ~ 1 kV) during evaporation.

In the majority of cases, the nucleation and growth of gold deposited by PVD on oxide supports is governed by the defect structure of the support (i.e. point defects (vacancies), steps, impurities, etc.) as a result of relatively weak interaction of gold with oxides and hence high surface diffusivity of gold ad-atoms.[17] Such effect may even be used for decorating surface defects to count them.[71]

Certainly, the substrate temperature can alter the growth processes. For example, the Au ad-atoms may escape from defect sites at elevated temperatures. In addition, the particles sinter at high temperatures. Very often, the size and the shape of the gold aggregates formed at low temperatures is kinetically limited as illustrated in Fig. 8.10 for Au deposited on the $Fe_3O_4(111)$ films.

(a)   (b)

Fig. 8.10. Room temperature STM images of Au particles as deposited by PVD on a $Fe_3O_4(111)$ film at 100 K (a) and after annealing to 500 K (b). The characteristic shapes of the annealed Au particles are shown in the inset.

In principle, a range of particle sizes between 1 and 10 nm with relatively narrow particle size distribution can be prepared by PVD. A degree of size control can be achieved by optimizing the coverage, substrate temperature and deposition flux.

8.6.2 *Cluster deposition*

In the range of very small sizes, a "soft-landing" of *mass-selected clusters* pre-formed in the gas phase was suggested as a highly controllable and flexible method,[77,78] albeit experimentally very sophisticated and much more "expensive" than PVD. The method allows, in principle, precise control over the cluster size and is primarily aimed at the so-called non-scalable regime of a gold particle size where each atom matters.[79] Note that when using this cluster deposition one has to take precautions against coalescence of deposited clusters, which implies low metal coverage.

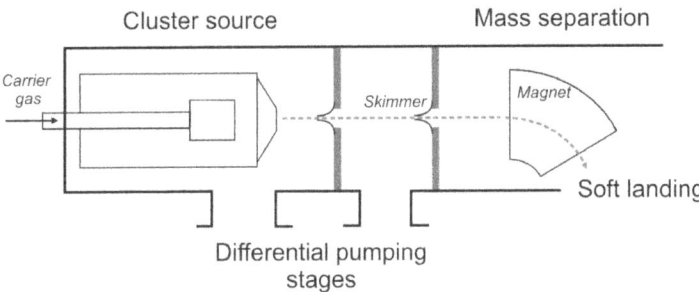

Fig. 8.11. Schematic representation of a mass-selected cluster deposition. Mass separation employs deflection of charged gold clusters in an electric or magnetic field. Here, only segmented magnet is shown, for simplicity.

Basically, the apparatus includes a cluster beam source, a mass filter, and facilities for soft landing deposition (Fig. 8.11). In the beam source, a metal vapour is ejected into a flow of a cooled inert gas, where the condensation of a supersaturated vapour into clusters occurs. The metal vapour can be generated by laser ablation, sputtering, a pulsed or continuous arc, or thermally.[77,80] Depending on the type of source, the gas pressure inside the condensation zone ranges from a few millibar to a few bar. The cluster size distribution depends on the time until the mixture of carrier gas and clusters exits the aperture into the vacuum region, which can be controlled, for example, by the gas flow rate. The expansion accelerates the clusters, which, in the limit of large pressure drop across the aperture ("free-jet expansion"), are thermally equilibrated with the bath gas and have the same velocity distribution irrespective of their size.

Mass separation and selection of the clusters is based on deflection of the charged species either in an electric or magnetic field. The so-called Wien filter applies orthogonal electric and magnetic fields to charged particles. Therefore, the equipment additionally requires a cluster ionizer (except of sputtering-based beam sources which contain substantial amounts of charged species). Some cluster sources even employ two mass filters; a high resolution instrument such as time-of-flight analyzer to measure the mass spectrum in the free beam and a lower resolution high-throughput device to narrow the native mass distribution prior to deposition. Mass separation of neutral clusters is still challenging (for more details, see Refs. 80,81).

Soft-landing of clusters implies that the clusters do not fragment upon collision with a substrate. The kinetic energy of clusters to ful-fil this regime is typically set to 0.1 eV per atom. Following the well-known Maxwell–Boltzmann velocity (v) distribution function, i.e. $f(v) \sim (m/2\pi kT)^{3/2} v^2 exp(-mv^2/2kT)$, the soft-landing regime may be achieved by properly choosing the mass m and temperature T of a carrier gas. For example, for Ar at 100 K, the velocity at the peak of the distribution is about 200 ms^{-1}, which corresponds to the impact energy of an Au cluster about 0.04 eV/atom. However, in the case of He at 300 K as the carrier gas, this would yield 1.5 eV/atom that falls into the range of medium impact energies (1–10 eV/atom).[81] In addition, soft landing can be achieved by decelerating the charged clusters by electric field while approaching a support (see, for example, Refs. 82,83).

Several experiments demonstrated that soft landing of metal clusters is possible.[79,84,85] STM characterization of Au_n^+ (n = 1 − 8) clusters deposited on $TiO_2(110)$ showed that, starting from dimers, Au_2^+, the clusters could be deposited intact, and no cluster agglomeration occurred, at least, at room temperature,[85] as shown in Fig. 8.12. Interestingly, single atom deposition (n = 1) revealed no small Au clusters on surface (see Fig. 8.12a). These observations indicate that Au monomers are highly mobile on the $TiO_2(110)$ surface, leading to aggregation into larger clusters, containing in the order of tens of atoms, on average.

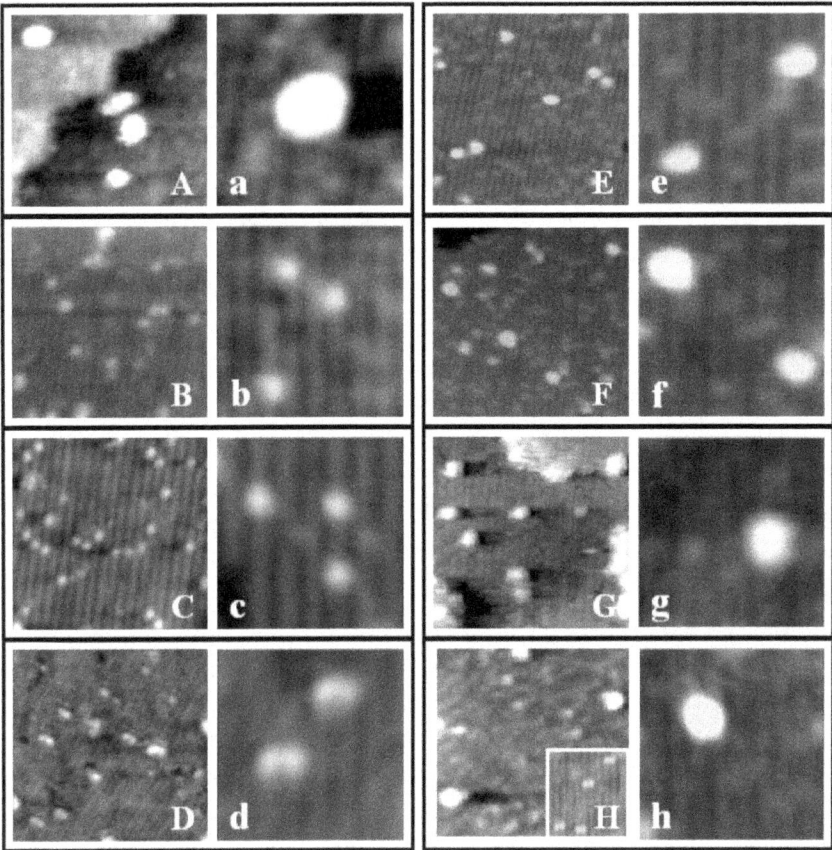

Fig. 8.12. STM images 14 nm × 14 nm (uppercase letters) and 5 nm × 5 nm (lowercase letters) of Au_n^+ clusters deposited on $TiO_2(110)$ at 300 K (n = 1 − 8 for (a–h), respectively). The bright protruding spots are the Au clusters, and the dim spots between the rows of 6.5 Å apart are bridging oxygen vacancies of $TiO_2(110)$. (Reprinted with permission from Tong *et al.*[85] Copyright (2005) American Chemical Society.)

Finally, the deposition onto a thin buffer layer of rare gas[86] (typically Xe) formed on a substrate cooled down to cryogenic temperatures (*ca.* 40 K) may also lead to the non-fragmented deposition of metal clusters. Subsequent heating to desorb the weakly bound buffer layer is accompanied by adsorption of the clusters onto a substrate.

8.6.3 *Reactive deposition methods*

The *buffer layer assisted deposition* just mentioned above was further extended to PVD of metals on amorphous solid water films or, simply, on ice formed by water adsorption on a planar substrate kept at $\sim$100 K.[87] After metal deposition, the sample is heated to the room temperature to desorb weakly bound water molecules. The formation and aggregation of particles on ice films is a complex process that is driven by the dewetting, islanding and sublimation of ice on heating and is yet not well understood.[88] Nonetheless, water assisted deposition of gold on silica films showed that one can independently control particle size and only change particle density by repeating the whole deposition procedure.[89]

Chemical Vapour Deposition (CVD) is based on the use of volatile complexes as precursors, which decompose on a target substrate kept at elevated temperatures. The CVD can be performed by direct sublimation of solid precursor in vacuum or in the carrier gas such as nitrogen or hydrogen, the latter additionally used as the reducing agent. Some preparations include bubbling the carrier gas through the precursor solution.

Most of the gold(I) and gold(III) precursors used to date are organometallic compounds such as alkyl(phosphine)gold(I) and dimethylgold(III) β-diketonates and their derivatives.[90-93] In general, synthesis of the gold precursors is often very complex and shows a low yield. In addition, many precursors are light and/or air-sensitive, and as such are difficult to handle. Dimethylgold(III) carboxylates were recently suggested as viable precursors due to their sufficient volatility and thermal stability.[94]

The principal advantage of CVD over PVD is the possibility to uniformly cover even rough surfaces without having shade zones.[91] However, the CVD method still suffers from the lack of control of cleanliness arising from the stripping of the ligands and decomposition at surface. It is noteworthy that this method is characterized by high deposition rates, thus resulting in granular films rather than isolated nanoparticles, and is, therefore, primarily aimed at deposition of gold thin films in microelectronics, metal coatings, etc. There are only a limited number of "surface-science" studies of CVD-prepared gold nanoparticles. For example, AFM measurements revealed significant sintering of Au particles on $TiO_2(110)$ under an ambient atmosphere at temperatures as low as 363 K.[95] However, pre-treatment of the surface with ultraviolet radiation before CVD prevented agglomeration of particles. Apparently, this behaviour resulted from the presence of hydroxyl groups which formed as a result of photo-induced dissociation of water on the titania surface. It was suggested that the hydroxyl groups react strongly with the gold(I)-phosphine precursor used and thereby brought about the formation of highly stable small gold particles, which showed limited agglomeration even at 493 K in air.

8.6.4 *Deposition of ordered particles*

Recently, the deposition methods have been developed which aim at a simultaneous control over particle size as well as distance between the Au particles. The latter parameter may be important for gold application in photonics and sensors as well as in catalysis of diffusion-controlled reactions.

One approach includes deposition of a monolayer of Au filled *micelles* with a different length of the polymeric ligand.[96] This preparation allows deposition of Au particles of the mean particle size ranged between 1 and 15 nm with very narrow particle size distribution. The micelles can be brought to the planar substrate by spin-coating techniques or, more preferably, by dip-coating, i.e. dipping the substrate into the micellar solution and pulling it slowly out.[97,98] The resulting systems showed hexagonally arranged pattern of Au nano-dots, with the average distance between Au nanoparticles being controlled by the total length of the diblock copolymers (see Fig. 8.13). The nature of substrate seems not to play a considerable role on particle spatial distribution.[96] Removal of the polymeric shell can be achieved by annealing in UHV or by treatment in oxygen plasma. However, this issue remains the most uncertain in these experiments.

Another approach for ordering metal particles on surfaces employs *electron beam lithography* (EBL) for production of a patterned template followed by metal deposition using PVD, as nicely demonstrated for Pd and Pt particles on various supports.[99–102] Briefly, a highly collimated electron beam is exposed to a thin layer of polymeric resist (such as poly(methyl methacrylate) spin-coated on a flat substrate (for example, a Si(100) wafer covered by the native oxide film or coated with thin film of alumina). The electron irradiation decomposes the polymer backbone, making it possible

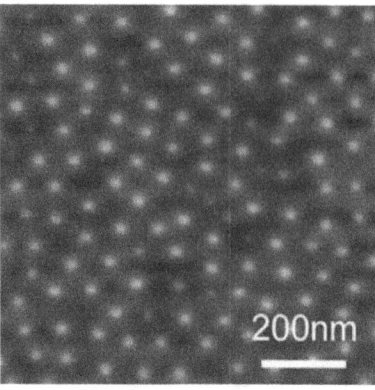

Fig. 8.13. AFM image of gold nanoparticles deposited by dip-coating of inverse micelles on ultra-thin TiC films.[98] The Au particles were synthesized using polystyrene-block-poly(2-vinylpyridine) as encapsulating agent. The mean particle height is around 2 nm. (Note that the lateral size of the particles appears much larger than 2 nm due to the tip-sample deconvolution effects.)

to dissolve the exposed polymer in a developing solution. Then a metal film is deposited on the surface by PVD, and the remaining polymer is removed with acetone. In the case of Pt, the resulting systems showed highly ordered arrays of metallic particles about 28 nm in lateral size with 100 nm periodicity.[99]

Although a combined EBL and PVD preparation was not yet applied to gold, the EBL was effectively used for substrate patterning and subsequent deposition of ligand protected gold particles, which selectively bound to the patterned substrate.[103] Another possibility is patterning of Langmuir-Blodgett films of colloidal gold.[104,105] More about *nanolithography* methods in application to gold can be found in the recent review.[106]

There are other gold deposition methods currently employed in research laboratories, which essentially combine those described above. The preferential choice in using one or another deposition method depends on many factors and is, of course, governed by the objectives of the study. On the one hand, all methods based on non-vacuum depositions have an inherent tendency to introduce contaminations into the system. On the other hand, the most clean, physical vapour and cluster depositions are hardly suited for preparations of technically relevant systems on a large scale.

8.7 "Surface Science" of Gold Nanoparticles: Case Studies

8.7.1 *Nucleation and growth*

To date, gold vapour deposited on $TiO_2(110)$ remains the most widely studied planar model system involving gold.[16–18] The (110) surface is the most stable surface of TiO_2 rutile and consists of alternating rows of titanium and oxygen atoms with half of the titanium atoms covered by so-called bridging oxygen. These oxygens are relatively weakly bound and can be removed upon high-temperature annealing (in order to form an electrically conducting support), ultimately creating oxygen vacancies, which can strongly influence the support's chemistry[107] (see also Fig. 8.12).

At very low coverage, PVD-deposited gold grows on $TiO_2(110)$ two-dimensionally (2D) at room temperature, which then subsequently changes to three-dimensional (3D) growth with increasing coverage.[108,109] At higher deposition temperatures, the growth mode is more 3D from the beginning, indicating that the 2D growth at low temperatures is a kinetically limited mode.[108–111] Very similar behaviour was observed for gold deposited on thin titania films grown on $Ru(0001)$[112] further supporting the concept that thin oxide films are, indeed, suitable supports for studying highly dispersed metal particles.

Early high resolution STM studies corroborated by density functional theory (DFT) calculations showed a direct relationship between gold particle nucleation and surface oxygen vacancies on $TiO_2(110)$.[113] Further studies,[114] however, have revealed that the $TiO_2(110)$ surface is, in fact, very sensitive to the traces of water in the UHV background leading to the formation of surface hydroxyl (OH) groups, which are likely to have influenced previous studies on gold deposition on this support. Indeed, hydroxyls were found to promote gold sintering and strongly affect the particle size distribution.[115]

When gold was deposited on $TiO_2(110)$ at 300 K and then annealed to 770 K the $(111)_{Au}//(110)_{TiO_2}$ orientation was preferred while deposition at 770 K preferentially gave rise to the $(112)_{Au}//(110)_{TiO_2}$ orientation.[16,116] Interestingly, the Au lattice does not appear to undergo any deformation in spite of the minimal strain that must be overcome to match the TiO_2 epitaxy, indicating that the interaction between gold and titania is rather weak. On the other hand, the relationship between titania and gold is strong

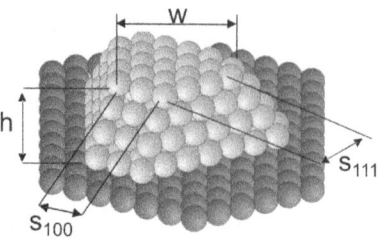

Fig. 8.14. Structural characteristics used for deriving the adhesion energy between a metal particle and a support.

enough such that particles can grow in a Au(111) epitaxy up to rather large sizes.[16]

In principle, one can derive the adhesion energy between a metal and a support, if the precise geometry of supported particles is known. For the metal particles, mainly exposing (100) and (111) facets, and contacting a substrate *via* the (111) plane (see Fig. 8.14), an analysis leads to the following expression for the adhesion energy[117]:

$$W_{adh} = \gamma_{111} \left(2 - \frac{3}{\sqrt{2}} \frac{h}{w} \frac{s+1}{2s+1} \right) \qquad (4)$$

where γ_{111} is the surface energy of the (111) surface, h (w) is height (width) of particles; s is the ratio of top-facet side lengths (s_{100}/s_{111}).

For example, STM images of the annealed gold particles on a $Fe_3O_4(111)$ thin film showed[118] well-faceted particles, exhibiting mostly hexagonal shape of top facets as shown in Fig. 8.10b. In addition, the histogram analysis of STM images revealed that the height of the particles was a multiple of ~2.5 Å (equal to the height of a monolayer of gold in the (111) orientation), thus indicating that particles grow by increasing the number of the atomic layers. Although atomic resolution of the top facets was not achieved, it seems plausible that the top facets show up the (111) surface owing to small (~3%) lattice mismatch between Au(111) and $Fe_3O_4(111)$ surfaces. Based on the structural parameters derived from STM and using the theoretical value of 1.28 J/m^2 for Au(111) as calculated by DFT,[119] Equation 4 yields around 2.3 J/m^2 for the adhesion energy between Au and the iron oxide. For comparison, Pd particles and Pt particles deposited on the same

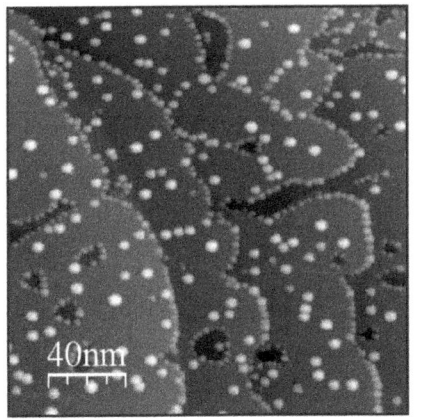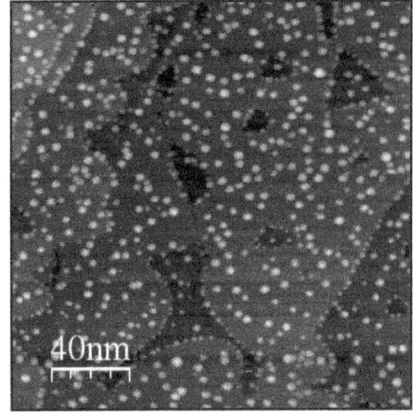

Fig. 8.15. STM images of Au particles deposited on fully oxidized (left) and partially reduced (right) $CeO_2(111)$ films.

support showed the energies of 3.1 J/m^2 and 3.8 J/m^2, respectively,[120] using the values of 1.92 J/m^2 for Pd(111) and 2.3 J/m^2 for Pt(111) from the same DFT calculations, for consistency.[119] It is evident that Au interacts with the iron oxide surface more weakly than do Pt and Pd.

On $CeO_2(111)$ thin films, gold preferentially nucleated on terraces at point defects (presumably, oxygen vacancies), which density could be varied by high-temperature annealing in UHV[121] (see Fig. 8.15). At increasing gold coverage, particles nucleated at the step edges exposing a large variety of low-coordinated sites. Only at high coverage, the Au particles grew homogeneously on the flat $CeO_2(111)$ terraces.[122] Gold exhibited a 3D growth mode from the onset resulting in nanoparticles, which were fairly stable towards sintering in vacuum at elevated temperatures. Increasing amounts of deposited gold essentially did not change the aspect (height/width) ratio of the particles, i.e. in agreement with thermodynamic considerations discussed in the Section 8.3.

Surprisingly, gold deposited onto crystalline alumina thin films did not show preferential nucleation as observed, for example, for Pd and Ag on the same support. The hemi-spherical particles were found randomly distributed on the surface.[118,123] This behaviour has been attributed to the particular structure of the thin alumina film grown on NiAl(110) which strongly interacts with the gold single atoms.[124]

In the above examples gold was deposited using PVD, which typically resulted in the mean particle size ca. 2–5 nm, except deposition at cryogenic temperatures where limited surface diffusion leads to the stabilization of single Au ad-atoms or the formation of very small aggregates.[21,125,126] For the preparation of monodispersed Au particles below 1 nm in size, deposition of mass-selected clusters seems to be the only method (see Section 8.6.2).

As mentioned above, the Au_n^+ (n = 2 − 8) clusters, deposited on TiO_2(110), showed no cluster agglomeration at room temperature[85] (see Fig. 8.12). In contrast, gold single atoms sintered rapidly and formed larger aggregates similarly to the samples prepared by PVD. In addition, the STM study revealed that supported Au_5 − Au_8 clusters all exhibited 3D-structures, albeit theory and gas-phase experiments indicated their planar configuration. These findings suggest that support has stabilization effect on particular structures of gold clusters. Interestingly, no direct evidence was found in these experiments for oxygen vacancies on TiO_2 to be required for binding the mass-selected clusters. Basically, the same conclusion has been drawn after revision of previous STM results of PVD-deposited Au particles.[115]

The role of oxygen vacancies (or more generally, point defects) on the structure of mass-selected gold clusters have been addressed using MgO(100) thin films.[79,127] Experimental and theoretical studies of Au_8^+ clusters deposited on perfect and defect-rich MgO films revealed that the so-called colour (or F-) centres play a crucial role in the reactivity of gold clusters in CO oxidation. DFT analysis showed that, although the Au_8 cluster adsorbed on F-centre is only slightly distorted as compared to the gas phase neutral cluster, it binds much more strongly to the defect, and the charge transfer from the oxygen vacancy to the Au_8 cluster occurs, as observed by IR spectroscopy of CO as a probe molecule.[79]

8.7.2 Particle size effects on the reactivity

The ability of the cluster deposition technique to vary a cluster size atom-by-atom allowed monitoring size dependent reactivity of the gold clusters, in particular in the CO oxidation. TPD and pulsed MB studies of supported Au_n clusters revealed non-scalable activity towards CO_2 formation

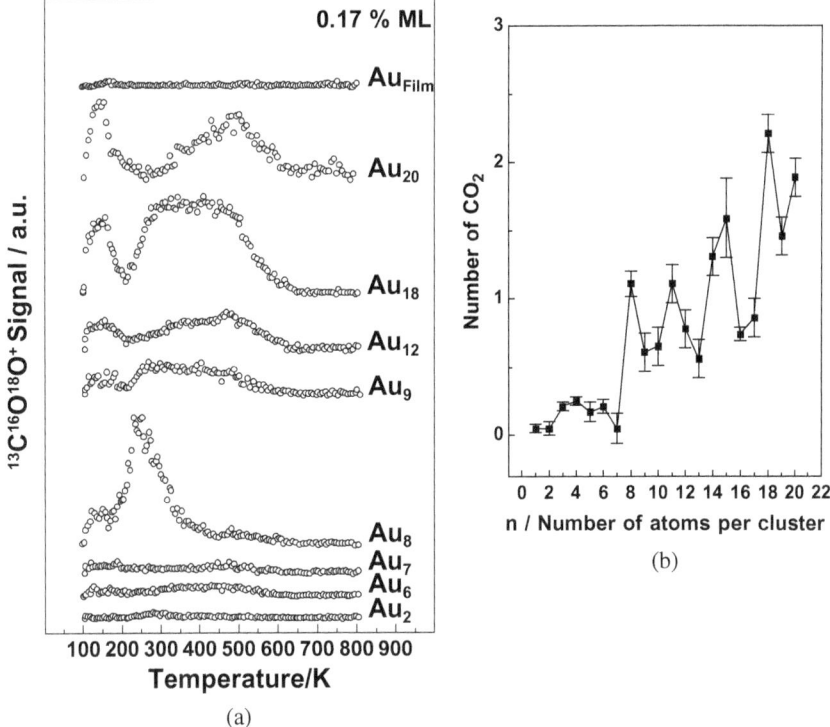

Fig. 8.16. (a) TPD signal of $^{13}C^{16}O^{18}O$ production upon sequential adsorption of $^{18}O_2$ and ^{13}CO at 90 K on Au_n clusters deposited onto defect-rich MgO thin films. The reactivity of Au_n is expressed in (b) as the number of formed CO_2 molecules per cluster. (Reproduced with permission from Arenz *et al.*[79] Copyright Wiley-VCH Verlag GmbH & Co. KGaA.)

in such that clusters below Au_8 on MgO[79,127] (see Fig. 8.16) and below Au_7 on $TiO_2(110)$[128,129] were essentially inert. At increasing cluster size somewhat oscillatory behaviour was observed as shown in Fig. 8.16b. Another important observation that came out from these studies is structural flexibility ("fluxionality"[79]) of the gold-based systems, which adopt the optimum structure in response to the gas adsorption and the most favourable reaction pathway.

Mass-selected Au clusters consisting of less than ten atoms were deposited on ion pre-sputtered HOPG surfaces.[23] STM inspection of Au_7 clusters suggested the absence of sintering upon deposition.[130] Only Au_8 can be significantly oxidized using atomic oxygen source and subsequently reduced by CO, as revealed by XPS. The results were somewhat different for Au clusters deposited on native silica films on Si(111), thus indicating importance of metal-support interactions for reactivity of gold.

TPD and IRAS studies of CO adsorption on gold nanoparticles deposited onto various oxide supports clearly showed a particle size effect in that small particles adsorb CO more strongly.[118,131] At the lowest Au coverage and low deposition temperatures, i.e. at the conditions providing the formation of small Au aggregates, CO was found desorbing at temperatures as high as 300 K. Such relatively strongly bound states have never been observed on Au single crystal surfaces[46,48,132,133] and, therefore, have been associated with highly uncoordinated gold atoms present on the surface of the smallest particles.[131] However, these states disappear upon annealing to 400–500 K, which is accompanied by strong reduction of the CO uptake due to gold sintering, in full agreement with STM results pointing out low thermal stability of gold clusters. Indeed, the spectra for annealed samples become very similar to those obtained on the stepped gold surfaces. Comparison of CO TPD profiles obtained for gold deposited on various oxide supports showed that, for a given nanoparticle size, the interaction of CO with Au particles is essentially identical.[118] These findings suggest that the support effects frequently reported in the literature for real catalytic systems (see Chapter 7), particularly in oxidation reactions, could be associated with defect structures of oxide supports that control the size distribution of gold nanoparticles and the gold/oxide interface. In addition, as shown above for the titania support, the presence of OH surface species either in the course of the catalyst preparation or under reaction conditions may also play a role in the support effects observed.

8.7.3 *Environmental effects*

To date, the majority of studies on the Au model catalysts were conducted under well-controlled but low-pressure conditions. Relatively unexplored are potential modifications of these systems in a reactive environment. In this respect, many fundamental questions still remain, including the possibility of reaction induced morphological changes in the system. From general bond conservation considerations, a weakening of the Au–Au bonds within the cluster may occur when an Au cluster strongly interacts with the reactive gas, ultimately leading to a disruption of the structure of the metal particle. Also, adsorption induced reshaping of gold particles may occur as recently shown by DFT calculations of an Au_{79} cluster ($\sim$1 nm in diameter) reacting with CO.[134] The strength of interaction between the gold nanoparticle and the support may play a significant role in determining the stability of these systems under realistic pressures and temperatures. However, structural *in situ* surface studies of gold nanoparticles under realistic reaction conditions are still scarce.

For example, no morphological changes of the Au particles deposited on thin FeO(111) films were observed in oxygen and hydrogen environments at pressures up to 2 mbar at room temperature as shown by STM *in situ*.[135] However, in CO and CO + O_2 (2:1) atmospheres, the destabilization of Au particles located at the step edges was observed even at $\sim$10^{-3} mbar leading to the formation of mobile Au species, which diffused across the surface. The results were rationalized in terms of a stronger interaction of gold with CO, as compared to O_2 and H_2, which significantly weakens the Au/support interaction for the smallest particles.

STM studies[136] of Au/TiO$_2$(110) revealed a form of Ostwald ripening (the growth of large particles at the expense of small particles) by exposing to 14 mbar of O_2 at 300 K. This process appeared to be even stronger in the stoichiometric mixture of O_2 and CO. Further *in situ* STM studies[137,138] revealed that an initial uniformly sized group of Au particles underwent severe Ostwald ripening at 450 K in 0.7 mbar of O_2. The presence of oxygen served to weaken Au–Au bonds, thereby promoting sintering. Note, however, that STM images taken under reaction conditions showed that the behaviour of particles that were initially of the same size could be quite different, as some particles decreased in size or even disappeared, while others seemed to remain stable.

The morphology of the supported Au particles on CeO$_2$(111) films was studied by STM *in situ* and *ex situ* in CO, O_2 and CO + O_2 environment at room temperature.[122] No visible changes were observed after exposing Au particles to pure O_2 up to $\sim$10 mbar. In the pure CO ambient, Ostwald ripening emerged above $\sim$1 mbar. Meanwhile, sintering of the Au particles was observed in CO + O_2 (1:1) mixture at much lower pressure ($\sim$10^{-3} mbar) that mainly occurred along the step edges. The results indicate that the structural stability of the Au/CeO$_2$ surfaces is intimately connected with its reactivity in the CO oxidation reaction. The results revealed both similarities and differences with the Au/TiO$_2$(110)[137] and Au/FeO(111)[135] systems, suggesting that Au/oxide interaction is directly involved in the structural stability of the supported Au nanoparticles.

Finally, STM studies were combined with reaction kinetics measurements of Au vapour deposited onto a TiO$_2$(110) single crystal and a titania thin film in the mbar pressure range.[139–141] Catalytic activity was found to be a function of the Au cluster size. Furthermore, maximum catalytic activity for these clusters was found to coincide with the metal-to-nonmetal transition occurring in clusters approximately 3 nm in diameter, as determined by tunneling spectroscopy. This cluster diameter also coincides with a cluster growth transition from the nucleation of two-dimensional islands to their agglomeration into hemispherical, three-dimensional structures.

8.8 Concluding Remarks

Numerous experimental and theoretical studies aimed at a better understanding of the unique properties of gold surfaces in the nanometer size regime are currently being performed. "Surface science" studies of model planar systems show that Au small aggregates exhibit electronic and chemical characteristics different from those of bulk gold. It seems plausible that the role of the oxide support is to stabilize particular atomic configurations, charge states or electronic properties of the ultra-small Au aggregates, which are in turn responsible for a distinct chemical behaviour.

It should be emphasized here that, in the case of the ultra-thin oxide films, the metal substrate lying underneath the film may affect the electronic and adsorption properties of Au species such that those depend on the film thickness. Recent studies of these systems, including single atoms, small clusters, and two-dimensional islands of gold, clearly demonstrated the complexity of the gold interaction with ultra-thin oxide films.[21]

Acknowledgements

I would like to thank all my coworkers, whose names appear in the references below, for their tremendous work in the laboratories of the group "Structure and Reactivity" at the Fritz-Haber Institute at Berlin. In addition, I am delighted to have been working for many years with Prof. Hajo Freund, who gave me the opportunity to head this group in the Department of Chemical Physics.

References

1. M. Haruta, *Chem. Rec.* **3** (2003) 75.
2. M. Haruta, *Gold Bull.* **37** (2004) 27.
3. G.C. Bond, D.T. Thompson and C. Louis, *Catalysis by Gold*, Imperial College Press, London, 2006.
4. A.S.K. Hashmi and G.J. Hutchings, *Angew. Chem. Int. Ed.* **45** (2006) 7896.
5. M.C. Kung, R.J. Davis and H.H. Kung, *J. Phys. Chem. C* **111** (2007) 11767.
6. D.W. Goodman, *Surf. Rev. Lett.* **2** (1995) 9.
7. D.W. Goodman, *J. Phys. Chem.* **100** (1996) 13090.
8. H.-J. Freund, *Angew. Chem. Int. Ed.* **36** (1997) 452.
9. C.T. Campbell, *Surf. Sci. Rep.* **27** (1997) 1.
10. P.L.J. Gunter, J.W. Niemantsverdriet, F. Ribeiro and G.A. Somorjai, *Catal. Rev. Sci. Eng.* **39** (1997) 77.
11. C.R. Henry, *Surf. Sci. Rep.* **31** (1998) 231.
12. D.R. Rainer and D.W. Goodman, *J. Mol. Catal. A*, **131** (1998) 259.
13. M. Bäumer and H.-J. Freund, *Progr. Surf. Sci.* **61** (1999) 127.
14. H.-J. Freund, *Surf. Sci.* **500** (2002) 271.
15. H.-J. Freund and G. Pacchioni, *Chem. Soc. Rev.* **37** (2008) 2224.
16. F. Cosandey and T. Madey, *Surf. Rev. Lett.* **8** (2001) 73.
17. R. Meyer, C. Lemire, S.K. Shaikhutdinov and H.-J. Freund, *Gold Bull.* **37** (2004) 72.
18. M. Chen and D.W. Goodman, *Acc. Chem Res.* **39** (2006) 739.
19. B.K. Min and C.M. Friend, *Chem. Rev.* **107** (2007) 2709.
20. J.A. Rodriguez, S. Ma, P. Liu, J. Hrbek, J. Evans and M. Perez, *Science* **318** (2007) 1757.
21. T. Risse, S. Shaikhutdinov, N. Nilius, M. Sterrer and H.-J. Freund, *Acc. Chem. Res.* **41** (2008) 949.
22. J. Gong and C.B. Mullins, *Acc. Chem. Res.* **42** (2009) 1063.
23. D.-C. Lim, C.-C. Hwang, G. Ganteför and Y.D. Kim, *Phys. Chem. Chem. Phys.* **12** (2010) 15172.
24. J. Libuda and H.-J. Freund, *Surf. Sci. Rep.* **57** (2005) 157.
25. G.A. Somorjai, *Appl. Surf. Sci.* **121/122** (1997) 1.
26. G. Rupprechter, *Catal. Today* **126** (2007) 3.
27. D.C. Meier and D.W. Goodman, *J. Am. Chem. Soc.* **126** (2004) 1892.
28. B.L.M. Hendriksen, S. Bobaru and J.W.M. Frenken, *Top. Catal.* **36** (2005) 43.
29. D. Tang, K. Hwang, M. Salmeron and G.A. Somorjai, *J. Phys. Chem. B* **108** (2004) 13300.
30. M. Salmeron and R. Schlögl, *Surf. Sci. Rep.* **63** (2008) 169.

31. J.V. Barth, H. Brune, G. Ertl and R.J. Behm, *Phys. Rev. B* **42** (1990) 93.
32. M.A. van Hove, R.J. Koestner, P.C. Stair, J.P. Bibérian, L.L. Kesmodel, I. Bartos and G. Somorjai, *Surf. Sci.* **103** (1981) 189.
33. U. Harten, A.M. Lahee, J.P. Toennies and C. Wöll, *Phys. Rev. Lett.* **54** (1985) 2619.
34. V. Heine and L.D. Marks, *Surf. Sci.* **165** (1986) 65.
35. J.K. Gimzewski, R. Berndt and R.R. Schlitter, *Phys. Rev. B* **45** (1992) 6844.
36. http://en.wikipedia.org/wiki/File:Atomic_resolution_Au100.JPG. (Accessed 1 February 2012.)
37. D.G. Fedak and N.A. Gjostein, *Acta Metall.* **15** (1967) 827.
38. D.G. Fedak and N.A. Gjostein, *Surf. Sci.* **8** (1967) 77.
39. H. Melle and E. Menzel, *Z. Natur.* **33a** (1978) 282.
40. K. Yamazaki, K. Takayanagi, Y. Tanishiro and K. Yagi, *Surf. Sci.* **199** (1988) 595.
41. N. Wang, Y. Uchida and G. Zehmpfuhl, *Surf. Sci.* **284** (1993) L419.
42. K. Takayanagi, Y. Tanishoro, K. Kobayashi, K. Akiyama and K. Yagi, *Jap. J. of Appl. Phys.* **26** (1987) L957.
43. A.R. Sandy, S.G.J. Mochrie, D.M. Zehner, K.G. Huang and D. Gibbs, *Phys. Rev. B* **43** (1991) 4667.
44. K.F. Peters, P. Steadman, H. Isern, J. Alvarez and S. Ferrer, *Surf. Sci.* **467** (2000) 1007.
45. Y. Jugnet, F.J. Cadete Santos Aires, C. Deranlot, L. Piccolo and J.C. Bertolini, *Surf. Sci.* **521** (2002) L639.
46. C. Ruggiero and P. Hollins, *Surf. Sci.* **377** (1997) 583.
47. M. Borbonus, R. Koch, O. Haase and K.H. Rieder, *Surf. Sci. Lett.* **249** (1991) L317.
48. C.J. Weststrate, E. Lundgren, J.N. Andersen, E.D.L. Rienks, A.C. Gluhoi, J.W. Bakker, I.M.N. Groot and B.E. Nieuwenhuys, *Surf. Sci.* **603** (2009) 2152.
49. G. Wulff, *Z. Kristal.* **34** (1901) 449.
50. W.L. Winterbottom, *Acta Metall.* **15** (1967) 303.
51. L.D. Marks, *Rep. Prog. Phys.* **57** (1994) 603.
52. C.L. Cleveland, U. Landman, T.G. Schaaff, M.N. Shafigullin, P.W. Stephens and R.L. Wetten, *Phys. Rev. Lett.* **79** (1997) 1873.
53. J.A. Ascencio, M. Perez and M. Jose-Yacaman, *Surf. Sci.* **447** (2000) 73.
54. C.L. Cleveland and U. Landman, *J. Chem. Phys.* **94** (1991) 7376.
55. P.M. Ajayan and L.D. Marks, *Phys. Rev. Lett.* **63** (1989) 279.
56. L. Mitome, Y. Tanishiro and K. Takayanagi, *Z. Phys. D* **12** (1989) 45.
57. S. Giorgio, C. Chapon, C.R. Henry, G. Nihoul and J.M. Penisson, *Phil. Magazin A* **64** (1991) 87.
58. T. Kizuka and N. Tanaka, *Phys. Rev. B* **56** (1997) R10079.
59. P. Hollins, *Surf. Sci. Rep.* **16** (1992) 51.
60. S.A. Chambers, *Surf. Sci. Rep.* **39** (2000) 105.
61. W. Weiss and W. Ranke, *Progr. Surf. Sci.* **70** (2002) 1.
62. J.G. Chen, J.E. Crowell and J.T. Yates Jr., *Surf. Sci.* **185** (1987) 373.
63. F. Rochet, S. Rigo, M. Froment, C. d'Anterroches, C. Maillot, H. Roulet and G. Dufour, *Adv. Phys.* **35** (1986) 237.
64. M. Bäumer, D. Cappus, H. Kuhlenbeck, H.-J. Freund, G. Wilhelmi, A. Brodde and H. Neddermeyer, *Surf. Sci.* **253** (1991) 116.
65. F. Rohr, M. Bäumer, H.-J. Freund, J.A. Meijas, V. Staemmler, S. Müller, L. Hammer, K. Heinz, *Surf. Sci.* **389** (1997) 391.

66. R.M. Jaeger, H. Kuhlenbeck, H.-J. Freund, M. Wuttig, W. Hofmann, R. Franchy and H. Ibach, *Surf. Sci.* **259** (1991) 235.
67. R. Franchy, J. Masuch and P. Gassmann, *Appl. Surf. Sci.* **93** (1996) 317.
68. C. Becker, J. Kandler, H. Raaf, R. Linke, T. Pelster, M. Dräger, M. Tanemura and K. Wandelt, *J. Vac. Sci. Technol. A*, **16** (1998) 1000.
69. J. Wollschläger, J. Viernow, C. Tegenkamp, D. Erdös, K.M. Schröder, H. Pfnür, *Appl. Surf. Sci.* **142** (1999) 129.
70. D.R. Mullins, P.V. Radulovic and S.H. Overbury, *Surf. Sci.* **429** (1999) 186.
71. J.-L. Lu, H.-J. Gao, S. Shaikhutdinov and H.-J. Freund, *Surf. Sci.* **600** (2006) 5004.
72. S. Surnev, M.G. Ramsey and F.P. Netzer, *Progr. Surf. Sci.* **73** (2003) 117.
73. S. Guimond, J.M. Sturm, D. Göbke, Y. Romanyshyn, M. Naschitzki, H. Kuhlenbeck and H.-J. Freund, *J. Phys. Chem. C* **112** (2008) 11835.
74. D. Löffler, J.J. Uhlrich, M. Baron, B. Yang, X. Yu, L. Lichtenstein, L. Heinke, C. Büchner, M. Heyde, S. Shaikhutdinov, H.-J. Freund, R. Wlodarczyk, M. Sierka and J. Sauer, *Phys. Rev. Lett.* **105** (2010) 146194.
75. H. Hövel and I. Barke, *Progr. Surf. Sci.* **81** (2006) 53.
76. D.R. Frankl and J.A. Venables, *Adv. Phys.* **19** (1970) 409.
77. P. Milani and S. Iannotta, *Cluster Beam Synthesis of Nanostructured Materials*, Springer, Berlin, 1999.
78. K. Wegner, P. Piseri, H. Vahedi Tafreshi and P. Milani, *J. Phys. D: Appl. Phys.* **39** (2006) R439.
79. M. Arenz, U. Landman and U. Heiz, *ChemPhysChem* **7** (2006) 1871.
80. C. Binns, *Surf. Sci. Rep.* **44** (2001) 1.
81. C. Binns, in *Metallic Nanoparticles* (ed J. Blackman), Elsevier, Amsterdam, 2009, vol. 6, p. 50.
82. H. Haberland, M. Mall, M. Moseler, Y. Quiang, T. Reiners and Y. Thurner, *J. Vac. Sci. Technol. A* **12** (1994) 2925.
83. D.A. Eastham, B. Hamilton and P.M. Denby, *Nanotechnology* **13** (2002) 51.
84. K. Bromann, H. Brune, C. Felix, W. Harbich, R. Monot, J. Buttet and K. Kern, *Surf. Sci.* **377** (1997) 1051.
85. X. Tong, L. Benz, P. Kemper, H. Metiu, M.T. Bowers and S.K. Burrato, *J. Am. Chem. Soc.* **127** (2005) 13516.
86. L. Huang, S.J. Chey and J.H. Weaver, *Phys. Rev. Lett.* **80** (1998) 4095.
87. E. Gross, Y. Horowitz and M. Assher, *Langmuir* **21** (2005) 8892.
88. J.S. Palmer, S. Sivaramkrishnan, P.S. Waggoner and J.H. Weaver, *Surf. Sci.* **602** (2008) 2278.
89. E. Gross, M. Assher, M. Lundwall and D.W. Goodman, *J. Phys. Chem. C* **111** (2007) 16197.
90. T.H. Baum and C.R. Jones, *J. Vac. Sci. Technol. B* **4** (1986) 1187.
91. E. Feurer and H. Suhr, *Appl. Phys. A* **44** (1987) 171.
92. E. Szlyk, P. Piszczck, I. Lakomska, A. Grodzicki, J. Szatkowski and T. Blaszczyk, *Chem Vap. Dep.* **6** (2003) 105.
93. P.D. Tran and P. Dopplet, *J. Electrochem. Soc.* **154** (2007) D520.
94. A.A. Bessonov, N.B. Morozova, N.V. Gelfond, P.P. Semyannokov, S.V. Trubin, Yu.V. Shevtsov, Yu.V. Shubin and I.K. Igumenov, *Surf. Coat. Technol.* **201** (2007) 9099.
95. K.-I. Fukui, S. Sugiyama and Y. Iwasawa, *Phys. Chem. Chem. Phys.* **3** (2001) 3871.

96. G. Kastle, H.G. Boyen, F. Weigla, G. Lengl, T. Herzog, P. Ziemann, S. Riethmüller, O. Mayer, C. Hartmann, J.P. Spatz, M. Möller, M. Ozawa, F. Banhart, M.G. Garnier and P. Oelhafen, *Adv. Funct. Mater.* **13** (2003) 853.

97. B. Roldan Cuenya, S.-H. Baeck, T.F. Jaramillo and E.W. McFahrland, *J. Am. Chem. Soc.* **125** (2003) 12928.

98. L.K. Ono and B. Roldan-Cuenya, *Catal. Lett.* **113** (2007) 86.

99. A.S. Eppler, J. Zhu, E.A. Anderson and G.A. Somorjai, *Top. Catal.* **13** (2000) 33.

100. J. Grunes, J. Zhu, M. Yang and G. Somorjai, *Catal. Lett.* **86** (2003) 157.

101. S. Johanson, E. Fridell and B. Kasemo, *J. Catal.* **200** (2001) 370.

102. K. Wong, S. Johanson and B. Kasemo, *Faraday Discussions* **105** (1996) 237.

103. P.M. Mendes, S. Jacke, K. Critchley, J. Plaza, Y. Chen, K. Nikitin, R.E. Palmer, J.A. Preece, S.D. Evans and D. Fitzmaurice, *Langmuir* **20** (2004) 3766.

104. M.H.V. Werts, M. Lambert, J.-P. Bourgoin and M. Brust, *Nano Letters* **2** (2002) 43.

105. M.-V. Meli and R.B. Lennox, *Langmuir* **19** (2003) 9097.

106. N. Stokes, A.M. McDonagh and M.B. Cortie, *Gold Bull.* **40** (2007) 310.

107. U. Diebold, *Surf. Sci. Rep.* **48** (2003) 53.

108. L. Zhang, R. Persaud and T.E. Madey, *Phys. Rev. B* **56** (1997) 10549.

109. S.C. Parker, A.W. Grant, V.A. Bondzie and C.T. Campbell, *Surf. Sci.* **441** (1999) 10.

110. A.K. Santra, A. Kolmakov, F. Yang and D.W. Goodman, *Japan. J. Appl. Phys.* **42** (2003) 4795.

111. N. Spiridis, J. Haber and J. Korecki, *Vacuum* **63** (2001) 99.

112. T. Diemant, H. Hartmann, J. Bansmann and R.J. Behm, *J. Catal.* **252** (2007) 171.

113. E. Wahlström, N. Lopez, R. Schaub, P. Thostrup, A. Rønnau, C. Africh, E. Laegsgaard, J.K. Nørskov and F. Besenbacher, *Phys. Rev. Lett.* **90** (2003) 026101.

114. O. Bikondoa, C.L. Pang, R. Ithnin, C.A. Muryn, H. Onishi and G. Thornton, *Nature Mat.* **5** (2006) 189.

115. D. Matthey, J.G. Wang, S. Wendt, J. Matthiesen, R. Schaub, E. Laegsgaard, B. Hammer and F. Besenbacher, *Science* **315** (2007) 1692.

116. F. Cosandey, L. Zhang and T.E. Madey, *Surf. Sci.* **474** (2001) 1.

117. K. Hojrup-Hansen, T. Worren, S. Stempel, E. Laegsgaard, M. Bäumer, H.-J. Freund, F. Besenbacher and I. Stensgaard, *Phys. Rev. Lett.* **83** (1999) 4120.

118. S. Shaikhutdinov, R. Meyer, M. Naschitzki, M. Bäumer and H.-J. Freund, *Catal. Lett.* **86** (2003) 211.

119. L. Vitos, A.V. Ruban, H.L. Skriver and J. Kollar, *Surf. Sci.* **411** (1998) 186.

120. Z.H. Qin, M. Lewandowski, Y.N. Sun, S. Shaikhutdinov and H.-J. Freund, *J. Phys. Chem. C* **112** (2008) 10209.

121. M. Baron, O. Bondarchuk, D. Stacchiola, S. Shaikhutdinov and H.-J. Freund, *J. Phys. Chem. C* **113** (2009) 6042.

122. J.-L. Lu, H.-J. Gao, S. Shaikhutdinov and H.-J. Freund, *Catal. Lett.* **114** (2007) 8.

123. C. Winkler, A. Carew, R. Raval, J. Ledieu and R. McGrath, *Surf. Rev. Lett.* **8** (2001) 693.

124. N. Nilius, M.V. Ganduglia-Pirovano, V. Brazdova, M. Kulawik, J. Sauer and H.-J. Freund, *Phys. Rev. Lett.* **100** (2008) 096802.

125. M. Yulikov, M. Sterrer, M. Heyde, H.-P. Rust, T. Risse, H.-J. Freund, G. Pacchioni and A. Scagnelli, *Phys. Rev. Lett.* **96** (2006) 146804.

126. H.M. Benia, X. Lin, H.-J. Gao, N. Nilius and H.-J. Freund, *J. Phys. Chem. C*, **111** (2007) 10528.
127. A. Sanchez, S. Abbet, U. Heiz, W.D. Schneider, H. Häkkinen, R.N. Barnett and U. Landman, *J. Phys. Chem. A* **103** (1999) 9573.
128. S.S. Lee, C.Y. Fan, T.P. Wu and S.L. Anderson, *J. Phys. Chem. B* **109** (2005) 11340.
129. S.S. Lee, C.Y. Fan, T.P. Wu and S.L. Anderson, *J. Am. Chem. Soc.* **126** (2004) 5682.
130. D.C. Lim, R. Dietsche, M. Bubek, T. Ketterer, G. Gantefor and Y.D. Kim, *Chem. Phys. Lett.* **439** (2007) 364.
131. C. Lemire, R. Meyer, S. Shaikhutdinov and H.-J. Freund, *Surf. Sci.* **552** (2004) 27.
132. D.A. Outka and R.J. Madix, *Surf. Sci.* **179** (1987) 351.
133. J.M. Gottfried, K.J. Schmidt, S.L.M. Schroeder and K. Christmann, *Surf. Sci.* **536** (2003) 206.
134. K.P. McKenna and A.L. Shluger, *J. Phys. Chem. Lett.* **111** (2007) 18848.
135. D.E. Starr, S. Shaikhutdinov and H.-J. Freund, *Top. Catal.* **36** (2005) 33.
136. X. Lai and D.W. Goodman, *J. Mol. Catal. A* **162** (2000) 33.
137. A. Kolmakov and D.W. Goodman, *Catal. Lett.* **70** (2000) 93.
138. A. Kolmakov and D.W. Goodman, *Surf. Sci.* **490** (2001) L597.
139. X. Lai, T.P. St. Clair, M. Valden and D.W. Goodman, *Prog. Surf. Sci.* **59** (1998) 25.
140. M. Valden, S. Pak, X. Lai and D.W. Goodman, *Catal. Lett.* **56** (1998) 7.
141. M. Valden, X. Lai and D.W. Goodman, *Science* **281** (1998) 1647.

Chapter 9

Theoretical Studies of Gold Nanoclusters in Various Chemical Environments: When the Size Matters

Hannu Häkkinen

University of Jyväskylä FI-40014 Jyväskylä, Finland.
Email: Hannu.J.Hakkinen@jyu.fi

9.1 Introduction

Gold nanoparticles (AuNPs) exhibit a rich array of interesting and important electronic, optical, chemical and catalytic properties, which has sparked a huge interest in gold-based systems in several interdisciplinary areas, leading to an explosive growth in the volume of both experimental and theoretical research.[1–3] A large variety of AuNPs differing by their size (1–100 nm), shape and surrounding can been synthesized (see Chapter 6). Among them the nanoparticles termed as clusters are made of a countable number of atoms (less than 150 atoms), which corresponds to particles smaller than 2 nm, and they offer a unique playground where the most elaborate *ab initio* calculation can accurately reproduce and interpret the experimental situations. This chapter contains an overview of the developments of structural determination of gold clusters in gas phase and as stabilized chemically by various ligands. Many gold particles with precise structure and properties can now be synthesized by wet chemistry and they can be handled as normal chemicals: stored, modified and functionalized for applications in medical therapy, biolabeling, sensing, nanoelectronics and catalysis. In

recent years, understanding of the stability, surface chemistry and functionalization of these interesting building blocks of nano-matter has taken a
quantum leap. This is facilitated by simultaneous breakthroughs in experimental and theoretical fronts concerning accurate structural determinations
of thiolate-stabilized gold clusters of 1–3 nm in diameter, in conjunction
with computational studies.

Computational studies on these systems are challenging for many reasons. The chemical and physical properties of gold can be understood only
if relativistic effects are taken properly into account (Chapter 2). This makes
gold a very peculiar metal that gives rise to specific properties.[2] The size of
many systems imposes a numerical burden for codes based on the density
functional theory. The systems are complex, since the ligand-passivated
nanocluster exhibits properties that call for an understanding of both the
surface (covalent) chemistry between metal atoms and molecules, and the
origins of the "metallic" properties of the gold core.

In dealing with the electronic properties, bulk gold is a good (6s) free-
electron metal. These properties result from electrons shared between an
infinite number of gold atoms and it is natural to ask up to what extent
this free-electron behaviour shows up in the electronic structure when the
number of atoms is reduced until forming nanocluster. (For an authoritative
review on cluster production techniques and analysis of electronic properties of simple metal clusters, see Ref. 4.) Nanoparticles larger than 2 nm are
known to be "metallic", i.e. they have an appreciable density of electron
states at the Fermi level (no HOMO-LUMO gap) and exhibit a typical surface plasmon at about 520 nm. Smaller particles will have a finite HOMO-
LUMO gap due to constriction of the delocalized Au(6s) derived states in
the finite volume. A rough but instructive estimate follows an elementary
discussion of a 3D free-electron metal.[5] For such a system, the density of
electron levels depends on the energy as $D(\varepsilon) \propto \sqrt{\varepsilon}$. At the Fermi energy ε_F
(energy of the highest occupied electron state), the mean spacing of electron
levels is $\delta(\varepsilon_F) = 1/D(\varepsilon_F) = 2\varepsilon_F/3Nz$ where N is the number of atoms and
z is the valence. Thus, decreasing the size of the particle (N) one increases
the energy gap at the Fermi energy. It is relevant to compare this energy gap
to thermal excitations that are of the order of kT (0.025 eV at room temperature). When the gap exceeds kT, quantum size effects due to discreteness

of the energy levels become dominant. For this rough estimate we can treat gold as monovalent, i.e. $z = 1$ and take a free electron value of $\varepsilon_F = 5.5\,\text{eV}$, which gives a limiting particle size of $N = 150$ atoms corresponding to a critical diameter $d = 1.7\,\text{nm}$. Below that size, gold clusters are expected to turn from "metallic" to "semiconducting".

This contribution deals with theoretical studies of gold clusters with a countable number of atoms (below 150). This regime is distinct from the larger nanoparticles that are discussed in most of the other chapters of this book. In the nanocluster regime, the atomic and electronic structures are intimately related, thus the computational results presented herein provide an overview of the achievements in understanding thermodynamic stability of gold clusters of increasing sizes: starting from a gold triangle with 3 atoms up to clusters made of 150 atoms. The relationship between structure and chemical properties, the transition from "semiconducting" to "metallic" electronic properties, the reactivity, the ligand-bond formation and the catalytic activity of these clusters are also tackled.

Given the huge body of published work in the area, this discussion is not meant to be an exhaustive review, rather it reflects personal views of the author who has had the privilege of being heavily involved in an extensive theoretical and experimental collaborative network since the late 1990s. At the same time it provides a glimpse of the timeline of development of ideas, which is hoped to be beneficial particularly for a novice entering this exciting field.

9.2 Clusters in Gas Phase

A few experimental methods have been developed to produce clusters in gas phase. They can be considered as model systems since they are made of pure gold but the challenge lies in the control of the actual size and the dispersity of the population being produced (see Chapters 6 and 8). This section deals with clusters made of 3 to 50 gold atoms and focuses on their thermodynamic stability and the evolution of their density of states as the number of atoms increases. Determination of absolute structures of metal nanoclusters in gas phase is experimentally a very challenging task. Cluster beams are controlled by electrostatic fields and most experimental techniques for structure analysis work conveniently only for beams of charged clusters.[4]

9.2.1 *Cationic clusters* Au_N^+

The first systematic structure determination for a range of cluster cations came from ion mobility experiments by the Karlsruhe group for $4 \leq N \leq 20$.[6] In the ion mobility experiment, an electrostatic field accelerates size-selected, charged clusters through a drift tube, filled by inert carrier gas (such as He, Ar). Collisions to the molecules of the carrier gas provide a drag force affecting the flight time through the tube. The flight time is measured and gives access to the collision cross sections. This latter is also obtained from theoretical calculations (density functional theory, DFT) of a set of structure candidates for a given cluster size. The comparison between measured and computed cross sections is crucial to conclude on the actual geometry of gold clusters.

The measured cross sections up to $N = 20$ are shown in Fig. 9.1 together with the calculated ones for the energetically most favourable cluster isomers up to $N = 13$. A close-up of the ground-state structures as well as of some close-lying isomers for $N = 6, 8, 10$ is shown in Fig. 9.2. Comparison to the theory shows that up to $N = 7$ the measured data points essentially coincide with the calculated values for the ground-state structures, all planar: a triangle, a rhombus, an "hour-glass shape", a triangle, and a centred hexagon for $N = 3$ to 7, respectively. All of these geometries are simply fragments of a close-packed hexagonal plane. This preference of gold for arranging into planar geometry is a good illustration of the achievements yielded by the combination between *ab initio* calculations and experimental observations. For $8 \leq N \leq 13$, an equally good match is observed for three-dimensional structures, which in many cases can be described as slightly relaxed fragments of fcc bulk structure, note e.g. the tetrahedral structure for $n = 10$ (isomer II in Fig. 9.2). For $N = 8, 10$ a close-lying isomer (8-II and 10-II) gave the best correspondence with the measured cross section. Here, the isomers were within 0.1 eV from the calculated ground state structure. Planar structures 8-I and 10-III, while predicted being energetically as good as the 3D structures 8-II and 10-II, were not observed in the experiment. The comparison shows that while theory is indispensable for

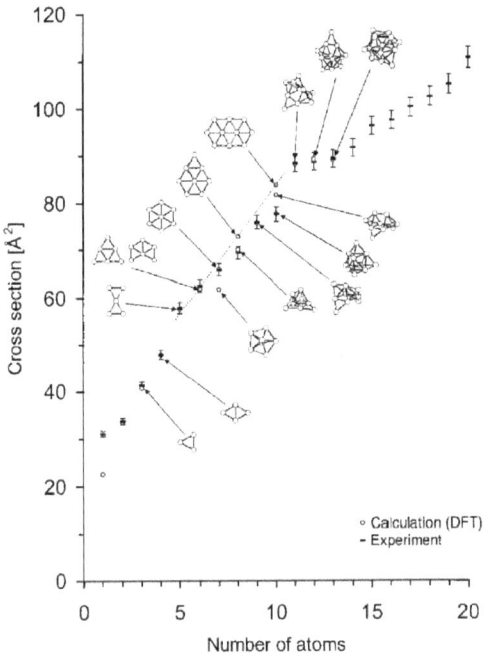

Fig. 9.1. Measured and predicted ion mobility cross sections for gold cluster cations. Reproduced from Ref. 6 by permission. Copyright 2002 American Institute of Physics.

interpretation of the experimental data, care should be exercised for taking theoretical structure prediction for granted, since little details in the ways the electron-electron interaction is treated in DFT may affect the sensitive energy balance between structure isomers of different dimensionality. The structural database of gas phase gold clusters (both cations and anions, to be discussed next) has served as a valuable benchmark for developing and testing reliable DFT methods for gold clusters for the past ten years.

9.2.2 *Anionic clusters Au_N^-*

Anionic clusters are amenable to structure determination by the ion mobility method as explained before. Besides cross sections, other experimental approaches such as photoelectron spectroscopy (PES) or trapped-ion electron diffraction (TIED), are used to confirm the structures. PES uses a UV or visible laser to produce photoelectrons whose detection provides

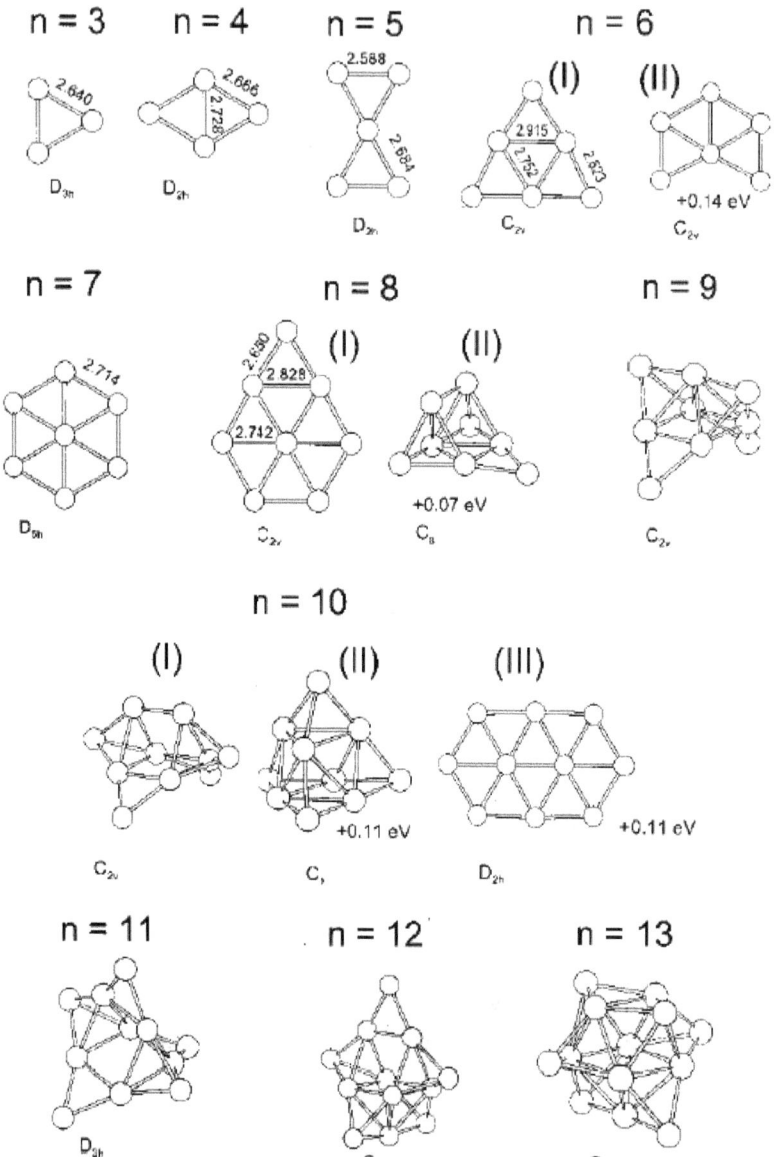

Fig. 9.2. Calculated of the most stable of gold cluster cations (Low-energy isomers). Up to seven atoms, the clusters arrange into planar geometry. Reproduced from Ref. 6 by permission. Copyright 2002 American Institute of Physics.

information on the detachment energies of electrons from negatively charged clusters (analogous to the well-known photoelectric effect on bulk metal surfaces). The Vertical Detachment Energy (VDE) is the energy required to remove the extra electron from the anion. This information can be compared to the structure of electron states obtained in a DFT calculation. In the TIED method, a "cloud" of size-selected, charged clusters is trapped by a set of electrostatic fields and is irradiated by a collimated electron beam. The scattering function of electrons can be measured, and the method can be considered to be analogous to powder X-ray diffraction. The experimental effort on anionic gold clusters intensified around 2001–2003 with several reports from ion mobility and PES experiments supported by DFT calculations.[6–8] A remarkable tendency of gold cluster anions to favour planar structures up to fairly large sizes was quickly discovered. Therefore an intriguing question is to determine at what size the 2D to 3D transition occurs when the number of atoms increases in gold clusters.

The Karlsruhe ion mobility experiment reported in 2002[7] convincingly set the 2D/3D transition size to $N = 12$. For $N = 12$, a bimodal arrival time distribution was observed and two distinct cross sections measured, allowing for a conclusion that for this cross-over size both planar and three-dimensional structure isomers were present in the beam. It was concluded that for all the other sizes only one isomer was observed. For most cases, the ground state structure predicted by DFT gave also the best fit to the experimental cross section; exceptions included 4-, 10- and 13-atom clusters where the first isomer above ground-state gave the best fit.

A complementary follow-up work involved high-resolution photoelectron spectroscopy (HR-PES) with the data interpreted via DFT calculations for the density of single-electron states (DOS) in the cluster anion for an extensive set of cluster isomers in the size range $4 \leq N \leq 14$.[8] In the single-particle interpretation, the measured PES data contains information about the distribution of binding energies of valence electrons detached from the cluster by a photon. These energies are also accessible with DFT calculations, including the energy of the weakest bound electron (vertical detachment energy, VDE, whose counterpart is the ionization potential for a finite neutral cluster and the work function for bulk metal surface). Comparing these "fingerprints" may thus provide a more sensitive measure to judge the presence of a given isomer or isomers in the beam, as compared to making structure assignments based on the collision cross section, which is a single number for a given cluster isomer.

The theoretical work reported in Ref. 8 confirmed the energetic stability of the earlier reported planar structures. In addition, comparison to the measured PES data (VDE values and spectral details) gave evidence of isomers that were present in the cluster beam for N = 4, 8, 12 and 13. As a prominent example, the theory predicts two close-lying planar structures for N = 10: a triangular and an elongated close-packed flake, see 10A and 10B in Fig. 9.3. The two independent DFT calculations[7,8] gave a consistent result by predicting the triangular structure 10A to be the ground state and the elongated isomer 10B to be 0.12 eV (Ref. 8) or 0.15 eV (Ref. 7) higher in energy. The geometrical cross section of the isomer 10B fits better with the measured collision cross section. The two structures deviate significantly from each other regarding the VDE value: for the ground state it is calculated to be 3.86 eV (Ref. 8) or 4.02 eV (Ref. 7) and for the higher-energy isomer it is 2.94 eV (Ref. 8) or 3.08 eV (Ref. 7). Early PES studies[9] assigned an experimental VDE of Au_{10}^- to be around 3 eV that would be consistent with the higher energy isomeric structure. However, the high-resolution experiment revealed that the low-energy feature in the PES is due to a minor isomer present in the beam and the experimental VDE for the dominant Au_{10}^- structure in the beam was determined to be 3.91 eV which is in an excellent

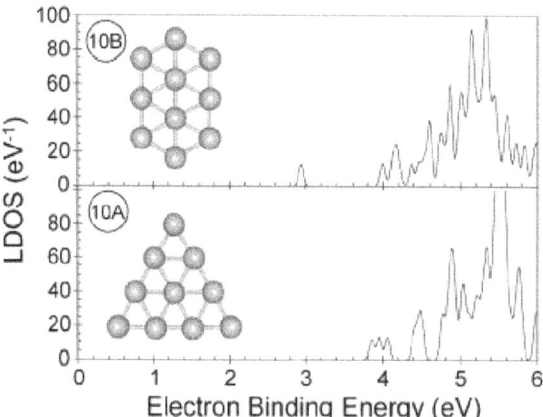

Fig. 9.3. The theoretical DOS of the two lowest-energy structures of Au_{10}^-. Adapted from Ref. 8 by permission. Copyright 2002 American Chemical Society.

agreement with the theoretical VDE values calculated for the ground state.[8] It has been shown recently that these isomers have different reactivities with molecular oxygen and their relative abundance can be controlled by source conditions.[10]

Not only does thermodynamic stability rule the cluster formation but kinetics may play an even more drastic role in enabling co-existence of cluster isomers of different dimensionality in the beam; the 2D/3D cross-over size $N = 12$ is one example. DFT calculations give a consistent large energy separation of about 0.5 eV for the most favourable 2D and 3D structures. This energy difference is far too large to be explained by thermal population and interconversion of structures under room temperature conditions in the beam. The most likely explanation is in the kinetics of formation of two structural "families", planar and three-dimensional, and their fast cooling in the beam. This was simulated via DFT-based tight-binding molecular dynamics simulations of the cooling process that showed that in the 2D/3D crossover region, supercooling to "wrong dimensionality" is possible.[11] It has to be noted that the standard DFT calculations probing energetics of isomers are strictly valid for $T = 0$ conditions only (no entropy effects included) and for that reason real molecular dynamics studies like the one reported in Ref. 11, using forces from semi-empirical theory such as tight-binding (or in some cases from DFT is the computer resources allow) should prove useful for probing thermal/entropy effects present in cluster beams.

Why does gold favour such large planar atomic structures? A systematic theoretical study of noble metal clusters Cu_7^-, Ag_7^- and Au_7^-, carried out in 2002, gave at least partial answers.[12] The stability of the planar structures was traced back to strong relativistic effects in bonding in gold,[2] which induce s-d hybridization, contraction of the Au-Au bond length and a significant overlap of d-orbitals.[12,13]

It is interesting to consider the effects of a particular feature of the electronic structure of $N = 12$ clusters that has been determined to be the 2D/3D crossover size. As mentioned in the Introduction, a subset of the full electronic structure of gold clusters consists of Au(6s) derived delocalized electron states.[14] Their global symmetries can be analyzed and related to a planar quantum-dot model (Fig. 9.4).[15] In that model, an electron number of 12 constitutes a shell closing. After the shell closure at 12 electrons, a new 8-electron shell opens with 2P and 1F symmetries, leading to the next

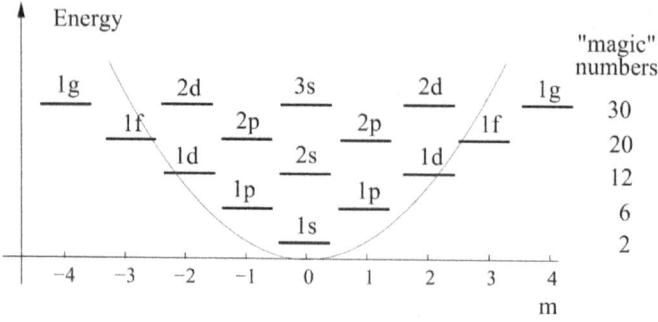

Fig. 9.4. A schematics of electron states in a planar harmonic quantum dot. Gold clusters containing a "magic number" of electrons have closed shell orbitals and exhibit a higher stability.

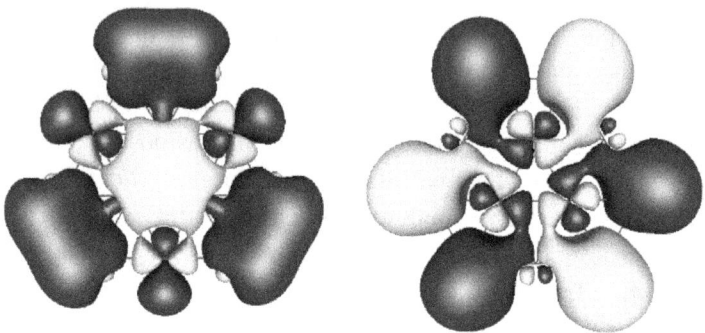

Fig. 9.5. Left: HOMO-1 state of D_{3h} Au_{12}^- cluster. Right: The HOMO state. 2S and 1F symmetries are seen.

"magic number" in a plane, 20. Figure 9.5 shows the two highest occupied orbitals for D_{3h} Au_{12}^-, for which the effective 6s-electron count is 13. Indeed the highest state, the singly-occupied molecular orbital (SOMO) shows a clear 1F symmetry and the state below that has a clear 2S symmetry of the 6-electron 2S1D shell. This shell closing is responsible for the large HOMO-LUMO gap of neutral Au_{12}, visible in the experimental photoelectron spectrum of the anion.[9]

9.2.3 *From flakes to cages to tubes: Anionic clusters with N = 13–24*

Two recent experimental investigations for larger anionic clusters, a PES study[14] and a trapped ion electron diffraction study,[16] confirmed the early conclusions from mobility experiments regarding the 2D/3D structural cross-over at the size Au_{12}^-. The experimental data were compared to the corresponding theoretical quantities using the same isomer database.[14] These studies confirmed the earlier photoelectron results for tetrahedral structures of Au_{16}^-[17] and Au_{20}^-.[18] Furthermore, these investigations showed the gradual transformation of the optimal structure from a near-planar, flat "cage" (N = 13, 14) to a tetrahedral cage, evolving finally to tubular structures for N > 20.

Figure 9.6 shows the computed energy-level diagram of electrons for cluster anions with N = 16, 18, 20, and visualization of selected orbitals for Au_{20}^- (the special Au_{16}^- case is discussed in the next section). Angular momentum analysis of the orbitals of Au_{20}^- shows that they can be generally divided into two major classes, the Au(5d) derived "band" of states in the energy range -3 eV to -8 eV, and to delocalized Au(6s) derived states above and below the 5d-band. A very important conclusion is obtained from these calculations: qualitatively, this medium-size cluster already shows all the characteristics of the bulk band of gold, where the free-electron-like 6s-derived band crosses the 5d-derived band and extends all the way from the gamma-point to the Fermi surface.[19] The upper edge of the 5d-band is

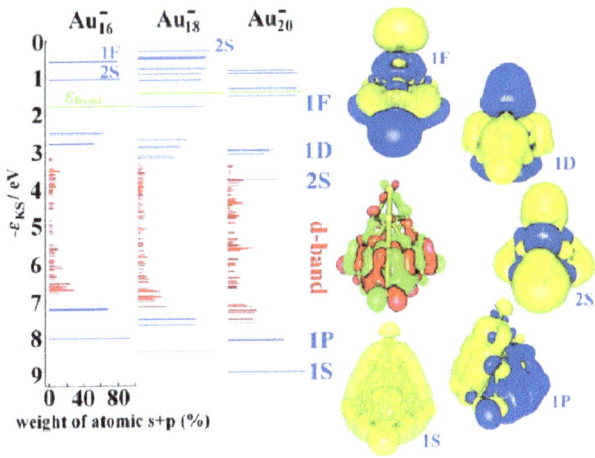

Fig. 9.6. Left: Angular-momentum analysis of Kohn-Sham eigenstates for Au_{16}^-, Au_{18}^- and Au_{20}^-. The states marked in blue have high local sp weight on atoms; globally they are free-electron-states over the cluster. A visualization of such states for Au_{20}^- is on the right. Reproduced from Ref. 14 by permission. Copyright 2007 Wiley.

about 2 eV below the Fermi surface, which also qualitatively matches the behaviour of the states in the Au_{20}^- cluster. The Au_{20}^- then shows the expected free-electron-like shell filling pattern of 21 electrons: $1S^2 1P^6 2S^2 1D^{10} 1F^1$ with $1F^1$ as SOMO, confirming that the large HOMO-LUMO gap of the neutral Au_{20} is the one between 20 and 21 electrons in the free-electron model. Many more examples of the free-electron behaviour of medium-size gold clusters can be analyzed, and some are given below.

9.2.4 Au_{16}^- : The smallest golden cage and the manifestation of shell closing of 18 delocalized electrons

In 2006, a combined photoelectron spectroscopy and density functional study[17] concluded that the experimentally observed isomer of the Au_{16}^- anion has tetrahedral symmetry and the geometry can be derived from the previously discovered[18] T_d-symmetric Au_{20}^- by removing the four vertex atoms and allowing for an outward relaxation of the 4 face-centred atoms, which yields a structure that was coined the "smallest golden cage" (see Fig. 9.7).

Since the early 1990s it had been known that Au_{16}^- exhibits an anomalously high VDE.[9] In a follow-up study it was shown that the high VDE of Au_{16}^- is due to a tendency to complete a shell of 18 delocalized electrons leading to a stability of the dianionic Au_{16}^{2-} cluster with a predicted HOMO-LUMO gap of 1.5 eV.[20] In the shell-model notation, the relevant shells are $1S^2 1P^6 1D^{10}$ with the 1D shell split by the symmetry. In terms of the T_d-symmetric crystal-field split molecular orbitals, the relevant configuration is $a_1^2 t_2^6 e^4 t_2^6$. The 2S shell is at high energies (and thus unoccupied) due to the fact that a radial node is not supported by the hollow cage. The cage maintains its robust geometry, with a minor Jahn-Teller deformation, over several charge states ($q = -1, 0, +1$), forming spin doublet, triplet and quadruplet states according to the Hund's rules. The cage is roomy enough that it can be doped endohedrally by guest atoms.

Despite the extensive experimental and theoretical work, one still cannot be fully conclusive about all the factors that drive the structural transitions as the size of gold cluster anions increases up to about N = 20. As discussed above, relativistic bonding effects are indicated from theory to be one factor in stabilizing the planar clusters up to N = 12. The evolution of "flat cages" or biplanar structures joint by edges for slightly larger sizes can also

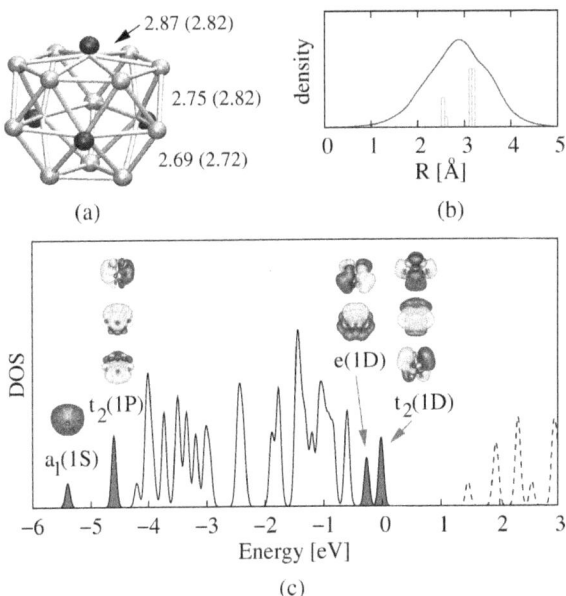

Fig. 9.7. Structure of Au_{16}^q. The numbers are values of the indicated Au–Au distances in the dianion, $q = -2$ (in parentheses for the cation, $q = +1$). The "stick" framework indicates the structure of the cation, and the "balls" are drawn from the coordinates of the dianion. Only the red face atoms move significantly during relaxation between different charge states. (b) Radial distribution of atoms (bars) and electrons (line) for the dianion. The radius of the cage is ~ 2.5 Å. (c) Density of Kohn-Sham molecular orbitals (DOS, folded by 0.05 eV Gaussians) of Au_{16}^{2-}. The HOMO state is at zero energy, and the empty states are denoted by a dashed line. The shaded and labelled peaks denote the delocalized, T_d crystal-field split states that are derived from a spherical cage confinement (angular momentum labeling in parentheses). These MO's are also visualized. t_2 is the 6-fold degenerate HOMO orbital (3 spatial orbitals × 2 for spin). Reproduced from Ref. 20 by permission of the PCCP Owner Societies 2007.

be thought of as caused by the tendency for quasi-planarity. As discussed above, Au_{16}^- is a 3D hollow cage and Au_{20}^- is a piece of fcc bulk structure. However, clusters that are slightly larger that $N = 20$ develop cage- or tube-like structures again and do not follow the motif of simply adding atoms to the fcc crystallite of $N = 20$.

9.2.5 *Anionic clusters with $N > 30$*

Since the early 1990s, the Au_{34}^- anion has been known to have a very large gap in the photoelectron spectrum,[9] reflecting a large HOMO-LUMO gap for the corresponding neutral cluster at 34 s-electrons in a configuration

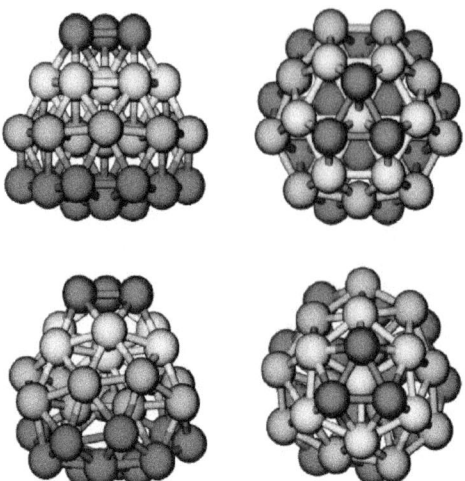

Fig. 9.8. Two views of predicted low-energy C_{3v} (Top) and C_3 (Bottom) structures of Au_{34}^-. The C_3 structure gives the best match with experimental electron diffraction and photoelectron data. Reproduced from Ref. 21 by permission. Copyright 2007 Wiley.

$1S^2 1P^6 1D^{10} 2S^2 1F^{14}$. A recent combination of TIED and UV-VIS PES methods in conjunction to density functional theory calculations shed light onto its most likely atomic structure.[21] The best fit to the measured TIED data was found for a C_3 structure that can be constructed from a more symmetric C_{3v} geometry via a twist (see Fig. 9.8), which increases the surface packing density, similar to what has been found for helical gold nanowires.[22]

In 2004, HR-PES investigation of cold mass-selected Cu_N^-, Ag_N^- and Au_N^- with $N = 53–58$ revealed interesting systematic behaviour.[23] In principle PES spectra are direct images of the electronic density of states. In the bulk form, Cu, Ag and Au have an electronic structure characterized by a half-filled and rather free-electron-like band, formed from the atomic s-orbitals. This band overlaps with the d-band, which lies some eV below the Fermi energy/level and is formed by the rather localized atomic d orbitals. It was observed that while the spectra of copper and silver clusters were practically identical in the upper valence band region, gold clusters exhibited a totally different spectral structure with only one band, although this band was heavily detailed. The series of highly degenerate peaks for Cu and Ag is a signature of high symmetry in the atomic structure. Indeed, by comparing the experimental PES to theoretical DOS calculated for several candidate structures, it was unambiguously concluded that copper and silver clusters in that size range have 55-icosahedron based ground-state structures (Fig. 9.9). On the other hand, several low-lying, low-symmetry isomers were found for Au_{55}^-. None of these structures produced DOS that would satisfactorily match the observed photo electron spectral features. It is possible that the true ground-state structure was not found or that in reality there are many low-energy structures present even at low temperatures (around 200 K). The preference for Au_{55}^- to have low-symmetry structures as opposed to symmetric ones was traced to relativistic bonding effects whose most dramatic effect is the shortening of the interatomic bond length and the increase of bulk modulus. Finally, one can remark that the highly *symmetric* Au_{55}^- metastable isomers, like an icosahedron, support degenerate 2P and 1G shell states (1G split by the I_h symmetry as for Ag_{55}^- and Cu_{55}^-). The large energy gap at around 5 eV electron binding energy in Fig. 9.9 for I_h Au_{55}^- can be identified as the "34 electron gap". This fact has relevance for discussion of symmetric cores of ligand-protected gold clusters discussed in the next section.

9.3 Ligand-Protected Nanocluster

9.3.1 *Synthesis of ligand-protected gold nanoparticles*

Several solution-phase preparation methods to grow and stabilize gold nanocluster with various ligands exist, starting from the classic work of Faraday in the mid 1800s (for a more detailed discussion see Chapters 1 and 5).[24-30] At the beginning of this section, two methods including stabilization by phosphines (Schmid) or thiols (Brust–Schiffrin) are briefly discussed as they relate to the systems where extensive computational work is available. Ligand-protected clusters are also often termed as monolayer-protected clusters (MPC).

In 1981, Schmid *et al.*[26] presented a method to stabilize 1–2 nm gold clusters with a narrow size-distribution by triphenylphosphines (TPP). The method involves reduction of the gold salt by a stream of diborane gas through benzene solution. This synthesis was a significant milestone that has spurred enormous research activities in characterization and utilization of thus-formed Au-TPP nanoparticles. A representative cluster of this

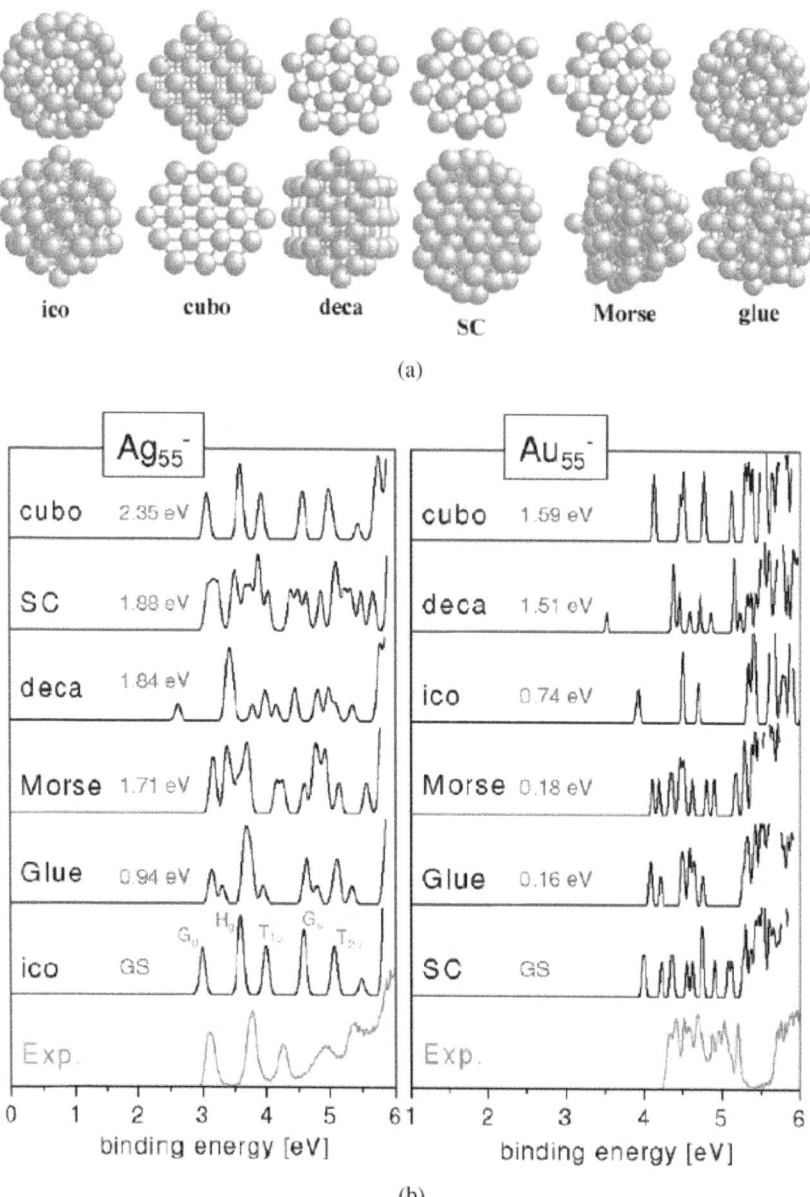

Fig. 9.9. Theoretical DOS for a number of structure isomers for Ag_{55}^- and Au_{55}^-. Comparison to the experimental PES shows that the observed Ag_{55}^- isomer is the icosahedral one. For gold, none of the structure candidates gives a perfect match. Reproduced from Ref. 23 by permission. Copyright 2004 American Physical Society.

nanoparticle material was labelled as $Au_{55}TPP_{12}Cl_6$ where the Au_{55} core was thought to have an fcc-type close-packed geometry. However, a later careful experimental analysis concluded that the cluster material produced by Schmid *et al.*'s method is heterogeneous in size (diameters ranging between 1 and 3 nm with a maximum at 1.4 nm) and structure and degrades upon storage in air.[27] In 1997, Hutchison *et al.* presented a refinement of the method where TOA+ (tetra octyl ammonium cation) is used to transfer $AuCl_4^-$(aq) ions to TPP-toluene solution.[25] Similarly, the reducing agent $NaBH_4$(aq) is transferred to the toluene solution by TOA+. This synthesis produces 1.4 nm (average diameter) Au-TPP particles. The particles can be stored in powder form and used for ligand-exchange reactions with various thiols, which further stabilizes them and enhances their functionality.[29] It is to be noted, however, that the detailed atomic compositions and structures of the dominant Au-TPP clusters formed by these methods remain unknown to date and the above assignments cannot be taken literally. In fact, a very recent DFT study suggested alternative compositions, slightly larger than the generic "Au55", that produce good stability both from the points of view of the electronic structure and steric protection.[28]

Brust *et al.* reported the preparation of thiol-stabilized gold clusters in 2–8 nm range in 1994 in a two-phase synthesis where $AuCl_4^-$(aq) ions were transferred to organic phase by TOA+ and reduced by $NaBH_4$ in the presence of dodecanethiol ($HSC_{12}H_{25}$).[30] This method has proven to be the most useful and versatile one to yield air-stable thiolate-protected clusters that can be handled much like ordinary chemicals. The method is amenable for use with a wide range of different thiols, including water-soluble ones. Later refinements to this method involved a one-phase modification and an improved control of thiol-to-gold ratio to yield truly monodispersed clusters in the 1–2 nm range. Three clusters are now atomically resolved via X-ray crystallography, namely $Au_{102}(SR)_{44}$, $Au_{25}(SR)_{18}^q$ ($q = -1, 0$) and $Au_{38}(SR)_{24}$. The impact of Brust *et al.*'s work is manifested by the number of citations it has drawn (over 2,700 at the time of writing in March 2011).

Although inert in the bulk phase, gold has rich complexation chemistry and is able to form chemical bonds in many oxidation states from $-I$ to $+V$.[1] Many complexes display Au^I–Au^I bonds whose length is in the 3.0–3.3 Å range; this interaction is generally termed aurophilic and the weak attraction between the filled $Au(d^{10})$ electron shells of two neighbouring atoms is of the order of a strong hydrogen bond. Gold is the metal where this effect is the strongest and its origin is to be found in the relativistic effects already mentioned (Chapter 2).[31] It is worth noting that even well before the breakthrough of total-structure-determination of thiolate-stabilized clusters, many sub-nm cluster complexes stabilized by a combination of phosphines and halides were crystallographically characterized.[32] Notable among them are various undecagold Au_{11} and tredecagold Au_{13} clusters that are inter-related via disproportionation reactions which is a special redox reaction where some gold atoms are oxidized and others are reduced. The largest crystallographically characterized phosphine-halide stabilized cluster to date is the complex $Au_{39}TPP_{14}Cl_6^q$.[33,34] These early achievements of synthesis and structure characterization are valuable from theoretical point of view; recent work[35] has found unifying principles to understand the stability of both phosphine-halide and thiolate-protected Au clusters on the basis of common features in the electronic structure, as discussed in Section 9.3.7.

9.3.2 *The noble metal-thiolate bond*

The affinity of organothiolate ligands (SR: sulfur bound to an organic radical R) to noble metal centres is a key interaction in diverse fields such as metal extraction from ores,[36] formation of metal-thiolate complexes that can act as therapeutic agents,[37] stabilization and functionalization of metal surfaces by self-assembled monolayers (SAMs)[38,39] and stabilizing nanoparticle growth by forming a "curved SAM" onto the metal core of the nanoparticle. The metal–SR bond is rather strong and competes with the metal–metal bond. For instance, DFT calculations with a generalized gradient correction (GGA) typically give a value 2.5 eV for the Au–SR bond, which is slightly larger than the experimentally known dissociation energy of 2.3 eV for the gas phase gold dimer Au_2. Using the simplest bond-counting argument, creating an atom vacancy on the principal crystal surfaces of gold costs energy in relation to the lost nearest neighbour Au–Au "bonds", scaled to the bulk cohesive energy (3.8 eV), as follows: 2.9 eV for Au(111), 2.5 eV for Au(100) and 2.2 eV for Au(110). These numbers are comparable to the Au–SR bond strength. In fact, *ab initio* molecular dynamics simulations have shown[40] that it is possible to pull an atomic Au

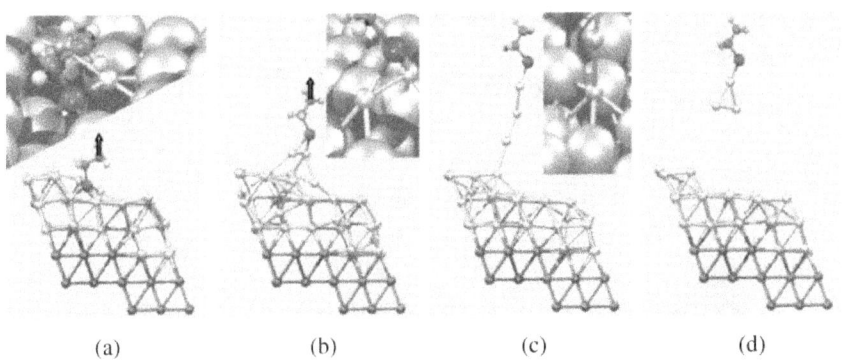

(a)   (b)   (c)   (d)

Fig. 9.10. Snapshots of different stages of pulling an ethylthiolate molecule from a gold surface. The strong Au–S bond acts as a clamp and an atomic gold chain is formed. Reproduced from Ref. 40 by permission. Copyright 2002 American Physical Society.

chain out of gold surface by using the thiolate-Au bond as a "clamp". The formed chain finally breaks at the Au-Au bond (Fig. 9.10).

Based on the above discussion, strong reconstruction of thiolate-SAM-covered noble metal surfaces can thus be expected. Due to the buried interface, it is only recently that a consistent picture of the RS adsorption configuration has emerged. It is now established that thiolates drive pronounced reconstruction of all noble metal surfaces.[39] Scanning tunnelling microscopy (STM) measurements of a monolayer of methyl thiolates (SMe) on Cu(111) show that the adsorbates occupy four-fold hollow positions on a pseudo-(100) reconstructed surface.[41] Low energy electron diffraction (LEED) measurements of SR adsorption on Ag(111) have revealed a ($\sqrt{7} \times \sqrt{7}$R19°) surface cell.[42] Several models for the structure of SR on Au(111) have been proposed based on experiments during the past few years: i) adsorption atop Au ad-atoms forming AuSR units,[43] ii) adsorption as RSAuSR complexes.[44] and iii) a combined model that comprises (AuSR)$_x$ polymers and SR adsorbed at surface point defects.[45] Among these three models, DFT calculations are in favour of the formation of RSAuSR complexes which have been shown to yield the observed c(4×2) superstructure on Au(111), energetically superior to earlier models.[46]

It is extremely interesting to note that similar RSAuSR complexes also protect gold cores when the curved SAM forms in thiolate-protected clusters. These complexes are polymeric and can be written generally as

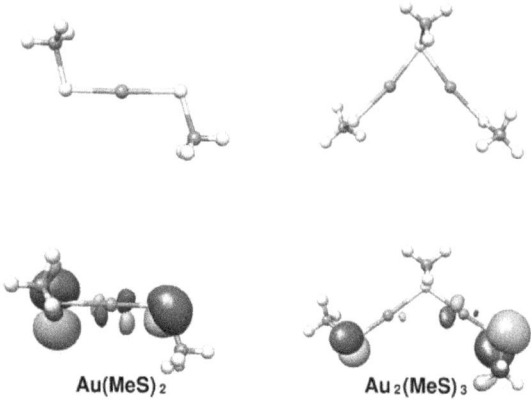

Fig. 9.11. Structures for neutral (Top) and anionic (Bottom) Au(SMe)$_2$ and Au$_2$SMe$_3$. The Kohn-Sham orbitals for the HOMO levels are shown for the charged systems. Reproduced from Ref. 47 by permission. Copyright 2010 American Chemical Society.

SR(AuSR)$_x$; the structures with $x = 1$ and $x = 2$ are shown for R = Methyl in Fig. 9.11 in neutral and anionic forms.[47] The Au–S bond in these complexes is of covalent type, with a very small polarization character (electron loss of 0.1 e to sulfur).[47,48] The capacity of MeSAuSMe complex to localize an electron is indicated by a large electron affinity (EA). The adiabatic EA is calculated to be 2.6 eV. The highest weights of the HOMO level in the anionic complex is found at the sulfur atoms. The properties of MeS(AuSMe)$_2$ are similar to that of MeSAuSMe.

9.3.3 *Early theoretical models*

Concentrating here on thiolate-protected clusters, the early theoretical models around the mid 1990s employed pre-parametrised potential functions for interatomic interactions and structures were explored via classical molecular dynamics methods, a considerable computational challenge at the time. A prevailing structural concept was the one with an atomically "smooth" Au/S interface and compact gold core. This paralleled the understanding of a similar smooth interface in the planar SAMs on the Au(111) surface. Much of the work of that era is highlighted in Refs. 49, 50.

Then, around 1998 the first electronic structure calculations of the protected clusters concentrated on the Au$_{38}$(SR)$_{24}$ cluster, where the structure motif for the Au_{38} core was drawn from classical simulations for medium-sized gold clusters.[50–52]

However, the most significant breakthrough occurred when one realized that the gold atoms of the core were different from those of the external shell. This gave rise to the "divide and protect" concept.

9.3.4 The "Divide and Protect" concept

In 2006, a novel "Divide and Protect" structural concept was introduced.[53] The new structural model emerged from density functional calculations with improved exchange-correlation functionals. In short, this approach revealed a peculiar composition of the gold cluster with a gold core protected by gold-thiolate tetra-units $(AuSR)_4$. It occurs through the "etching" of Au atoms from the Au_{38} core by sulfur, as shown by refined DFT calculations and it leads to the formation of six square-like $(AuSR)_4$ units (Fig. 9.12). Consequently, the composition $Au_{38}(SR)_{24}$ could be written as $Au_{14}[(AuSR)_4]_6$. The gold atoms in the cluster were found to be in two distinct chemical states, the ones in Au_{14} core essentially neutral ("metallic") and those inside the "rings" oxidized (formally Au^I). The binding energy of one $(AuSR)_4$ units to the Au_{14} core was found to be quite weak, about 1 eV. A follow-up systematic study on the $(AuSR)_x$ units showed that they are polymeric for $x \geq 4$, i.e. the binding energy per one AuSR unit saturates.[48]

Comparison to the earlier alternative structure model with a disordered gold core[52] showed that the calculations modified also that structure and a tendency to form distinct $(AuSR)_4$ units was observable (Fig. 9.12), signalling an energetic competition between two driving factors: to optimize the number of metallic Au–Au interactions and to optimize the number of covalent Au–S interactions. The energetic competition is due to the improved description of the Au–Au interaction strength (it has been known that the earlier DFT calculations overestimated that interaction significantly). Structurally, the new concept provided also an attractive model for optimized packing in the ligand shell, since steric repulsion among long (such as C_{12}) or bulky (such as glutathione GSH) ligands could be avoided.

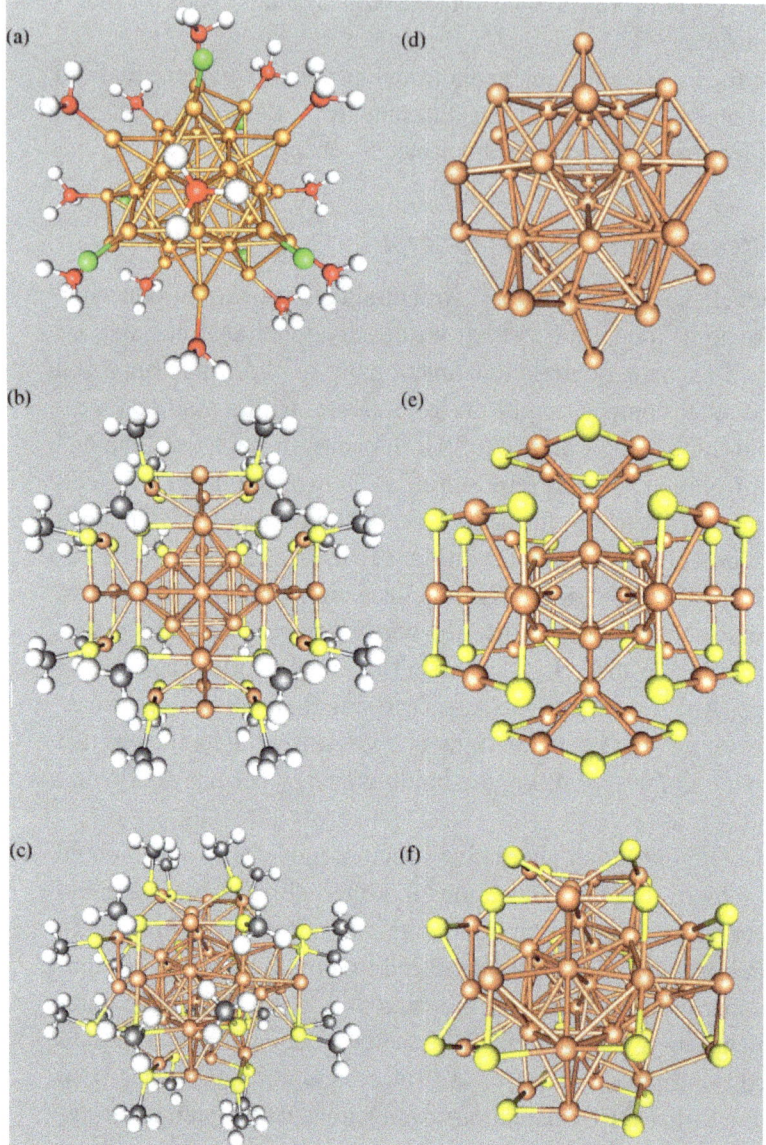

Fig. 9.12. (a) A phosphine/chloride passivated $Au_{39}(PH_3)_{14}Cl_6^-$ cluster, (b, c) two isomers of $Au_{38}(SR)_{24}$. (d) Au_{39} core, shown by a 90 degree rotation about the horizontal axis on the left; (e) Au–S framework of B, with 45 degree rotation about the vertical axis on the left; and (f) Au–S framework of C. Au: orange-brown, S: yellow, P: red, Cl: green, C: dark grey, H: white. Reproduced from Ref. 48 by permission. Copyright 2006 American Chemical Society.

In 2007, a related model ("core-in-cage") was introduced for the $Au_{25}(SR)_{18}$ cluster consisting of a Au_7 core protected by two $(AuSR)_3$ and one $(AuSR)_{12}$ units.[54]

9.3.5 The experimental breakthroughs: X-ray crystallography for all-thiolate protected Au_{102} and Au_{25} clusters and the success of the superatom model

In 2007, the first ever total-structure-determination of an all-thiol protected gold nanoparticle was published by the group of R.D. Kornberg.[55] based on X-ray diffraction at 1.1 Å resolution from single crystals containing a distinct compound with 21 kDa Au core mass, protected by p-MBA ligands (p-MBA = para mercapto benzoic acid, SC_6H_4COOH). Specifically, the composition was determined as $Au_{102}(p$-$MBA)_{44}$ and the crystal unit cell was observed to contain an enantiomeric pair of these clusters.

A subsequent thorough analysis of the atomic structure and full density functional treatment of the electronic structure of the $Au_{102}(p$-$MBA)_{44}$ cluster (with all of its 762 atoms and 3366 valence electrons) resulted in a clear picture of the identity of the protecting gold-thiolate ligands and the gold core, and the underlying reasons for the thermodynamic stability of this compound.[35] It was found that the atomic structure of the $Au_{102}(p$-$MBA)_{44}$ compound (Fig. 9.13) consists of an approximately D_{5h}-symmetric Au_{79} metallic core with a protective gold-thiolate layer of composition $Au_{23}(p$-$MBA)_{44}$. The $Au_{23}(p$-$MBA)_{44}$ layer can further be decomposed into $RS(AuSR)_x$ units with 19 units for $x = 1$ and 2 units for $x = 2$, which are anchored to the core via sulfur in atop positions. The Au_{core}-S-Au_{ligand} angle is close to 90 degrees and the Au_{ligand} atoms are linearly coordinated with two sulfurs. Hence $Au_{102}(p$-$MBA)_{44}$ is more accurately described in the formulation $Au_{79}[p$-$MBA(Au\ p$-$MBA)]_{19}$ $[p$-$MBA(Au\ p$-$MBA)_2]_2$. The gold atoms in the cluster are in two distinct chemical states: the 79 core Au atoms (Au_{core}) are in a metallic (Au^0) state whereas the 23 Au atoms (Au_{ligand}) that belong to protecting RS-$(AuSR)_x$ units are oxidized. Consequently, the composition evokes the predicted "divide and protect" structure motif.[53] The total number of $RS(AuSR)_x$ units, 21, is intimately related to the electronic stability of the particle.

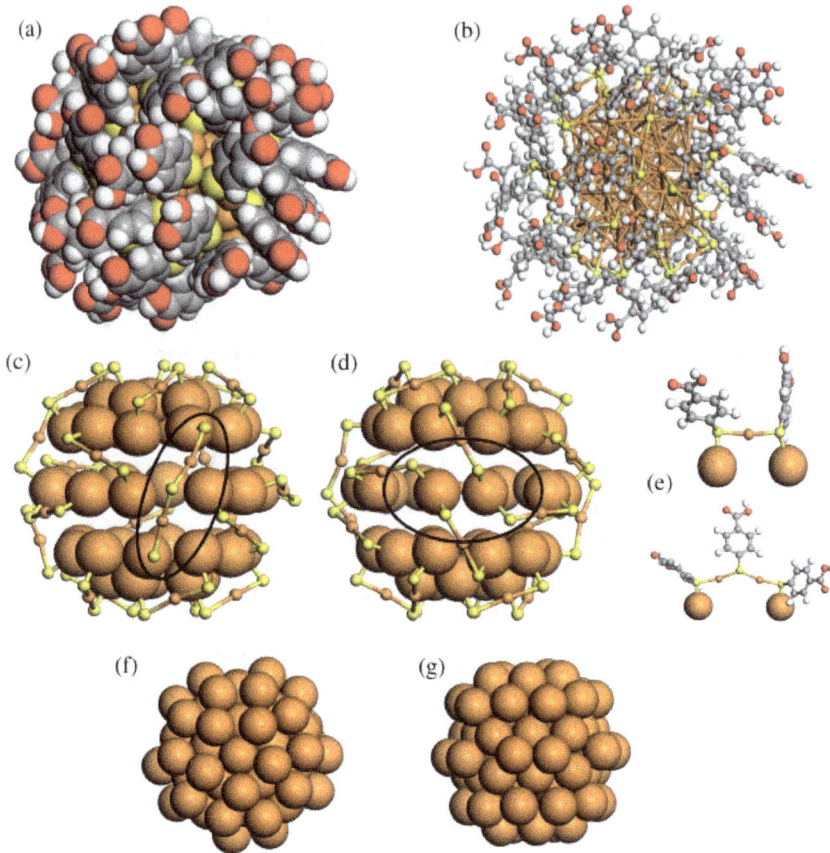

Fig. 9.13. Core-shell structure of the $Au_{102}(p\text{-MBA})_{44}$ cluster. (a) Space-filling and (b) ball-stick representations of the $Au_{102}(p\text{-MBA})_{44}$ nanoparticle. Au: orange, S: yellow, C: grey, O: red, H: white. (c, d) Two views of the 40-atom surface of the Au_{79} core, together with the passivating $Au_{23}(p\text{-MBA})_{44}$ ligand shell. The Au atoms in the ligand shell are depicted by the smaller orange spheres. The "structure defects" at the core-mantle interface (two Au atoms with two Au-S bonds, and a long RS-$(AuSR)_2$ unit) are highlighted. (e) Close-up of the protecting RS-$(AuSR)_x$ unit with $x = 1,2$. (f, g) Two views of the Au_{79} core, which has a symmetry of D_{5h} (within 0.4 Å tolerance). To associate the atoms to grey scale, please see the subfigure e). Reproduced from Ref. 35 by permission. Copyright 2008 National Academy of Sciences.

A confirmation of the metallic character of the Au_{79} core came through analysis of radial difference in the cumulative induced charge when the $Au_{102}(p\text{-MBA})_{44}$ compound was made either cationic or anionic (that is, remove or add one electron, re-calculate the electron density, and analyse the radial density difference, see Fig. 9.14). In both cases, the major

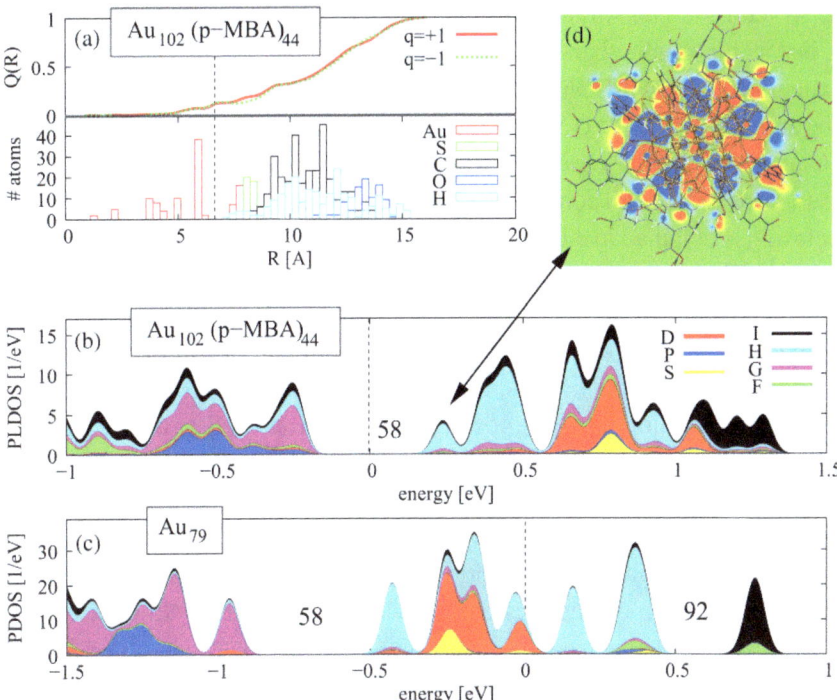

Fig. 9.14. Electronic structure analysis of the $Au_{102}(p\text{-MBA})_{44}$ cluster. (a) The radial dependence of the integrated induced charge Q(R) upon removing (Red curve) and adding (Green curve) one electron to the neutral $Au_{102}(p\text{-MBA})_{44}$ cluster (top panel), and the radial distribution of atoms (Lower panel). The dashed line indicates a midpoint between the surface of Au_{79} core and the Au-thiolate layer. $Q(R) = 4\pi \int^R \Delta\rho(r)r^2 \, dr$ where $\Delta\rho(r) = \rho^0(r) - \rho^q(r)$ is the induced charge-difference from two DFT calculations for the neutral and charge particle. (b) The angular-momentum-projected local electron density of states (PLDOS) (projection up to the I-symmetry, i.e., $l = 6$) for the Au_{79} core in $Au_{102}(p\text{-MBA})_{44}$. (c) The same for the bare Au_{79} without the Au-thiolate layer. d) A cut-plane visualization of the LUMO-state of the $Au_{102}(p\text{-MBA})_{44}$ cluster. Note the H-symmetry (10 angular nodes) at the interface between the Au_{79} core and the gold-thiolate layer. In b, the zero energy corresponds to the middle of the HOMO-LUMO gap, while in c the zero energy is at the HOMO level (Dashed lines). Shell-closing electron numbers are indicated in **b, c**. Reproduced from Ref. 35 by permission. Copyright 2008 National Academy of Sciences.

portion (90%) of the induced charge was found in the $Au_{23}(p\text{-MBA})_{44}$ shell. Virtually no change was observed inside a radius of 5 Å and only 10% of the induced charge resides at the interface between the Au_{79} core and the $Au_{23}(p\text{-MBA})_{44}$ protective layer (5 Å $<$ R $<$ 7 Å). Since a metallic cluster accepts charge only at its surface, it could be directly concluded that the electronic structure of the Au_{79} should feature delocalized-electron shell structure, just as the smaller bare metallic Au clusters.

The calculated HOMO-LUMO gap of Au_{102}(p-MBA)$_{44}$ cluster was found to be appreciable, about 0.5 eV, indicating a major electron shell closing. The analysis of angular momentum character of the electron states of the bare Au_{79} core and the full compound revealed the exact mechanism how this shell closing is obtained in the protected cluster (Fig. 9.14). The stability of the Au_{102}(*p*-MBA)$_{44}$ particle is due to several co-existing factors: (i) formation of a compact, symmetric-enough metal core that can support clear electron-shell structure, (ii) complete chemical protection (passivation) of the core surface by RS-(AuSR)$_x$ units, the number of which has to be "just right" so that (iii) a major gap is exposed in the electron shell structure. The resulting *superatom* is a thermodynamically stable species at ambient conditions just as the ordinary atoms in the Periodic Table. Very recently, a combined experimental-theoretical investigation on the NIR absorption by Au_{102}(p-MBA)$_{44}$ in solution and solid phases documented a full spectroscopic characterization of this cluster in a wide mid-IR–NIR–VIS–UV range.[56]

The existence of two different protective units in Au_{102}(*p*-MBA)$_{44}$, RSAuSR and RS(AuSR)$_2$ may seem surprising. However, polymeric "zig-zag" chains or rings of such units are known (see Refs. 34, 52 and references therein). The RS(AuSR)$_2$ unit may exist (at least) in two conformations, a sharp-angle or wide-angle "V" with the central Au–S–Au angle around 100 or 124 degrees, respectively. While the latter one was found in Au_{102}(*p*-MBA)$_{44}$, the "sharp-angle V" unit was found to offer an ideal building block to protect a much smaller particle Au_{25}(SR)$_{18}$ where the composition could be written as Au_{13}[RS(AuSR)$_2$]$_6$.[57] The central Au_{13} core is a slightly distorted icosahedron with the 6 RS(AuSR)$_2$ ligands octahedrally arranged around the core (Fig. 9.15). Taking the cluster to be anionic (q = −1) renders the system as an 8-electron "superatom" (the 6 SR(AuSR)$_2$ units localize in total 6 electrons from the 14-electron (6s) shell structure of Au_{13}^- and in spherical harmonics notation, the 6s-derived shells are $1S^2 1P^6$).

Remarkably, *two* simultaneous and independent experiments[58,59] confirmed the structural prediction for Au_{25}(SR)$_{18}^-$. The cluster was passivated by using the SCH$_2$CH$_2$Ph ligand. The unit cell of the crystal was found to contain also a TOA$^+$ counterion (tetra octyl ammonium, a phase-transfer

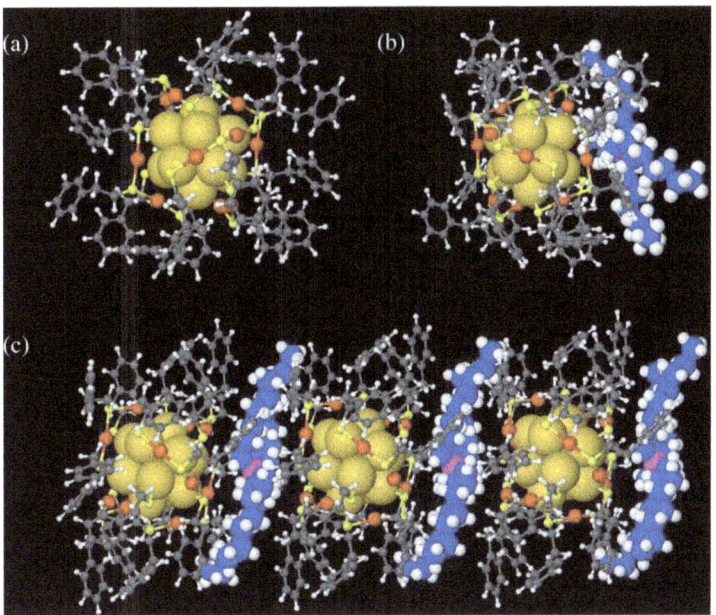

Fig. 9.15. (a) Geometry of the $Au_{25}(SR)_{18}^{-1}$ in the gas phase and (b) with the TOA+ counterion in the crystal. SR = SEtPh. (c) shows the unit cell in (b), replicated three times along the c-axis of the crystal. Reproduced from Ref. 61 by permission. Copyright 2010 American Chemical Society.

agent) which confirmed the anionic state of the gold cluster (Fig. 9.15). The theoretical analysis revealed that the close-to-icosahedral Au_{13} is also quite rigid with respect to charge q = 0,1, confirming in part the earlier experimental results that indicated robustness of optical spectra with respect to charges q = −1,0,1.[60] The robustness can be understood straightforwardly from the superatom model: it was observed that the Au_{13} core remains rigid for different charge states, hence the major optical transitions of q = 0,1 clusters are still over the "HOMO-LUMO gap" of the q = −1 cluster, since transitions inside the same angular-symmetric shell (1P→1P) are forbidden by the dipole selection rule.[57,61] Soon afterwards, a more detailed analysis of the optical excitations of a model cluster $Au_{25}(SR)_{18}^{-}$ followed.[59,62] The observation of $Au_{25}(SR)_{18}^{-}$ added yet another member to the "family" of ligand-protected clusters with 8 delocalized electrons (see below), a family that was born by early predictions[63] and synthesis[64] of $[Au_{13}(PMe_2Ph)_{10}Cl_2]^{3+}[(PF_6)_3]^{3-}$ complex.

9.3.6 *Phosphine-stabilized Au$_{11}$ and Au$_{39}$ clusters: superatoms with 8 and 34 electrons*

Not only thiol moiety can play a stabilizing role for gold clusters, but also phosphine provides interesting bonding capability with gold. Various Au$_{11}$ and Au$_{13}$-based phosphine-halide passivated clusters have been characterized in solid state by X-ray diffraction since the late 1970s (for review, see Ref. 32). The undecagold compounds generally have the formula Au$_{11}$(PR$_3$)$_7$X$_3$ where X = halide or thiolate, and the gold skeleton often has an approximate C$_{3v}$ symmetry. A recent investigation dealt with the electronic structure of clusters Au$_{11}$(PH$_3$)$_7$(SMe)$_3$ and Au$_{11}$(PH$_3$)$_7$Cl$_3$ which are homologous models for a recently reported thiolate-stabilized cluster Au$_{11}$(S-4-NC$_5$H$_4$)$_3$(PPh$_3$)$_7$.[65]

The calculated HOMO-LUMO gaps of these compounds are 1.5 eV for X = SMe and 2.1 eV for X = Cl.[35] The dominant angular momentum character of the states around the gap was found to change from P-symmetry to D-symmetry. In the delocalized electron model this corresponds to closing of the 8 electron (in configuration 1S21P6) gap. This gap exposure is due to the fact that the three halide or thiolate ligands localize one electron each out of the eleven conduction electrons from the gold core. It is interesting to note that a halide and a thiolate ligand act here in analogous roles, although the character of the Au–Cl bond is more "iono-covalent" than that of the Au–SR bond. The seven phosphine ligands act as weak surfactants in both systems, without modifying the electron shell structure of the gold core.

A tredecagold compound [Au$_{13}$(PMe$_2$Ph)$_{10}$Cl$_2$][PF$_6$]$_3$ was experimentally characterized in 1981,[64] confirming earlier theoretical predictions of stable ligand-protected icosahedral gold clusters.[63] The three hexafluorophosphate anions stabilize the triple-cationic gold compound in the crystal structure. The calculated HOMO-LUMO gap for the homologous relaxed Au$_{13}$(PH$_3$)$_{10}$Cl$_2^{3+}$ compound is 1.8 eV, very similar to the undecagold compounds.

In 1992, the Au$_{39}$(PPh$_3$)$_{14}$Cl$_6^q$ (q is the total charge) compound was isolated and crystallized, and for fifteen years remained the largest "soluble" cluster with an unambiguously determined structure.[33] The geometrical arrangement of the Au$_{39}$ core of this cluster is close to D$_3$ symmetry, and can be also described as two hexagonal close-packed (hcp) crystallites, joined together by 30 degree twist (Fig. 9.19). There is only one fully coordinated gold atom in the centre of a hexagonal antiprismatic cage. The calculated HOMO-LUMO gap was found to be large, 0.8 eV, for the anionic compound (q = −1).[35] The angular momentum analysis of the electron states around the gap showed that the gap closes a band of states that have dominantly F-character while the states above the gap have a major G-character. The F-shell closing indicates an effective conduction electron count of 34 in the gold core. This is consistent with the fact that there are six iono-covalent AuCl bonds at the surface, thereby reducing the effective count of delocalized electrons from 40 to 34.

9.3.7 *The unifying superatom concept*

The above analysis of precisely known compositions and structures of all-(mono)thiolate, phosphine-halide or phospine-(mono)thiolate protected gold clusters suggests that all these compounds can be expressed by a formula.[35]

$$(L_s \bullet Au_N X_M)^q \tag{1}$$

where the gold cluster (core size N) is protected by M electron-withdrawing ligands X and s "weak" ligands that do not affect the effective free-electron-count of the gold core but they complete the sterical protection (typically L = phosphine). The compound may have an overall charge q. It is essential to recognize the identity of the electron-withdrawing ligands X in the "Divide and Protect" concept. For instance, in the $Au25(SR)18 = Au13[RS(AuSR)2]6$ cluster $X = RS(AuSR)2$ ($N = 13$ and $M = 6$) and $Au102(SR)44$ has two types of X ligands, 19 of RSAuSR type and 2 of RS(AuSR)2 type as discussed in Section 9.3.5 (thus $N = 79$ and $M = 21$). In phosphine-halide protected clusters, X = halide. All the shell closing numbers (hence the electronic stability) n_e can be evaluated with an "effective gold valence" $v_A = 1$ from

$$n_e = N v_A - M - q \tag{2}$$

The result of this analysis is summarized in Table 9.1. A compound having a closed electron shell and a complete chemical protection of the metal core can be called a "noble-gas superatom".

Table 9.1. Experimentally determined band gaps for free gas-phase gold cluster anions from photoelectron spectroscopy vs. theoretical DFT values (PBE functional) for HOMO-LUMO gaps of passivated gold cluster compounds that correspond to $n_e = 8, 34$, and 58 conduction-electron shell closings. N_{core} is the number of gold atoms in the cluster core. For details see Ref. 35.

	Experiment			Theory (Ref. 38)		
Shell closing, n_e	Cluster	Gap (eV)		Cluster compound	N_{core}	Gap (eV)
8e ($1S^2 1P^6$)				$Au_{11}(PH_3)_7(SMe)_3$	11	1.5
8e				$Au_{11}(PH_3)_7Cl_3$	11	2.1
8e				$Au_{13}(PH_3)_{10}Cl_2^{3+}$	13	1.8
8e				$Au_{25}(SMe)_{18}^-$	13	1.2
34e (8e + $1D^{10}2S^2 1F^{14}$)	Au_{34}^- (Refs. 9, 21)	1.0		$Au_{39}Cl_6(PH_3)_{14}^-$	39	0.8
58e (34e + $2P^6 1G^{18}$)	Au_{58}^- (Refs. 21, 23)	0.6		$Au_{102}(p\text{-MBA})_{44}$	79	0.5
58e				$Au_{102}(SMe)_{44}$	79	0.5

9.3.8 Use of the superatom concept to understand the reactivity of gold clusters: dioxygen activation and CO oxidation

Surfaces of bulk gold are chemically inert, but finely dispersed gold particles with size below a few nanometres are catalytically active species for O–O, C–C and C–H bond activation.[66–71] Notably, when gold particles are catalytically active for oxidation reactions (essentially CO oxidation), they function well at ambient temperature and pressure (see Chapter 7). This propensity makes gold-based nanoscale catalysts interesting systems to exploit for green chemistry. Intensive research over two decades has concluded that several factors are contributing to this behaviour. Many active gold catalysts are prepared on reducible oxides, and strong interactions between the support and the particle may create active sites at the periphery of the particle/support interface. These interactions may also induce charge transfer to or from the particle. From purely geometric arguments, small particles have always a high ratio of low-coordinated edge and corner atoms to well-coordinated atoms in the middle of facets, which may increase the reactivity provided that the low-coordinated gold atoms bind the reactants more effectively. Room-temperature vibrations of atoms in very small clusters can be soft, leading to fluctuations of the shape and structure which may further lower reaction barriers.[68,72] Less attention has been paid to quantum size effects that are known to dominate the physical and chemical properties of small isolated metal clusters in vacuum ($N \leq 150$ atoms, $d \leq 1.7$ nm). Metal clusters with such few atoms have molecule-like distinct electronic states, and all the chemistry then arises from the interactions between fairly few frontier orbitals of the cluster and the reactant molecules.

First experimental and theoretical indications on the catalytic activity of the smallest substrate-supported gold clusters, just a few atoms in size, were observed over a decade ago.[67] Temperature-programmed-desorption experiments performed on model catalysts prepared by soft-landing mass-selected gold clusters on a magnesia surface concluded that an eight-atom cluster is the smallest one that catalyses CO oxidation. Accompanying density functional calculations showed that a model cluster Au_8 with two atomic layers can indeed bind both reactants, CO and O_2, and activate O_2 by transferring electron charge to the $2\pi^*$ orbital of di-oxygen.[67,68] The charging mechanism was later verified by Fourier-transform infrared absorption study of CO stretch frequencies at the Au_8/MgO reaction center.[69] Recently, two independent experiments gave strong indications on the catalytic activity of ~ 1 nm or even sub-nanometre supported Au clusters. Gold catalysts prepared on inert supports from ligand-protected clusters containing initially about 50 Au atoms were shown to be effective for partial oxidation of styrene.[83] The catalytic particles had a size distribution that was peaked around 1.5 nm. On the other hand, a aberration-corrected scanning tunnelling electron microscopy study concluded that the most active species catalyzing CO oxidation on iron-oxide support are bi-layer 0.5 nm clusters containing only about 10 Au atoms.[70] Although this claim has recently been challenged,[71] these observations clearly call for theoretical understanding on the size-dependence of binding and activation of oxygen at gold nanoclusters and the possible role of the quantum-size effects.

Ligand-protected and chemically passivated Au clusters provide an ideal testing ground for these studies, since the composition and structure of several clusters are by now known precisely and their electronic structure is well understood as discussed above. A recent study considered clusters with the overall size (including the ligand shell) extending up to 2.4 nm and demonstrated by density functional calculations that quantum size effects are instrumental for binding of dioxygen to the nano-catalyst.[73] Partial removal of the protective phosphine-halide or gold-thiolate layer activates the cluster with an occupied electron state over an energy gap that originates from the HOMO-LUMO gap of the fully protected cluster. Transfer of this electron to dioxygen $2\pi^*$ orbital stabilizes dioxygen adsorption in an activated superoxo O_2^- form. A clear correlation between the energy gap and the binding energy of the O_2^- species is found. Surprisingly, only the smallest clusters with an overall size of 1.2–1.8 nm (Au core size 0.9 to 1.5 nm) and energy gaps larger than 0.5 eV are able to bind O_2^* appreciably (Table 9.2). Comparison to O_2 adsorption on bulk Au(111) surface shows that the favourable binding of O_2^* to the smallest clusters requires proper alignment of $2\pi^*$ orbitals in the energy gap of the cluster to facilitate electron transfer from the cluster to O_2. Hybridization of $2\pi^*$ orbitals with the gold 5d-derived band weakens the bonding, and this mechanism is responsible for the weaker binding of O_2 to larger clusters. On the smallest clusters the oxidation reaction $2CO + O_2 \rightarrow 2CO_2$ proceeds effectively via the Langmuir-Hinshelwood mechanism with reaction barriers that are below 0.7 eV, indicating low-temperature activity.

The effect of the electron counting rule (Eq. 2) for understanding the reactivity of the partially protected gold cluster can be demonstrated here in case of cluster 1 in Table 9.2. Removal of a single halogen ligand increases the electron count to $n^* + 1$, thus a new electron state in the gold core is occupied. This state is located over an energy gap, which originates from the HOMO-LUMO gap of the fully protected parent cluster. This mechanism turns the partially protected cluster electropositive and thus reactive towards adsorption of electronegative O_2. The fully protected cluster 1 has 8 electrons in the gold core in a configuration $1S^2 1P^6$ of centre-of-mass spherical harmonics and removing one Cl from the ligand shell modifies the electronic structure to $1S^2 1P^6 1D^1$, i.e. the gold core has nine itinerant electrons. The ninth electron occupies the D-symmetric state. Upon O_2

Table 9.2. Dioxygen adsorption on various partially protected gold clusters. n_e is the electron count from equation (2). Removal of the ligand(s) changes the electron count to n_e^*. BE(O_2) is the binding energy of O_2 to the activated cluster. d(Au–O) is the distance between the adsorbed molecule and the closest Au atom. d(O–O) is the intramolecular bond length. R = methyl. From Ref. 73.

Fully protected cluster	Total diameter (nm)	Core diameter (nm)	n_e	HOMO-LUMO gap (eV)	Ligand removed	n_e^*	BE(O_2) (eV)	d(Au–O) (Å)	d(O–O) (Å)
1. $Au_{11}(PH_3)_7Cl_3$	1.2	0.9	8	2.03	Cl	9	0.95	2.17	1.31
2. $Au_{25}(SR)_{18}^{-1}$	1.6	0.9	8	1.25	$Au_2(SR)_3$	9	0.72	2.24	1.31
3. $Au_{25}(SR)_{18}^{-1}$	1.6	0.9	8	1.25	$2 \times Au_2(SR)_3$	10	0.62	2.36	1.30
								2.20	1.32
4. $Au_{39}(PH_3)_{14}Cl_6^{-1}$	1.8	1.5	34	0.85	Cl	35	0.59	2.25	1.32
5. $Au_{102}(SR)_{44}$	2.2	1.5	58	0.53	$Au(SR)_2$	59	0.08	2.19	1.30
6. $Au_{144}(SR)_{60}$	2.4	1.7	84	0.08	$Au(SR)_2$	85	−0.15	2.22	1.29
7. Au(111) surface				0			−0.54	2.26	1.24

adsorption this state depletes completely and the electron is transferred to one of the $2\pi^*$ orbitals of O_2, initially empty in the gas-phase triplet O_2. Occupation of this orbital and stretch of the O–O bond length (1.31 Å vs. 1.24 Å for the calculated O–O bond in the neutral triplet O_2) verifies that the dioxygen is activated to the superoxo O_2^- state. When the size of the gold cluster increases, the HOMO-LUMO gap of the fully protected cluster decreases (Table 9.2), and thus the energy gain of transferring charge to the $2\pi^*$ orbital of O_2 decreases. Concomitantly, the degree of hybridization of the molecular O_2 states with the Au states increases. These two factors play a role in the weakening of the Au–O bond. Cluster 6, which has a very small energy gap (<0.1 eV), functions already at the bulk limit of metallic Au(111) surface 7, displaying a "zero-order" Au–O bond and a metastable O_2 molecular adsorption.

9.3.9 *Outlook and challenges for theory*

Traditionally, the "phosphine chemistry" and the "thiolate chemistry" have been regarded as separate branches to prepare ligand-protected gold nanoparticles; no general, unifying theoretical concepts have been available to understand and classify the wealth of experimental information on the well-defined, discrete compounds. The recent experimental and theoretical advances discussed in this contribution provide now certain guiding principles for molecular-precision synthesis and functionalization of these exciting building blocks of nano-materials that are finding applications in diverse fields of biolabeling, photonics, sensing and nanocatalysis. The early suggestions to use "magic" metal clusters as "superatoms" to build novel materials may perhaps now be realized by ligand-protected gold clusters. This is obviously a huge open field for high-level theory and computations. However, theory always needs concrete contact points to ongoing experiments. To close this contribution, a few "burning questions" are briefly mentioned.

As yet unknown cluster compositions and structures. The definite total-structure determinations discussed here have indicated spherical electron shell closings of 8 ($Au_{11}PR_3X_3$, $Au_{25}(SR)_{18}^{-1}$), 34 ($Au_{39}PR_{14}Cl_6^{-1}$) and 58 ($Au_{102}(SR)_{44}$) electrons, as well as the 14-electron shell closing in a nano-rod-shaped metal core ($Au_{38}(SR)_{24}$). Many other shell closings and the corresponding compositions are waiting to be discovered. On the small side, possibilities include 2, 18 and 20 electrons for roughly spherical metal cores and 16 electrons for a strongly oblate core. It is interesting to note that compositions $Au_{39}(SR)_{23}$[74] and $Au_{40}(SR)_{24}$[75] have been determined from high-resolution mass spectrometry; those would correspond to 16 electron systems. On the larger side, a composition of $Au_{68}(SR)_{34}$ that would correspond to the 34-electron shell closing (if the cluster is neutral) has been reported.[76] One can also ponder the upper limit of the cluster size where the electron shell-closing effects would still be significant for stabilization of the cluster via opening a HOMO-LUMO gap. An interesting case is that of the so-called 29 kDa Au cluster (140 to 150 gold atoms and 50 to 60 thiolates): recent high-resolution mass-spectrometry investigations have reported compositions of $Au_{144}(SR)_{59}$,[77] $Au_{144}(SR)_{60}$ and $Au_{146}(SR)_{59}$[78] and $Au_{144}(SR)_{60}$.[79] A theoretical model, fitting very well with the reported powder X-ray data, has been presented for $Au_{144}(SR)_{60}$,[80] and the stability of this cluster is assigned to the high geometrical symmetry of the atomic arrangement (the itinerant electron count of the cluster is 84, nowhere near any electronic shell closing).

Ligand-exchange reactions. The detailed structural knowledge of several clusters now facilitates computational studies of the mechanisms of place-exchange reactions of thiolates, routinely used for functionalization of the nanoparticle surface. This is a sub-field of great interest where no theoretical insight is currently available. An interesting question is whether that the obtained cluster stabilities and geometries can be retained by functionalizing ligands such that, e.g. a toxic phosphine-stabilized gold cluster could be made non-toxic via exchange to amine ligands that are more amenable to biological applications.

Ligand-protected clusters as nanocatalysts. As discussed in Section 9.3.8, activation of ligand-protected gold clusters by partial removal of the protective layer can make them robust electro-positive species that can bind and reduce dioxygen and catalyze ambient CO oxidation reactions. Quantum size effects, characterized by the magnitude of the HOMO-LUMO gap of the parent cluster, determine the binding energy of dioxygen to the activated sub-nanometer gold cluster. By "ligand-engineering", it seems possible to tune the number of occupied electron states in the metal core and thus control the catalytic activity. Several routes for achieving this could be explored. Synthesis of ligand-protected gold clusters has now reached a stage where exploration of techniques to fabricate robust material systems from these building blocks for catalysis has started.[81,82] Concerning thiolate-protected clusters, the recent study[89] indicated that although sulfur has traditionally been considered as poison for a catalyst, controlled extent of thiolate-protection can be beneficial since it aids achieving near-monodispersity and can be used to limit the particle size to the active sub-2nm region. Partial removal of the thiolate layer could be achieved by suitable thermal treatment.[81] In solution-phase, dynamical ligand-exchange reactions may be the key to understand the recently reported activity of $Au_{25}(SR)_{18}$ cluster for selective hydrogenation of unsaturated ketones and aldehydes[83] and the electrochemical reduction of oxygen[84] Gold has potential as an effective catalyst for low-temperature activation of O=O, C=O, C=C and C–H bonds. A better understanding of ligand-protected gold nanocluster could pave a way to explore catalytic systems made out of these particles for practical applications in green chemistry.

Acknowledgements

I wish to thank all the following scientists whose contributions to various collaborations, some of them tracing back to the late 1990s, have each taught me some new aspects about the physics and chemistry of bare, supported and passivated gold clusters and nanoparticles: C.J. Ackerson, C.M. Aikens, J. Akola, R.N. Barnett, T. Bernhardt, G. Calero, C.L. Cleveland, P. Frondelius, H. Grönbeck, U. Heiz, K. Honkala, E. Hulkko, B. v. Issendorff, P. Jadzinsky, K.A. Kacprzak, J. Koivisto, R.D. Kornberg, P. Koskinen, U. Landman, O. Lopez-Acevedo, K. Manninen, M. Moseler, R.W. Murray, M. Pettersson, P. Pyykkö, T. Tsukuda, M. Walter, L.-S. Wang, R. L. Whetten and L. Wöste. The author's work is supported by the Academy of Finland and the Finnish IT Center for Science (CSC).

References

1. H. Schmidbaur, *Gold: Progress in Chemistry, Biochemistry and Technology*, Wiley, Chichester, 1999.
2. P. Pyykkö, *Chem. Rev.* **88** (1988) 563.

3. G. Schmid, *Chem. Rev.* **92** (1992) 1709; M.-C. Daniel and D. Astruc, *Chem. Rev.* **104** (2004) 293; P. Pyykkö, *Angew. Chem. Int. Ed.* **43** (2004) 4412; P. Pyykkö, *Inorg. Chim. Acta* **358** (2005) 4113; R.W. Murray, *Chem. Rev.* **108** (2008) 2688; H. Häkkinen, *Chem. Soc. Rev.* **37** (2008) 1847; C.M. Aikens, *J. Phys. Chem. Letters* **2** (2011) 99; R. Sardar, A.M. Funston, P. Mulvaney, and R.W. Murray, *Langmuir* **25** (2009) 13840; J.F. Parker, C.A. Fields-Zinna and R.W. Murray, *Acc. Chem. Res.* **43** (2010) 1289.

4. W.A. de Heer, *Rev. Mod. Phys.* **65** (1993) 611.

5. C. Kittel, *Introduction to Solid State Physics*, Wiley & Son, Hoboken, New Jersey, 1996, pp. 146–151.

6. S. Gilb, P. Weis, F. Furche, R. Ahlrichs and M.M. Kappes, *J. Chem. Phys.* **116** (2002) 4094.

7. F. Furche, R. Ahlrichs, P. Weis, C. Jacob, S. Gilb, T. Bierweiler and M.M. Kappes, *J. Chem. Phys.* **117** (2002) 6982.

8. H. Häkkinen, B. Yoon, U. Landman, X. Li, H.-J. Zhai and L.-S Wang, *J. Phys. Chem. A.* **107** (2003) 6168.

9. K.J. Taylor, C.L. Pettiette-Hall, O. Cheshnovsky, and R.E. Smalley, *J. Chem. Phys.* **96** (1992) 3319.

10. W. Huang and L.S. Wang, *Phys. Chem. Chem. Phys.* **11** (2009) 2663.

11. P. Koskinen, H. Häkkinen, B. Huber, B. von Issendorff and M. Moseler, *Phys. Rev. Lett.* **98** (2007) 15701.

12. H. Häkkinen, M. Moseler and U. Landman, *Phys. Rev. Lett.* **89** (2002) 33401.

13. H. Grönbeck and P. Broqvist, *Phys. Rev. B* **71** (2005) 73408.

14. B. Yoon, P. Koskinen, B. Huber, O. Kostko, B. von Issendorff, H. Häkkinen, M. Moseler and U. Landman, *Chem. Phys. Chem.* **8** (2007) 157.

15. E. Janssens, H. Tanaka, S. Neukermans, R.E. Silverans and P. Lievens, *New J. Phys.* **5** (2003) 46.

16. X. Xing, B. Yoon, U. Landman, and J.H. Parks, *Phys. Rev. B* **74** (2006) 165423.

17. S. Bulusu, X. Li, L.-S. Wang, and X.C. Zeng, *Proc. Nat. Acad. Sci. USA* **103** (2006) 8326.

18. J. Li, X. Li, H.-J. Zhai, and L.-S. Wang, *Science* **299** (2003) 864.

19. I. Opahle, PhD Thesis, Technical University of Dresden, 2001.

20. M. Walter and H. Häkkinen, *Phys. Chem. Chem. Phys.* **8** (2006) 5407.

21. A. Lechtken, D. Schooss, J.R. Stairs, M.N. Blom, F. Furche, N. Morgner, O. Kostko, B. von Issendorff and M.M. Kappes, *Angew. Chem. Int. Ed.* **46** (2007) 2944.

22. Y. Kondo and K. Takayanagi, *Science* **289** (2000) 606.

23. H. Häkkinen, M. Moseler, O. Kostko, N. Morgner, M.A. Hoffmann and B. von Issendorff, *Phys. Rev. Lett.* **93** (2004) 093401.

24. M. Faraday, *Philos. Trans. R. Soc. London* **147** (1857) 145; D. Thompson, *Gold Bull.* **40** (2007) 4; J. Turkevich, *Gold Bull.* **18** (1985) 86; J. Turkevich, P. C. Stevenson and J. Hillier, *Discuss. Faraday Soc.* **11** (1951) 55; J. Turkevich, P.C. Stevenson and J. Hillier, *J. Phys. Chem.* **57** (1953) 670.

25. J.A. Dahl, B.L.S. Maddux and J.E. Hutchison, *Chem. Rev.* **107** (2007) 2228.

26. G. Schmid, R. Pfeil, R. Boese, F. Bandermann, S. Meyer, G.H.M. Calis and J.W.A. van der Velden, *Chem. Ber.* **114** (1981) 3634.

27. D.H. Rapoport, W. Vogel, H. Cölfen, and R. Schlögl, *J. Phys. Chem.* **101** (1997) 4175.

28. M. Walter, M. Moseler, R.L. Whetten and H. Häkkinen, *Chem. Sci.* **2** (2011) 1583.
29. L.O. Brown and J.E. Hutchison, *J. Am. Chem. Soc.* **119** (1997) 12384.
30. M. Brust, M. Walker, D. Bethell, D.J. Schiffrin and R. Whyman, *J. Chem. Soc. Chem. Comm.* (1994) 801.
31. P. Pyykkö, *Chem. Rev.* **97** (1997) 597.
32. H. Schmidbaur, *Gold: Progress in Chemistry, Biochemistry and Technology* Wiley, Chichester, 1999, p. 511.
33. B.K. Teo, X. Shi and H. Zhang, *J. Am. Chem. Soc.* **114** (1992) 2743.
34. B.K. Teo and H. Zhang, *Coord. Chem. Rev.* **143** (1995) 611.
35. M. Walter, J. Akola, O. Lopez-Acevedo, P.D. Jadzinsky, G. Calero, C.J. Ackerson, R.L Whetten, H. Grönbeck and H. Häkkinen, *Proc. Acad. Natl. Sci. USA* **105** (2008) 9157.
36. D.E. Rawlings, *Ann. Rev. Microbiol.* **41** (2002) 279.
37. C.F. Shaw, *Chem. Rev.* **99** (1999) 2589.
38. A. Ulman, *Chem. Rev.* **96** (1996) 1533; J.C. Love, L.A. Estroff, J.K. Kriebel, R.G. Nuzzo and G.M. Whitesides, *Chem. Rev.* **105** (2005) 1103.
39. D.P. Woodruff, *Phys. Chem. Chem. Phys.* **10** (2008) 7211.
40. D. Kruger, H. Fuchs, R. Rousseau, D. Marx and M. Parrinello, *Phys. Rev. Lett.* **89** (2002) 186402.
41. S.M. Driver and D.P. Woodruff, *Surf. Sci.* **457** (2000) 11.
42. A.L. Harris, L. Rothberg, L.H. Dubois, N.J. Levinos and L. Dhar, *Phys. Rev. Lett.* **64** (1990) 2086.
43. M.G. Roper, M.P. Skegg, C.J. Fisher, J.J. Lee, D.P. Woodruff and R.G. Jones, *Chem. Phys. Lett.* **389** (2004) 87.
44. P. Maksymovych, D.C. Sorescu and J.T. Yates, Jr., *Phys. Rev. Lett.* **97** (2006) 146103.
45. A. Cossaro, R. Mazzarello, R. Rousseau, L. Casalis, A. Verdini, A. Kohlmeyer, L. Floreano, S. Scandolo, A. Morgante, M.L. Klein and G. Scoles, *Science* **321** (2008) 943.
46. H. Grönbeck, H. Häkkinen and R.L. Whetten, *J. Phys. Chem. C* **112** (2008) 15940.
47. K.A. Kacprzak, O. Lopez-Acevedo, H. Häkkinen, and H. Grönbeck, *J. Phys. Chem. C* **114** (2010) 13571.
48. H. Grönbeck, M. Walter and H. Häkkinen, *J. Am. Chem. Soc.* **128** (2006) 10268.
49. R.L. Whetten, J.T. Khoury, M.M. Alvarez, S. Murthy, I. Vezmar, Z.L. Wang, P.W. Stephens, C.L. Cleveland, W.D. Luedtke and U. Landman, *Adv. Mater.* **8** (1996) 428.
50. C.L. Cleveland, U. Landman, T.G. Schaaff, M.N. Shafigullin, P.W. Stephens and R.L. Whetten, *Phys. Rev. Lett.* **79** (1997) 1873.
51. H. Häkkinen, R.N. Barnett and U. Landman, *Phys. Rev. Lett.* **82** (1999) 3264; R. N. Barnett, C.L. Cleveland, H. Häkkinen, W.D. Luedtke, C. Yannouleas and U. Landman, *Eur. Phys. J D* **9** (1999) 95; H. Grönbeck and W. Reoni, *Int. J. Quantum Chem.* **80** (2000) 598; I.L. Garzón, K. Michaelian, M.R. Beltrán, A. Posada-Amarillas, P. Ordejón, E. Artacho, D. Sánchez-Portal and J.M. Soler, *Phys. Rev. Lett.* **81** (1998) 1600.
52. I.L. Garzon, C. Rovira, K. Michaelian, M.R. Beltrán, P. Ordejón, J. Junquera, D. Sánchez-Portal, E. Artacho and J.M. Soler, *Phys. Rev. Lett.* **85** (2000) 5250.
53. H. Häkkinen, M. Walter and H. Grönbeck, *J. Phys. Chem. B* **110** (2006) 9927.
54. T. Iwasa and K. Nobusada, *J. Phys. Chem. C* **111** (2007) 45.
55. P.D. Jadzinsky, G. Calero, C.J. Ackerson, D.A. Bushnell and R.D. Kornberg, *Science* **318** (2007) 430.

56. E. Hulkko, O. Lopez-Acevedo, J. Koivisto, Y. Levi-Kalisman, R.D. Kornberg, M. Pettersson and H. Häkkinen, *J. Am. Chem. Soc.* **133** (2011) 3752.

57. J. Akola, M. Walter, R.L. Whetten, H. Häkkinen and H. Grönbeck, *J. Am. Chem. Soc.* **130** (2008) 3756.

58. M.W. Heaven, A. Dass, P.S. White, K.M. Holt and R.W. Murray, *J. Am. Chem. Soc.* **130** (2008) 3754.

59. M. Zhu, C.M. Aikens, F.J. Hollander, G.C. Schatz and R. Jin, *J. Am. Chem. Soc.* **130** (2008) 5883.

60. Y. Negishi, N. K. Chaki, Y. Shichibu, R.L. Whetten and T. Tsukuda, *J. Am. Chem. Soc.* **129** (2007) 11322.

61. J. Akola, K.A. Kacprzak, O. Lopez-Acevedo, M. Walter, H. Grönbeck and H. Häkkinen, *J. Phys. Chem C* **114** (2010) 15986.

62. C.M. Aikens, *J. Phys. Chem. C* **112** (2008) 19797.

63. D.M.P. Mingos, *J. Chem Soc. Dalton Trans.* **13** (1976) 1163–1169.

64. C.E. Briant, B.R.C. Theobald, J.W. White, L.K. Bell, M.P. Mingos and A.J. Welch, *J. Chem. Soc. Chem. Comm.* **5** (1981) 201–202.

65. K. Nunokawa, S. Onaka, M. Ito, M. Horibe, T. Yonezawa, H. Nishihara, T. Ozeki, H. Chiba, S. Watase and M. Nakamoto, *Organomet. Chem.* **691** (2006) 638.

66. G.J. Hutchings and R. Joffe, *Appl. Catal.* **20** (1986) 215; M. Haruta, T. Kobayashi, H. Sano, and N. Yamada, *Chem. Lett.* **2** (1987) 405; M. Haruta, *Catal. Today* **36** (1997) 153; G.C. Bond and D. Thompson, *Cat. Rev. Sci. Eng.* **41** (1999) 319; M. Valden, X. Lai and W. Goodman, *Science* **281** (1998) 1647; M.D. Hughes, Y.J. Xu, P. Jenkins, P. McMorn, P. landon, D.I. Enache, A.F. Carley, G.A. Attard, G.J. Hutchings, F. King, E.H. Stitt, P. Johnston, K. Griffin and C.J. Kelly, *Nature* **437** (2005) 1132; M. Turner, V.B. Golovko, O.P.H. Vaughan, P. Abdulkin, A. Berenguer-Murcia, M.S. Tikhov, B.F.G. Johnson and R.M. Lambert, *Nature* **454** (2008) 981; T. Ishida and M. Haruta, *Angew. Chem. Int. Ed.* **46** (2008) 7154; R. Meyer, C. Lemire, S. Shaikhutdinov and H.J. Freund, *Gold Bull.* **37** (2004) 72; N. Lopez, T.V.W. Janssens, B.S. Clausen, Y. Xu, M. Mavrikakis, T. Bligaard and J.K. Nørskov, *J. Catal.* **223** (2004) 232.

67. A. Sanchez, S. Abbet, U. Heiz, W.-D. Schneider, H. Häkkinen, R.N. Barnett and U. Landman, *J. Phys. Chem. A* **103** (1999) 9573.

68. H. Häkkinen, S. Abbet, A. Sanchez, U. Heiz and U. Landman, *Angew. Chem. Int. Ed.* **42** (2003) 1297.

69. B. Yoon, H. Häkkinen, U. Landman, A.S. Wörz, J.-M. Antonietti, S. Abbet, K. Judai and U. Heiz, *Science* **307** (2005) 403.

70. A.A. Herzing, C.J. Kiely, A.F. Carley, P. landon and G.J. Hutchings, *Science* **321** (2008) 1331.

71. Y. Liu, C.-J. Jia, J. Yamasaki, O. Terasaki and F. Schüth, *Angew. Chemie Int. Ed.* **49** (2010) 5771.

72. K.A. Kacprzak, J. Akola and H. Häkkinen, *Phys. Chem. Chem. Phys.* **11** (2009) 6359.

73. O. Lopez-Acevedo, K. A. Kacprzak, J. Akola and H. Häkkinen, *Nature Chem.* **2** (2010) 329.

74. Y. Negishi, K. Nobusada and T. Tsukuda, *J. Am. Chem. Soc.* **127** (2005) 5261.

75. H. Qian, Y. Zhu and R.C. Jin, *J. Am. Chem.* **132** (2010) 4583.

76. A. Dass, *J. Am. Chem. Soc.* **131** (2009) 11666.

77. N.K. Chaki, Y. Negishi, H. Tsunoyama, Y. Shichubu and T. Tsukuda, *J. Am. Chem. Soc.* **130** (2008) 8608.

78. C.A. Fields-Zinna, R. Sardar, C.A. Beasley and R.W. Murray, *J. Am. Chem. Soc.* **131** (2009) 16266.

79. H. Qian and R.C. Jin, *Nano Letters* **9** (2009) 4083.

80. O. Lopez-Acevedo, J. Akola, R.L. Whetten, H. Grönbeck and H. Häkkinen, *J. Phys. Chem. C* **113** (2009) 5035.

81. N. Zheng and G.D. Stucky, *J. Am. Chem. Soc.* **128** (2006) 14278.

82. Y. Liu, H. Tsunoyama, T. Akita and T. Tsukuda, *Chem. Comm.* **46** (2010) 550; H. Tsunoyama, N. Ichikuni, H. Sakurai and T. Tsukuda, *J. Am. Chem. Soc.* **131** (2009) 7086.

83. Y. Zhu, H. Qian, B.A. Drake and R.C. Jin, *Angew. Chem. Int. Ed.* **49** (2010) 1295.

84. W. Chen and S.W. Chen, *Angew. Chemie Int. Ed.* **48** (2009) 4386.

Chapter 10
Optical and Thermal Properties of Gold Nanoparticles for Biology and Medicine

Romain Quidant

Plasmon nano-optics group, ICFO — The Institute of Photonic Sciences, Av. Carl Friedrich Gauss, No. 3, 08860 Castelldefels (Barcelona), Spain. Email: romain.quidant@icfo.es

10.1 Introduction

Light plays an increasing role in life science especially with the recent developments of new optical techniques that enable both imaging biological processes down to the molecular level and treating patients. In parallel, recent advances in nanotechnologies have brought new tools to medicine, both to diagnose and cure diseases. In this chapter, the aim is to discuss recent research that sits at the convergence of optics, nanotechnology and health. This research is based on the extraordinary optical properties of gold nanoparticles (AuNPs) supporting Localized Surface Plasmon (LSP) Resonances. We discuss how plasmonic properties can turn AuNPs into efficient nano-sources of either light or heat for biological and medical applications.

LSP resonances of NPs are associated with a dramatic increase of both their absorption and scattering cross-sections. For gold such resonances occur in the visible range of the spectrum and their features (central wavelength and position) are determined by the particle geometry and environment.[1] Upon suitable illumination matching the resonance conditions, the light is efficiently coupled to the nanoparticle. Part of the coupled light is efficiently scattered: (i) in the near field, leading to an enhanced field at the particle surface, and (ii) to the far field, the NP acting as an efficient optical antenna. The remaining part of the energy is absorbed and dissipated into heat, creating an increase of the metal temperature. While the optical and thermal response of AuNPs are intrinsically connected, it has been shown that AuNPs can be engineered to act more specifically as efficient point-like sources of either light or heat[2,3] (see Chapter 3). In this chapter, we discuss how these properties open an extraordinary potential in life sciences. The chapter is organized into three sections. The first section introduces the use of plasmonic nanostuctures as efficient biosensors for biomolecule detection. In particular we discuss what governs their sensitivity and show that lithographically prepared gold nanostructures can be designed to achieve enhanced performances. We will see in the second section that AuNPs offer great opportunities as contrast agents in bioimaging towards efficient optical diagnosis. Finally, we review in the last section the use of AuNPs as point-like source of heat for photothermal cancer therapy.

10.2 Gold Nanoparticles for Biomolecule Sensing

10.2.1 *LSPR sensing: concept and motivation*

Since surface plasmon modes are bound to the metal, their resonance features are naturally strongly sensitive to a local change of the refractive index. This dependence to the surrounding dielectric function that explicitly appears in the expression of the dispersion relations of both extended (SPR) and localized surface plasmon resonances (LSPR) (see Chapter 3), is the foundation of the use of surface plasmons for optical sensing. Plasmon sensors based on SPR supported by extended flat gold films have been widely studied and have led to several commercial devices that are broadly used as tabletop systems by chemists and biologists in the detection of biomolecules and study of they specific binding. SPR sensors usually monitor changes in the resonance condition (via the incident angle) associated to the modification of the surrounding refractive index. They have been applied to many different contexts including unraveling biological mechanism, drug design, virology, etc. Like SPR at extended flat metal films, LSPR supported by

AuNPs can be used for sensing. In this case, a typical sensing experiment consists in monitoring the frequency shift in the LSPR resonance.

The use of 3D AuNPs instead of homogeneous gold films is motivated by at least three advantages: (i) Firstly, unlike SPR, LSPR supported by particles can be directly coupled to free space light without needing any bulk prism or grating, which limit integration. (ii) Along the same lines, the use of nano-sized particles offers possibilities of integrating a large number of sensing sites on a chip for parallel assays. (iii) Finally, while the SPR configuration is expected to be more sensitive to bulk refractive index changes, i.e. a homogeneous change of refraction index over the whole dielectric superstrate in contact with the metal film (as for liquid and gas sensing), LSPR sensors have the potential to be more sensitive to shallow changes of refractive index as those induced by the binding of small molecules at the metal surface.[4] This can be easily understood when considering that the magnitude of the resonance shift is directly related to the spatial overlap between the plasmonic mode and the volume occupied by the target molecules when binding to the metal (see details in next subsection). Indeed, NPs offer much more flexibility than films to tailor the spatial distribution of the electromagnetic mode and therefore of the sensing volume. In the next subsection, we will discuss how exploiting the electromagnetic coupling between adjacent AuNPs can enhance the sensing sensitivity.

10.2.2 *Sensitivity of LSPR sensors*

The maximum resonance shift of LSPR sensors upon binding of molecules is determined by various parameters related to the sensor itself, its surrounding medium and the target molecule. It has been established that such dependence can be formalized into the following equation, adopted from exact theory for planar SPR sensing, and accepted to some extent as a valid approximation for LSPR.

$$\Delta\lambda \approx m(n_{analyte} - n_{medium})(1 - e^{(-2d/L_d)}) \tag{1}$$

where m is the bulk sensitivity in Refractive Index Unit (RIU) change per nm, $n_{analyte}$ and n_{medium} are the refractive indexes of the adsorbing molecule and medium surrounding the system, d is the thickness of the adsorbed layer and L_d is the electromagnetic field decay length. Equation 10.1 suggests that there are quite a number of parameters that influence the change in the resonance. In the case in which the system is adopted for particular analyte in a given medium ($n_{analyte}$, n_{medium} and d are then fixed), to increase resonance shift one can tune m and L_d. m and L_d, are strongly correlated, accounting for the near-field enhancement and modal distribution, respectively. Equation 10.1 shows that the resonance shift and thus the sensor sensitivity can be increased by concentrating the plasmonic mode into a nanoscale volume and maximizing its spatial overlap with the volume occupied by the analyte.

While simple and intuitive, the 2D model of Equation 10.1 fails in accounting for the inhomogeneous spatial 3D distribution of the near field in LSPR sensing as well as the non-uniform assembly of molecules due to steric hindrance, etc. Indeed, it supposes a uniform molecular coverage, and consequently does not apply to single or few molecule-binding events.

One of the more rigorous approaches, based on perturbation theory, yields a more complicated expression, attempting to address the important differences between LSP and SPP. Derived initially from dielectric non-lossy resonators, Equation 10.2 gives more insight into the contributions of spatial near-field distribution around 3D nanoparticles and the location and size of the perturbation:

$$\Delta\lambda \approx \lambda \frac{(n_{analyte} - n_{medium}) \int_{V_{analyte}} \varepsilon_{medium} |\vec{E}^2| dV}{n_{medium} \int_{V_{nearfield}} \varepsilon_{metal} |\vec{E}^2| dV} = \lambda \frac{\Delta n}{n_{medium}} C$$

(2)

in which $V_{analyte}$ and $V_{nearfield}$ are the volumes occupied by the analyte and the plasmonic mode, respectively. Equation 10.2 clearly highlights that the resonance wavelength shift is directly proportional to the mode-analyte overlap, C.

10.2.3 State of the art in LSPR sensing: from single particle to engineered architectures

The simplest implementation of LSPR sensing is based on large random ensembles of AuNPs.[5] The particles are conjugated with a receptor that has specific binding affinity with the target molecule to be detected. By monitoring the LSPR resonance shift induced by the antigen binding one gets direct information about its concentration and binding coefficient.[6-8]

Motivated by getting rid of size dispersion and agglomeration that tend to broaden the resonance bandwidth and thus limit the shift read out, Raschke and coworkers demonstrated LSPR sensing on a single gold nanosphere immobilized on glass.[9] In their experiment, scattering spectroscopy using dark-field microscopy was used to monitor the specific binding between biotin and streptavidin in the micromolar range. While single particle sensing is *a priori* very attractive for the integration of a large number of sensors for parallel assays it is in practice incompatible with simple detection schemes. However, fair signal over noise ratio can be achieved with a moderate number of particles, maintaining an overall smaller size as compared to typical SPR pads.

As the magnitude of the resonance shift to a given change of the dielectric environment is strongly dependent on the optical near-field distribution associated with the LSPR, the geometry of the AuNP has a strong influence on the sensor sensitivity. Recent studies have demonstrated that particle shapes with sharp edges, such as nanorices,[10] nanostars[11] and bipyramids[12] that leads to intense *hot spots* feature up to several folds higher sensitivities as compared to spherical AuNPs. Alternatively, other geometries as for instance gold nanorings,[13] prepared by projection lithography (also known as colloidal lithography), show remarkable sensing properties.[14]

Beyond isolated particles and random ensembles, further control of the sensitivity can be achieved by accurately engineering the electromagnetic coupling between adjacent AuNPs. The concept of plasmon mode engineering for improving plasmonic sensing was first suggested in 2004, in a theoretical proposal by Enoch and coworkers.[15] The proposed configuration consists of a periodic ensemble of plasmonic dimers formed by two adjacent gold cylinders separated by a nanometer-sized air gap (Fig. 10.1). Upon illumination linearly polarized along the particle alignment, a strong gradient of surface charges is created across the gap, leading to a concentration of light within the gap region.[16] The results showed that an array of gold dimers is about five times more sensitive as compared to an array of isolated particles. Similar calculations performed more recently by using another numerical method have confirmed the enhanced sensitivity of gold dimers.[17]

Based on these numerical predictions, arrays of dimers with different gap sizes were fabricated by e-beam lithography on a glass substrate.[18] In order to maximize the level of reproducibility in the gap size all over

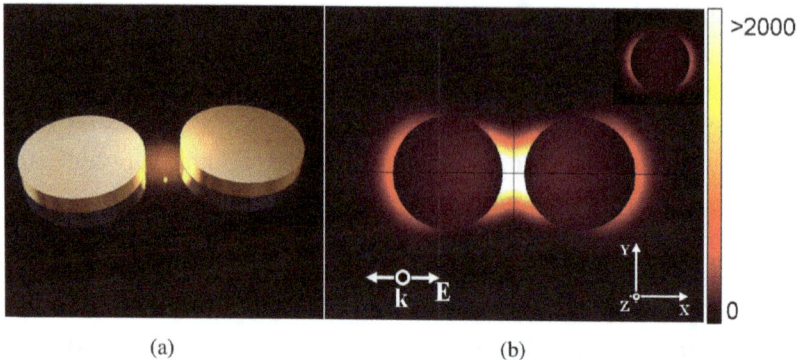

>2000

0

Fig. 10.1. (a) Plasmonic dimer formed by two gold cylinders. (b) FEM calculations of the normalized electric field intensity distribution at resonance (659 nm), (inset) single gold disk at resonance.

each of the arrays, the conventional positive resist process combined with lift-off, usually used in plasmonics, was substituted by an alternative process based on negative resist combined with reactive ion etching. Using this process, arrays of gold cylinder dimers with gaps as small as 10 nm were successfully fabricated. The sensing properties of the fabricated structures were tested by measuring the extinction resonance shifts after binding BSA (Bovine Serum Albumin) to a self-assembled monolayer of mercaptoundecanoic acid (MUA). The experimental data are summarized in Fig. 10.2 in which the resonance shift induced by BSA binding is plotted as a function of the gap size. Two different regimes can be identified. For gap sizes greater than 60 nm, the resonance shift is small and nearly constant. Under these conditions, the weak near-field coupling between the adjacent particles forming the dimers behave similarly to isolated particles. A drastically different regime is observed while decreasing the gap size from 60 nm to contact. The shift increases exponentially until reaching a maximum for a gap of about 30 nm. For this gap size, the sensitivity to the BSA binding is about five times larger than with isolated particles, in good agreement with the predictions of Ref. 15. Further decrease of the gap leads to a dramatic drop of the shift followed by a second maximum. In order to understand this discrete evolution, one needs to consider the geometry and the binding properties of the BSA molecule. BSA is an elongated 14 nm molecule that tends to bind perpendicularly to the MUA layer. The maximum resonance shift thus corresponds to a gap size for which two BSA molecules bounded

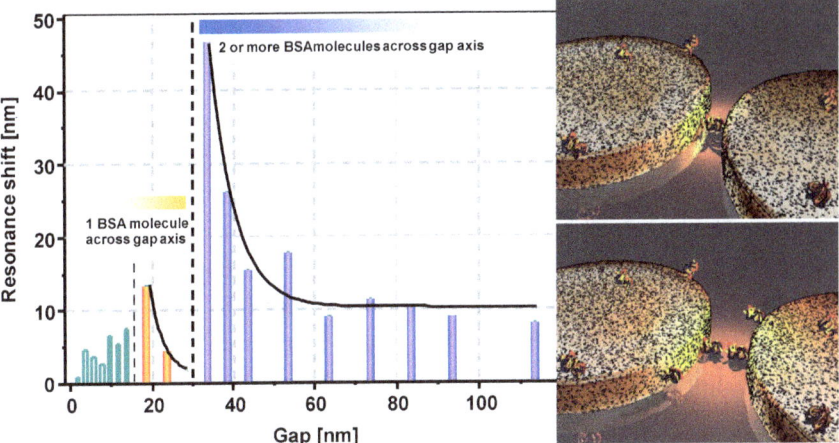

Fig. 10.2. Enhanced detection of biomolecules in an array of gold dimers: (Left) Evolution with the dimer gap size of the resonance shift induced by BSA binding. (Right) Artistic view illustrating both binding regimes.

across the gap fill in the whole sensing volume. For slightly shorter gaps (of about 25 nm), only one molecule can fit across the gap decreasing the spatial overlap with the dimer mode and subsequently leading to a large drop of the resonance shift. Further decreasing the gap size leads to a second maximum corresponding to a single molecule filling in the gap.

Exploiting this same concept of plasmon mode engineering, several other approaches have been considered to increase the sensitivity of LSP-based plasmonic sensors.[19–21] References 19 and 20 propose the use of coupled plasmonic geometries with low symmetry to exploit the optical equivalent to Electromagnetically Induced Transparency (EIT) in atom physics. This coherent phenomenon leads to a sharp transparent window in the LSPR extinction peak that can be used to improve the resonance shift read-out. Along the same strategy, Liu and coworkers have recently used EIT in planar metamaterials to achieve narrow resonance line widths featuring enhanced sensitivity to the surrounding refractive index.[21]

10.2.4 *Towards integrated biosensing platforms*

While recent studies have mostly focused on investigating new configurations with enhanced sensitivity, and testing them with model molecular systems, the use of LSPR sensing to address concrete biological problems is only at an early stage. The increasing interactions between physicists, chemists and medical doctors render possible testing LSPR sensing with pertinent biological molecules. For instance, lots of effort is currently focused on detecting very low concentrations of cancer markers in blood patients, as a novel strategy to achieve earlier diagnosis and treatment monitoring of cancer. Reliable tracking of biomolecules will though require the integration of LSPR sensors into a microfluidic platform[22] that could enable multiple assays in a short time. The current extensive research in this direction should soon provide compact and turnkey analytical devices to be used in biomedical discipline.

10.3 Gold Nanoparticles as Contrast Agents for Bio-imaging: Application to Cancer Diagnosis

Bioimaging is an important discipline that has the potential to advance biology and help in early detection and treatment monitoring of disease. Imaging of biological samples is usually based on the use of contrast agents designed to target to specific biomarkers and monitor biological processes down to the molecular level.

While quantum dots and fluorescent molecules have been widely used as contrast agents, AuNPs offer several key advantages:

(i) The light absorbed and scattered by AuNPs can be strong at their LSPR resonance, despite their small size.
(ii) AuNPs do not exhibit limitations related to the power of the incident laser such as photobleaching and blinking, making imaging over a long period of time possible.
(iii) The dimensions and geometry of AuNPs can be changed to tune their LSPR resonance into the near infrared region of the spectrum, in which optical penetration into biological samples is longer.
(iv) Unlike semiconductors quantum dots or most organic dyes, AuNPs are fully biocompatible and nontoxic.

The main optical imaging techniques using AuNPs as a contrast agent can be classified as three main families: Linear microscopy, nonlinear microscopy and photo-acoustic imaging. In this section, we discuss the concept and specificities of each approach and review some of their respective recent advances.

10.3.1 *Linear imaging techniques*

Among the different techniques that can strongly benefit from the use of AuNPs as contrast agent, we start by discussing the first family of imaging modalities that are based on linear optical processes.

10.3.1.1 *Reflectance microscopy*

Advanced reflectance-based optical methods for *in vivo* imaging of superficial tissue (OCT, RCM etc.) have recently been proposed as non-invasive clinical-imaging techniques, in particular for mapping structural changes associated with the development of cancer. Such techniques are based on the change in the reflective properties of affected tissues. Despite their promise to perform real-time optical biopsies, the modest contrast of regions of early malignancy is usually too low to be of significant clinical value. Here AuNPs show great potential to enhance the reflection contrast. In that case, each nanoparticle is used as a nano-reflector. By proper functionalization and conjugation of the metal surface (see Chapter 10), the particle can selectively target the region of interest, making it more reflective.

Combining confocal reflectance microscopy with AuNP labeling, cancer cervical cells, both isolated and 3D-agglomerated into a phantom mimicking epithelial tissues, were successfully imaged by Sokolov and coworkers.[24] Exploiting the over expression of EGFR (Epithelial Growth factor Receptor), cancer epithelial cells were targeted with 12 nm gold nanoparticles bioconjugated with monoclonal anti-EGFR antibodies (see Fig. 10.3). Firstly, data show that the gold nanospheres preferentially bind to the surface of the cells. Moreover, the actual distribution of EGFR markers within the phantom can be assessed showing a huge potential for *in vivo* cancer detection and treatment monitoring. More recently, other studies have applied AuNP-enhanced reflectance to imaging of other cancer markers like prostate-specific membrane antigen (PSMA) using AuNPs conjugated with aptamers.[25]

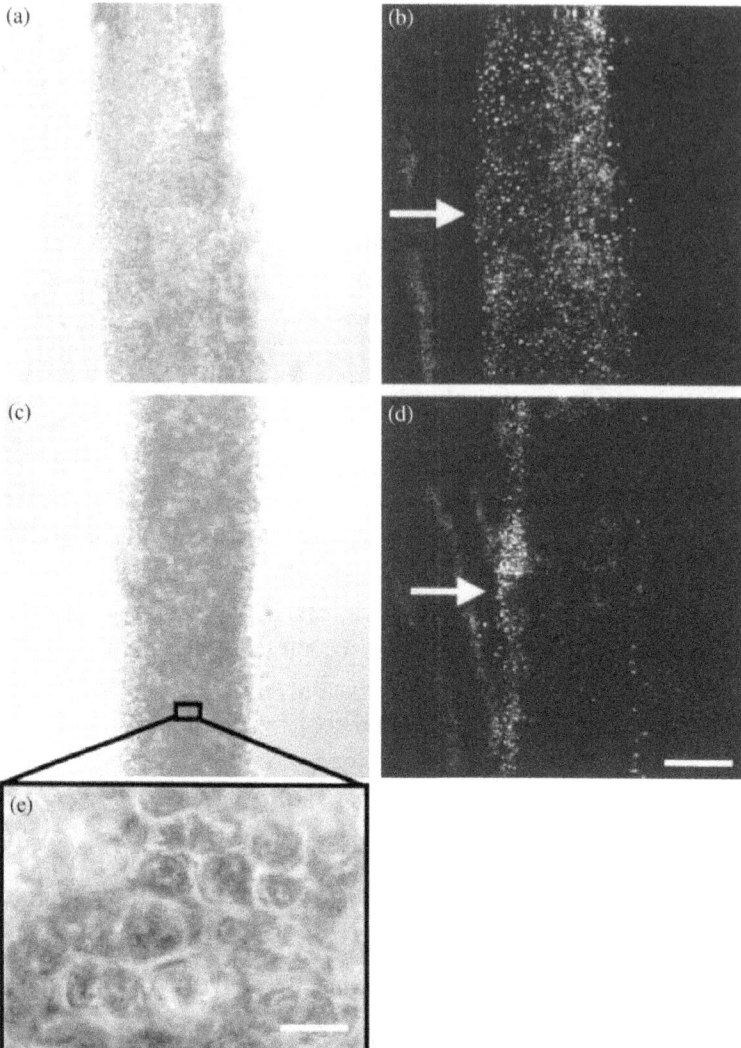

Fig. 10.3. Transmittance (a, c, and e) and reflectance (b and d) images of engineered tissue constructs labeled with anti-EGFR/gold conjugates. The tissue constructs consist of densely packed, multiple layers of cervical cancer (SiHa) cells. The contrast agents were added on top of the tissue phantoms in 10% PVP solution in PBS (a and b) or in pure PBS (c and d). After incubation for 30 min at room temperature, the phantoms were transversely sectioned with a Krumdieck tissue slicer, and the sections were imaged using the Zeiss Leica inverted laser scanning confocal microscope with X10 (a–d) objective. A small spot on a tissue construct was imaged using X40 oil immersion objective to show high density of the epithelial cells in the phantom (e). Reflectance images were obtained with 647 nm excitation. Arrows show the surfaces exposed to the contrast agents. The scale bars are (a–d) 200 μm and (e) 20 μm (From Ref. 24).

10.3.1.2 *Dark-field microscopy*

Beyond the high reflectivity of AuNPs, one can exploit their extraordinary optical properties associated to their LSPR resonances. A simple way to use AuNPs as a contrast agent is based on their ability to efficiently scatter light with frequencies within their plasmon band. El Sayed and coworkers first suggested fast-screening cancer cells targeted with AuNPs using dark-field microscopy.[26] Unlike conventional (bright) transmission microscopy, in a dark-field microscope, the numerical aperture (NA) of the illumination condenser does not overlap with the NA of the collection objective lens. In other words, in absence of any scattering centers, the incident light rays are not collected and lead to a black (dark) background. This technique thus enables us to image tiny objects with a large signal over noise. For instance, a dark field was used to perform the first scattering spectroscopy of individual plasmonic nanoparticles.[27]

The data of El Sayed *et al.* on epithelial cancer cell lines targeted with gold nanospheres conjugated to anti-EGFR show that this non-scanning, fast and simple imaging technique enables us to discriminate between cancer cells and healthy cells after incubation within a solution of the conjugated AuNPs. Additionally, when combined with linear scattering spectroscopy, the level of agglomeration of the AuNPs can be assessed. However, the significant direct scattering from the cell can be a major drawback when dealing with low concentrations of markers, as it strongly limits the minimum number of AuNPs that can be detected.

10.3.1.3 *Enhanced-fluorescence microscopy*

Another approach relies on exploiting the ability of AuNPs to enhance the fluorescence yield of fluorophores located at their vicinity. Through the adjustment of the particle resonance and the fluorophore-metal distance, one can substantially affect the excitation and decay channels of the molecule.[28–30] For very short distances, typically shorter than 10 nm, the fluorescence gets quenched as the nonradiative decay channels prevail over the radiative ones. For larger distances the balance is inverted and fluorescence is enhanced as the result of a combination of processes including enhanced light absorption, enhanced radiative decay and enhanced reemission of fluorescence to the far field (antenna effect).

While near-infrared fluorophore-like ICG (Indocyanine Green) has been extensively used for molecular imaging, it usually suffers from low quantum yields in aqueous media (typically of about 1%) that limit the imaging sensitivity. Recently, it was shown that the emission of weak NIR fluorophores, such as ICG, could be dramatically enhanced to about 80% by gold nanoshells consisting of spherical dielectric core coated with a thin gold layer.[31,32] ICG molecules were positioned at an optimum distance of about 10 nm by growing a silica layer around the nanoparticle surface. Interestingly, the spacer layer can also incorporate some Fe_3O_4 magnetic nanoparticles that enables using the same complex for both fluorescence imaging and Magnetic Resonance Imaging (MRI). Despite its lower spatial resolution, MRI is complementary to fluorescence since it penetrates tissues to depth of several centimeters. Bardhan and coworkers first demonstrated the suitability of their fluorescent–magnetic contrast agent conjugated to anti-HER2 (Human Epidermal growth factor 2) for *in vitro* imaging breast cancer cells that over-express HER2.[33] Very recently, the same authors extended their approach to a mice model[34] to track the circulation of the complex through the mice body over several days. *In vivo* fluorescent imaging was used to monitor the concentration of complexes after their injection in the blood flow (Fig. 10.4). Their data nicely show that due to their bioconjugation, the complexes remain within the tumor region for longer than in healthy tissues. There is thus a time window of several hours during which the fluorescent map enables detection and/or treatment (see Section 10.2) of cancer tissues. It was also observed that the complexes were fully eliminated from the mice organism after 72 hours.

10.3.2 *Nonlinear imaging techniques*

Another attractive aspect of plasmonic AuNPs is their ability to dramatically enhance, through their intense local fields, weak nonlinear optical processes occurring either in the particle itself or in its vicinity.

10.3.2.1 *Multiphoton imaging*

Among the nonlinear imaging modalities that can benefit from enhanced plasmonic fields, let's first mention Second harmonic generation (SHG)

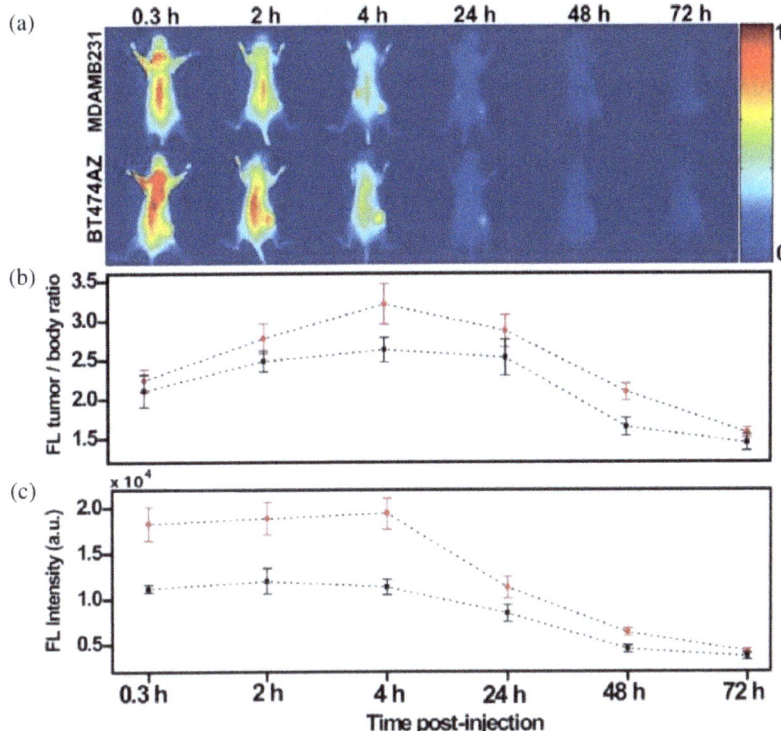

Fig. 10.4. *In vivo* Fluorescence monitoring of gold nanoshell/ICG complexes in mice: (a) NIR images of mice with HER2 low expressing MDAMB231 xenografts (Top) and HER2 overexpressing BT474AZ xenografts (Bottom) at 0.3, 2, 4, 24, 48, and 72 h after injection of nanocomplexes. (b) Fluorescence (FL) intensity of tumor-to-body ratio at different time points of mice with BT474AZ xenografts (n) 6) and MDAMB231 xenografts (n) 3) and showing maximum fluorescence at 4 h. (c) Fluorescence intensity comparison of tumors only between BT474AZ (n) 6) and MDAMB231 (n) 3) showing 71.5% increase in signal at 4 h in BT474AZ tumors compared to MDAMB231 tumors, p) 0.003 across tumor types. (From Ref. 34).

microscopy. SHG (also known as frequency doubling) is a nonlinear optical process, in which photons from a pulsed laser source interacting with a nonlinear material are effectively "combined" to form new photons with twice the energy, and therefore half the wavelength of the incident photons.[35] A unique feature of SHG is that it requires the presence of an asymmetric distribution of the second harmonic sources (non-centrosymmetry). Thus cell membranes, their proteins and their crucial contribution to cellular physiology are ideally suited to interrogation with such technique. However,

SHG is a process with poor efficiency that, in practice, requires advanced detection schemes. It was shown that complexes formed by nonlinear dyes attached to AuNPs could be used to strongly enhance SHG at the membrane of cells.[36]

Similarly to SHG, Third harmonic generation (THG) microscopy, in which incident photons at frequency ω lead to photons at 3ω, can be boosted by plasmon nanooptics. The absence of asymmetry requirements, combined with the high third-order susceptibility $\chi^{(3)}$ of gold,[37] makes AuNPs excellent examples of efficient sources of THG for cell imaging, without any need for additional nonlinear molecules. In the experiment by Yelin and coworkers, AuNPs were grown into the cells from tiny gold seeds to a size at which they were resonant with the near infrared light from a femto-second Titanium-Sapphire laser source.[38]

10.3.2.2 SERS imaging

Another imaging modality that can strongly benefit from AuNPs is Raman imaging. Raman spectroscopy is a spectroscopic technique used to study vibrational, rotational, and other low-frequency modes in a system. It relies on inelastic scattering, or Raman scattering, of monochromatic light, usually from a laser in the visible, near infrared, or near ultraviolet range. The laser light interacts with molecular vibrations, phonons or other excitations in the system, resulting in the energy of the laser photons being shifted up or down. Consequently, the shift in energy gives information about the phonon modes in the system and hence is very powerful at identifying the presence of given specie via its structural fingerprint. However Raman signal is very weak (typically 1 out of 10^7 photons) and the integration times needed render imaging very tedious in practice. Nevertheless, it was shown that metallic nanostructures could dramatically enhance the Raman cross section by more than ten orders of magnitude[39,40] to reach levels of signal comparable to fluorescence and enable single molecule measurements. The exact mechanism of the so-called Surface Enhanced Raman Scattering (SERS) has, however, been a matter of debate for many years between two main theories, based on electromagnetic and chemical mechanisms, respectively. In the electromagnetic theory, the local field enhancement near resonant AuNPs augments both the excitation field experienced by the molecules and their emitted Raman signal. While the electromagnetic theory of enhancement can be applied regardless of the molecule being studied, it does not fully explain the magnitude of the enhancement observed in many systems. The chemical mechanism involves charge transfer between the chemisorbed species and the metal surface. It only applies to specific cases though, where the molecules have formed a chemical bond with the surface. After many years of debate, nowadays it is pretty well accepted that in many SERS experiments, both mechanisms may coexist.[40]

While SERS was discovered in the 1970s, it has only recently been exploited for bioimaging.[41] The approach that has been considered by several groups is based on preparing a complex consisting of a gold nanoparticle combined to an efficient Raman active molecule. After a proper bioconjugation, the complex can be used to target cancer cells to map the distribution of cancer markers.[42-44] Lately, the technique has been extended to *in vivo* imaging on a mouse model.[45] The main advantage of SERS imaging over other approaches is the possibility to simultaneously detect multiplex analytes by exploiting the sharper bandwidth of Raman peaks as compared to fluorescence peaks.

10.3.2.3 *Two-photon induced Luminescence*

In addition to frequency generation and Raman scattering, there has also been a growing interest in exploiting a $\chi^{(3)}$-based phenomenon known as two-photon absorption. Such process is based on the simultaneous absorption of two photons with the same energy that excites the molecules into higher energy state. In this context, it was shown that the strong local field resulting from the agglomeration of AuNPs at the surface of cells could be used to dramatically enhance the two-photon absorption and thus the subsequent fluorescence of adjacent molecules of the cell membrane.[38] Alternatively, one can directly exploit the two-photon induced luminescence (TPL) from AuNPs. Such a process, first reported by Mooradian in the 1960s,[46] is slightly different from two-photon absorption in molecules, which requires simultaneous absorption of two coherent photons. TPL-based imaging has recently received lot of attention in the plasmon nano-optics community as a powerful technique to probe the near-field optical response of plasmonic nanostructures.[47-50] Recently, Durr and

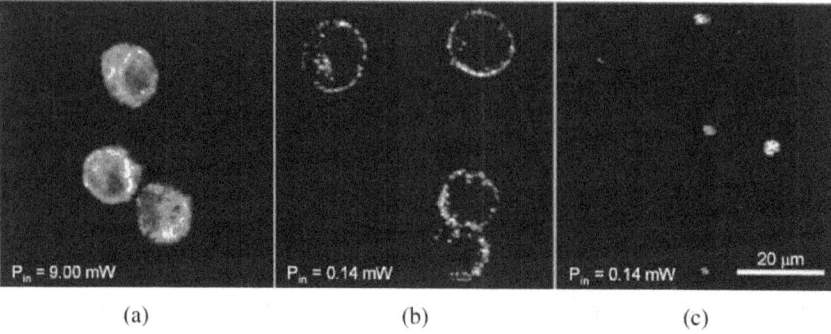

Fig. 10.5. Two-photon luminescence (TPL) images of cancer cells placed on a coverslip from a cell suspension. (a) TPL image of unlabeled cells. (b) TPL image of nanorod-labeled cells. Imaging required 9 mW of excitation power in unlabeled cells to get same signal level obtained with only 140 μW for nanorod labeled cells, indicating that TPL from nanorods can be more than 4000 times brighter than TPAF from intrinsic fluorophores. (c) TPL image of nonspecifically labeled cells. (From Ref. 51.)

coworkers used TPL microscopy to perform targeted imaging of cancer cells in three dimensional tissue phantoms.[51] In their experiment, gold nanorods designed to be resonant at 760 nm were bioconjugated with EGFR to target skin cancer cells (Fig. 10.5). Their data show the distribution of the nanorods at the cells membrane. Discrete bright spots within the cytoplasm are indicative of endosomal uptake of EGFR receptors with nanorods inside the cells. More recently, the same approach has been combined with 3D imaging *in vivo* to characterize the intestinal blood vessels of a mouse.[52]

10.3.3 *Photo-acoustic imaging*

To close this section on the use of AuNPs as contrast agent in bioimaging, we review recent advances in photo-acoustic microscopy for *in vivo* imaging. Photo-acoustic imaging is based on the photo-acoustic effect: upon illumination with laser pulses at optical frequencies, biological tissues adsorb part of the delivered energy and convert it into heat, leading to transient thermoelastic expansion and thus wideband (e.g., MHz) ultrasonic emission. The generated ultrasonic waves are then detected by ultrasonic transducers to form an image with a spatial resolution down to tens of micrometers. Upon near infrared illumination, the penetration depth is on the order of centimeters making this modality very well suited for *in vivo* imaging. The optical absorption in biological tissues can be due to endogenous molecules such

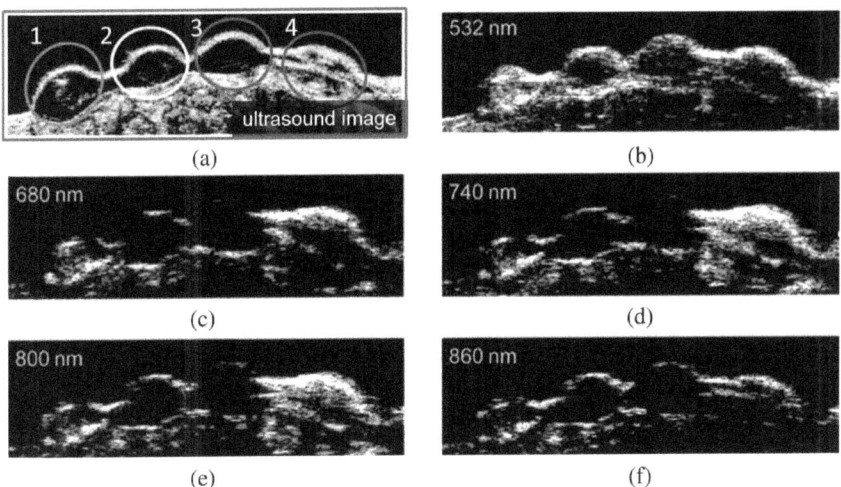

Fig. 10.6. Ultrasound (a) and photoacoustic (b–f) images of gelatin implants in mouse tissue *ex-vivo* at laser illumination wavelength 532, 680, 740, 800, and 860 nm, respectively. The gelatin implants containing the cells with targeted AuNPs (1), control A431 cells (2), the A431 cells mixed with mPEG-SH coated AuNPs (3), and NIR dye (4) are indicated on the ultrasound image. Tuning the incident wavelength enables discriminating agglomerated AuNPs targeted to cancer cells from the non-targeted AuNPs. The images measure 44 mm laterally and 11 mm axially (From Ref. 57).

as hemoglobin or melanin. Since blood usually has much larger absorption levels than surrounding tissues, there is sufficient endogenous contrast for photo-acoustic imaging to visualize blood vessels. Recent studies have shown that photo-acoustic imaging can be used *in vivo* for tumor angiogenesis monitoring, blood oxygenation mapping, functional brain imaging, and skin melanoma detection. However, when aiming at imaging regions with low absorption, the use of contrast agents is required to provide sufficient signal over noise. In this context, the enhanced absorption of gold nanoparticles makes them very attractive candidates to boost the ultrasonic signal.[53–55] As a practical example, Wang and coworkers have recently demonstrated efficient intravascular imaging of macrophages using agglomeration of gold nanospheres as a novel strategy towards monitoring of cardiovascular diseases.[56] Interestingly, multi-wavelength imaging (changing the wavelength of the optical illumination) is an efficient way to discriminate the signal from the agglomerated AuNPs targeted to cancer cells from the non-targeted AuNPs.[57]

In this section, we have reviewed some of the main bio-imaging modalities in which AuNPs can behave as an efficient contrast agent. Lately there has been a clear tendency towards multimodal plasmonic imaging in which a single nanoprobe could be used by multiple imaging methods; for instance, to confirm the development of a disease.

Following this trend, in the following section we will discuss how the same complexes used for imaging can lead to promising novel cancer therapy based on local photo-heating.

10.4 Photothermal Properties of Gold Nanoparticles and their Application to Photothermal Cancer Therapy

In vivo local delivery of heat has raised a growing interest in particular for local tissue ablation. Conventional methods include laser-induced therapy, Microwave and Radio Frequency ablation, and magnetic and focused ultrasound ablation. However, all of these approaches suffer from a common limitation, which arises from the fact that heating is nonspecific, hence it leads to the damage of healthy tissues. Alternatively, the use of magnetic fields to heat magnetic particles targeted to tissues was suggested. Similarly, in 2003 Hirsch and coworkers proposed to use AuNPs as local heat sources controllable by an external laser illumination.[58] As well as ablation, it has also been suggested to exploit a more moderate temperature increase in gold complexes for local drug delivery. Interestingly, photothermal therapy is fully compatible with molecular targeting used for diagnosis and the same gold complex could thus be used to detect the disease and treat it. In this last section we first discuss the optimization of heat generation in AuNPs before reviewing the latest advances in photothermal cancer therapy and thermal-induced drug delivery.

10.4.1 *Optimizing heat generation in gold nanoparticles*

When a metal nanoparticle is illuminated, part of the intercepted light is scattered in the surroundings while the other part gets absorbed and ultimately dissipated into heat.[59] The efficiency of each of these processes can be characterized by σ_{sca} and σ_{abs}, the elastic scattering and the absorption cross-sections, respectively. The sum of these two processes leads to light attenuation characterized by the extinction cross-section σ_{ext}:

$$\sigma_{ext} = \sigma_{sca} + \sigma_{abs} \tag{3}$$

Depending on the size and the shape of the nanoparticle, the balance between scattering and absorption can vary substantially.[59–62] For instance, while small gold spheres (<10 nm in diameter) mainly act as invisible nano-sources of heat,[63,64] scattering processes dominate for diameters larger than ∼50 nm.[59] Here, we focus on the absorption processes and the subsequent heat generation. The general expression of the absorption cross section for a nanoparticle illuminated by a plane wave is (in mks units):

$$\sigma_{abs} = \frac{k}{\varepsilon_0 |E_0|^2} \int_V \mathfrak{I}(\varepsilon_\omega) |E(r)|^2 \, dr \tag{4}$$

where $k = 2\pi n/\lambda_0 = n\omega/c$ is the wave vector, n is the refractive index of the surrounding medium, ε_ω is the permittivity of the nanoparticle material at frequency ω, E_0 the electric field amplitude of the incoming light considered as a plane wave and $E(r)$ the total electric field amplitude. $\mathfrak{I}(\varepsilon_\omega)$ denotes the imaginary part of the dielectric function. The integral is calculated over the nanoparticle volume V. The power of heat generation Q inside the nanostructure is directly proportional to σ_{abs}:

$$Q = \sigma I = \sigma c \varepsilon_0 |E|^2 \tag{5}$$

where $I = n c \varepsilon_0 |E_0|^2$ is the intensity of the incoming light and using Equation 10.4 we obtain:

$$Q = \frac{n^2 \omega}{2} \mathfrak{I}(\varepsilon_\omega) \int_V |E(r)|^2 \, dr = \int_V q(r) \, dr \tag{6}$$

where $q(r) = (n^2\omega/2)\mathfrak{I}(\varepsilon_\omega)|E(r)|^2$ is the volumetric power density of heat generation. This expression shows that the quantity of generated heat is governed by the electric field intensity within the metal. Consequently, the drastic influence of the particle geometry on the plasmon mode distribution offer some degree of control for designing efficient nano heat sources, remotely controllable by laser illumination.

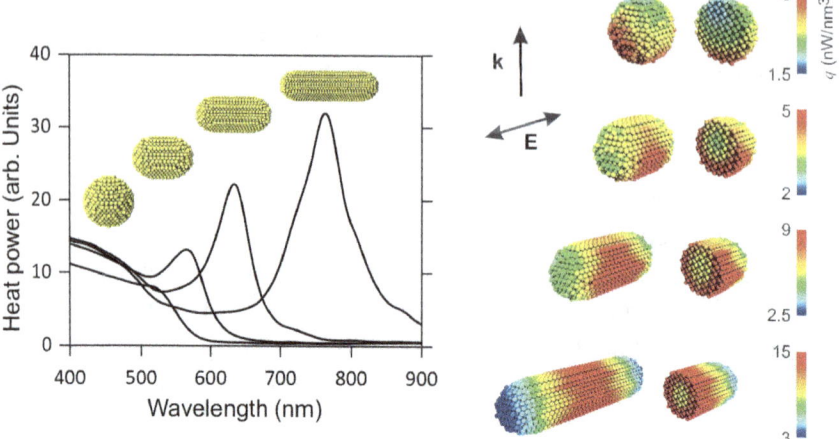

Fig. 10.7. Heat generation in gold nanoparticles: (Left) Evolution of the heat power spectrum with the particle aspect ratio (at constant gold volume) (Right) 3D mapping of the heat power density computed for the four nanoparticle shapes at their respective plasmon resonances.

Recently the Green Dyadic Method (GDM)[12,13] has been used to quantify the influence of the geometry of a gold AuNPs on its heating efficiency. The GDM makes it possible to map the spatial distribution of the heat power density inside the nanoparticles, providing further insight into the influence of the particle shape and the illumination conditions on the origin of heat. Figure 10.7 displays calculations of heat power spectra $Q(\omega)$ for different geometries of gold nanoparticles surrounded by water and illuminated by a plane wave. We fix the intensity of the incoming light at $1\,\text{mW} \cdot \mu m^{-2} = 10^{5}\text{W} \cdot \text{cm}^{-2}$.

To illustrate the influence of the particle geometry, the heat generation of a sphere progressively elongating into a rod-like structure at a constant volume ($4\pi r_{\text{eff}}/3$, where $r_{\text{eff}} = 25\,\text{nm}$) is shown in Fig. 10.7 (left panel). The successive nanorods aspect ratios are 1:1 (sphere), 1.4:1, 2:1 and 3:1. Two major features arise from the calculations. First the LSP resonance markedly depends on the nanoparticle shape. A redshift is indeed expected for nanorods compared with spheres.

Beyond the resonance redshift, a substantial increase of the heating efficiency is observed, by a factor 5 from the sphere to the 3:1 nanorod. The GDM can be efficiently employed to understand this feature. Figure 10.7

(right panel) represents the distribution of the heat power density $q(r)$ around and across each of the geometries. Interestingly, for a sphere excited at the LSP resonance, the heat generation arises mainly from the outer part of the particles facing the incoming light. Consequently, the major part of the nanoparticle sees weak electric field intensity and thus does not contribute to heating. However, for elongated nanorods, the field can further penetrate the inner part of the particle, thus making the whole metal volume more efficiently involved in the heating process. It should be underlined that the heat generation mainly arises from the central part of the nanorods because the extremities undergo charge accumulation that leads to a weaker electric field inside the structure. Using another numerical method, systematic temperature calculations have been performed on other particles geometries.[65]

The temperature increase induced by gold nanorods randomly agglomerated on a glass surface has recently been measured experimentally by using Fluorescence Polarization Anisotropy (FPA).[66,67]

10.4.2 *Photothermal therapy (Thermal ablative therapy)*

In the original work by Hirsch and coworkers,[58] gold nanoshells, formed by a dielectric core surrounded by a thin gold layer, were tuned to be resonant in the near infrared where the transmission through tissues is optimal. Human breast carcinoma cells incubated with nanoshells *in vitro* were found to have undergone photothermally-induced death upon exposure to NIR light (820 nm, 35 W/cm^2). Conversely, cells without nanoshells displayed no loss in viability after the same periods and conditions of NIR illumination. Moreover, exposure to low doses of NIR light (820 nm, 4 W/cm^2) in solid tumors treated with metal nanoshells reached average maximum temperatures capable of inducing irreversible tissue damage ($\Delta T = 37.4 \pm 6.6°C$) within 4–6 min. Importantly, controls treated without nanoshells demonstrated a significantly lower average temperature increase on exposure to NIR light ($\Delta T < 10°C$). Shortly afterwards, the feasibility of this approach was successfully tested *in vivo* on a mice model.[68] Gold nanoshells coated with Polyethylene glycol (PEG) were intravenously injected into the mice. The tumor was then illuminated with a diode laser during three minute sessions. After ten days of treatment, complete resorption of the tumor was observed.

	No nanoshells	Non-specific antibody	Anti-HER2

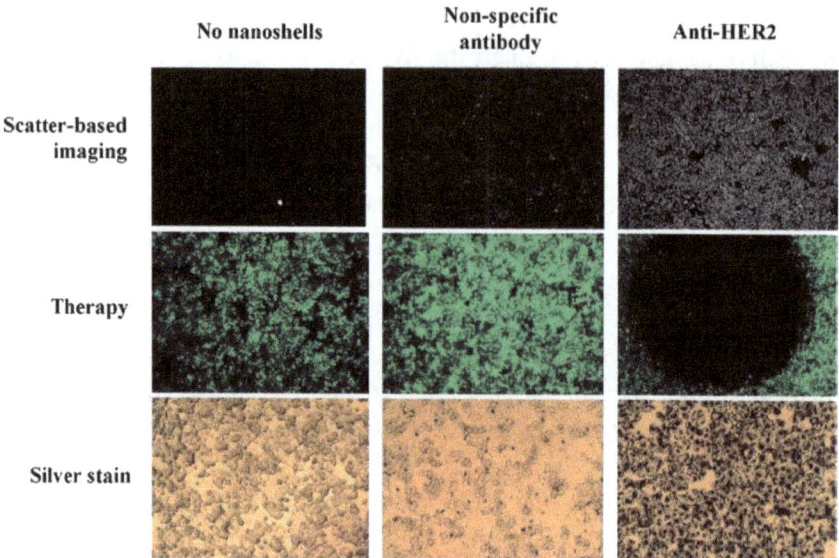

Scatter-based imaging / Therapy / Silver stain

Fig. 10.8. Combined imaging and therapy of SKBr3 breast cancer cells using HER2-targeted nanoshells. Scatter-based dark-field imaging of HER2 expression (Top row), cell viability assessed via calcein staining (Middle row), and silver stain assessment of nanoshell binding (Bottom row). Cytotoxicity was observed in cells treated with a NIR-emitting laser following exposure and imaging of cells targeted with anti-HER2 nanoshells only. Note increased contrast (Top row, right column) and cytotoxicity (Dark spot) in cells treated with a NIR- emitting laser following nanoshell exposure (Middle row, right column) compared to controls (Left and middle columns). (From Ref. 69.)

More than 90 days after the treatment, all treated mice remained healthy and free of tumors.

Since then, numerous studies have been carried out to push this initial proposal towards clinical trials. Special attention has been given to investigating new particle geometries and their specific targeting to cancer cells (see Section 10.2). In recent years, the company Nanospectra[70] has started clinical tests on head and neck cancer using gold nanoshells.

10.4.3 *Drug delivery*

Beyond tissue ablation, moderate temperature increases below the damage threshold can be used for the control delivery of molecules, which is in high demand, especially for drug delivery and gene therapy. A first step towards this goal is to exploit photothermal heating of AuNPs to control dehybridization and release of DNA in cells. In the recent experiment of

Lee and coworkers,[71] thiol-modified sense oligonucleotides were attached to gold nanorods before hybridizing antisense oligonucleotides. When the minimum temperature on the complex reaches the melting temperature of the short duplex, the short double-stranded oligonucleotides denature. The antisense oligonucleotides are released and are allowed to bind to a portion of the corresponding mRNA into the cell. Lately this approach has been extended to deliver other types of molecules. In the approach by Huschka and coworkers[72] DAPI (4', 6-diamidino-2-phenylindole), a water-soluble blue fluorescent dye, is intercalated into the dsDNA of a gold nanoshell/ daDNA complex. The nanoshell-dsDNA-DAPI complexes were incubated with H1299 lung cancer cells. Upon illumination with an 800 nm CW laser, corresponding to the peak resonant wavelength of the nanoshell complexes, the DAPI molecules were released from the nanoshell complexes. Subsequent to release, the DAPI diffused through the cytoplasm and into the cell nucleus, where it preferentially bound and stained the nuclear DNA. *A priori*, this strategy could be extended to a multitude of other guest molecules that associate with the host dsDNA carrier including small organic fluorophores, steroid hormones, and therapeutic molecules. For example, the quest to find dsDNA intercalators that inhibit the uncontrollable replication of tumor cells comprises an entire field of cancer research.

10.5 Conclusion

Throughout this chapter we have seen that the extraordinary optical properties of AuNPs offer a set of new opportunities to improve diagnosis and therapy of diseases such as cancer. Enhanced optical fields near lithographically prepared gold nanostructures can be exploited for sensing purposes to detect low concentrations of biomolecules. Their integration into microfluidic chip may soon lead to novel compact analytical platforms able to track cancer markers in the blood of patients. Also, conjugated colloidal AuNPs act as bright, stable and biocompatible contrast agents for *in vivo* biomolecule tracking in biological tissues, opening new horizons in optical diagnosis. Remarkably, one can exploit the ability of these same conjugated AuNPs to heat up upon suitable illumination to either destroy cancer cells or control intra-cellular drug delivery.

Despite their great potential, most of the above mentioned approaches based on AuNPs are still being developed in research laboratories. Prior to becoming new tools for physicians with positive repercussion on health care, further optimization is required and some issues need to be addressed. For instance, for applications in which AuNPs need to be injected into the patient's body, extensive toxicity tests still need to be performed to assess potential accumulation in the patient organs.

References

1. U. Kreibig and M. Vollmer, *Optical Properties of Metal Clusters*, Springer Series in Materials Science, New York, 1995.
2. G. Baffou, R. Quidant and C. Girard, *Appl. Phys. Lett.* **94** (2009) 153109.
3. G. Baffou, C. Girard and R. Quidant, *Phys. Rev. Lett.* **104** (2010) 136805.
4. M. Svedendahl, S. Chen, A. Dmitriev and M. Käll, *Nano Lett.* **9** (2009) 4428–4433.
5. T. Okamoto, I. Yamaguchi and T. Kobayashi, *Optics Letters* **25** (2000) 372–374.
6. P. Englebienne, *Analyst* **123** (1998) 1599–1603.
7. P. Englebienne, M. Verhas and A. Van Hoonacker, *Analyst* **126** (2001) 1645–1651.
8. M.P. Kreuzer, R. Quidant, G. Badenes, M. -Pilar Marco, *Biosensors and Bioelectronics* **21** (2006) 1345–1349.
9. G. Raschke, S. Kowarik, T. Franzl, C. Sönnichsen, T.A. Klar, J. Feldmann, A. Nichtl and K. Kürzinger, *Nano Lett.* **3** (2003) 935–938.
10. H. Wang, D.W. Brandl, F. Le, P. Nordlander and N.J. Halas, *Nano Lett.* **6** (2006) 827–832.
11. C.L. Nehl, H. Liao and J.H. Hafner, *Nano Lett.* **6** (2006) 683–688.
12. K.M. Mayer, F. Hao, S. Lee, P. Nordlander and J.H. Hafner, *Nanotechnology* **21** (2010) 255503.
13. J. Aizpurua, P. Hanarp, D.S. Sutherland, M. Käll, Garnett W. Bryant, and F.J. García de Abajo, *Phys. Rev. Lett.* **90** (2003) 057401.
14. E.M. Larsson, J. Alegret, M. Käll, and D.S. Sutherland, *Nano Lett.* **7** (2007) 1256–1263.
15. S. Enoch, R. Quidant and G. Badenes, *Opt. Express* **12** (2004) 3422–3427.
16. J. Kottmann and O. Martin, *Opt. Express* **8** (2001) 655–663.
17. P.K. Jain and M.A. El-Sayed, *Nano Lett.* **8** (2008) 4347–4352.
18. S.S. Aćimović, M.P. Kreuzer, M.U. González and R. Quidant, *ACS Nano* **3** (2009) 1231–1237.
19. N. Verellen, Y. Sonnefraud, H. Sobhani, F. Hao, V.V. Moshchalkov, P. Van Dorpe, P. Nordlander and S. Maier, *Nano Lett.* **9** (2009) 1663.
20. N. Liu, T. Weiss, M. Mesch, L. Langguth, U. Eigenthaler, M. Hirscher, C. Sonnichsen and H. Giessen, *Nano Lett.* **10** (2010) 1103.
21. A.B. Evlyukhin, S.I. Bozhevolnyi, A. Pors, M.G. Nielsen, I.P. Radko, M. Willatzen and O. Albrektsen, *Nano Lett.* **10** (2010) 4571.

22. T. Thorsen, S.J. Maerkl and S.R. Quake, *Science* **298** (2002) 580–584.
23. D. Gerion and G. Day, *BioPharm International* **2323** (2010) 38–45.
24. K. Sokolov, M. Follen, J. Aaaron, I. Pavlona, A. Malpica, R. Lotan and R. Richard-Kortum, *Cancer Research* **63** (2003) 1999–2004.
25. D.J. Javier, N. Nitrin, M. Levy, A. Ellington and R. Richards-Kortum, *Bioconjugate Chem.* **19** (2008) 1309–1312.
26. I.H. El-Sayed, X. Huang and M.A. El-Sayed, *NanoLetters* **5** (2005) 829–834.
27. T. Klar, M. Perner, S. Grosse, G. von Plessen, W. Spirkl, and J. Feldmann, *Phys. Rev. Lett.* **80** (1998) 4249.
28. E. Dulkeith, A.C. Morteani, T. Niedereichholz, T.A. Klar, J. Feldmann, S.A. Levi, F.C.J.M. van Veggel, D.N. Reinhoudt, M. Möller and D.I. Gittins, *Phys. Rev. Lett.* **89** (2002) 203002.
29. P. Anger, P. Bharadwaj and L. Novotny, *Phys. Rev. Lett.* **96** (2006) 113002.
30. S. Kühn, U. Håkanson, L. Rogobete and V. Sandoghdar, *Phys. Rev. Lett.* **97** (2006) 017402.
31. R. Bardhan, N.K. Grady and N.J. Halas, *SMALL* **4** (2008) 1716–1722.
32. R. Bardhan, N.K. Grady, J.R. Cole, A. Joshi and N.J. Halas, *ACS Nano* **3** (2009) 744–752.
33. R. Bardhan, W. Chen, C. Perez-Torres, M. Bartels, R.M. Hschka, L.L. Zhao, E. Morosan, R.G. Pautler, A. Joshi and N.J. Halas, *Advanced Materials*, **19** (2009) 3901–3909.
34. R. Bardhan, W. Chen, M. Bartels, C. Perez-Torres, M.F. Botero, R. Ward MacAninch, A. Contreras, R. Schiff, R.G. Pautler, N.J. Halas and A. Joshi, *Nano Lett.* **10** (2010) 4920–4928.
35. R. W. Boyd, *Nonlinear Optics*, 2nd edn, Academic Press; Elsevier, New York, 2003.
36. G. Peleg, A. Lewis, O. Bouevitch, L. Loew, D. Parnas and M. Linial, *Bioimaging* **4** (1996) 215–224.
37. J. Renger, R. Quidant, N.F. van Hulst and L. Novotny, *Phys. Rev. Lett.* **104** (2010) 046803.
38. D. Yelin, D. Oron, S. Thiberge, E. Moses and Y. Silberberg, *Optics Express* **11** (2003) 1385–1391.
39. S. Nie and S.R. Emory, *Science* **275** (1997) 1102–1106.
40. E.C. Le Ru, E. Blackie, M. Meyer and P.G. Etchegoin, *The Journal of Physical Chemistry C* **111** (2007) 13794–13803.
41. J. Kneipp, H. Kneipp, W.L. Rice and K. Kneipp, *Anal. Chem.* **77** (2005) 2381–238.
42. S. Lee, S. Kim, J. Choo, S. Young Shin, Y. Han Lee, H. Young Choi, S. Ha, K. Kang, and C. Hwan Oh, *Anal. Chemistry* **79** (2007) 916–922.
43. S. Lee, H. Chon, M. Lee, J. Choo, S. Young Shin, Y. Han Lee, I. Joo Rhyu, S. Wook Son and C. Hwan Oh, *Biosensors and Bioelectronics* **24** (2009) 2260–2263.
44. H. Park, S. Lee, L. Chen, E. Kyu Lee, S. Young Shin, Y. Han Lee, S. Wook Son, C. Hwan Oh, J. Myong Song, S. Ho Kang and J. Choo, *Phys. Chem. Chem. Phys.* **11** (2009) 7444–7449.
45. M.V. Yigit, L. Zhu, M.A. Ifediba, Y. Zhang, K. Carr, A. Moore and Z. Medarova, *ACS Nano* **5** (2011) 1056–1066.
46. A. Mooradian, *Phys. Rev. Lett.* **22** (1969) 185–187.

47. M.R. Beversluis, A. Bouhelier and L. Novotny, *Phys. Rev. B* **68** (2003) 115433.
48. P.J. Schuck, D.P. Fromm, A. Sundaramurthy, G.S. Kino and W.E. Moerner, *Phys. Rev. Lett.* **94** (2005) 017402.
49. P. Mühlschlegel, H.-J. Eisler, O.J.F. Martin, B. Hecht and D.W. Pohl, *Science* **308** (2005) 1607–1609.
50. P. Ghenuche, S. Cherukulappurath, T.H. Taminiau, N.F. van Hulst and R. Quidant, *Phys. Rev. Lett.* **101** (2008) 116805.
51. N.J. Durr, T. Larson, D.K. Smith, B.A. Korgel, K. Sokolov and A. Ben-Yakar, *Nano Lett.* **4** (2007) 941–945.
52. S-C. Tang, Ya-Y. Fu, W-F. Lo, T-E. Hua and H-Y. Tuan, *ACS Nano* **4** (2010) 6278–628.
53. X. Yang, S.E. Skrabalak, Z.-Y. Li, Y. Xia and L.V. Wang, *Nano Lett.* **7** (2007) 3798–3802.
54. M. Eghtedari, A. Oraevsky, J.A. Copland, N.A. Kotov, A. Conjusteau and M. Motamedi, *Nano Lett.* **7** (2007) 1914–1918.
55. S. Mallidi, T. Larson, J. Aaron, K. Sokolov and S. Emelianov, *Optics Express* **15** (2007) 6583.
56. B. Wang, E. Yantsen, T. Larson, A.B. Karpiouk, S. Sethuraman, J.L. Su, K. Sokolov and S.Y. Emelianov, *Nano Lett.* **9** (2009) 2212–2217.
57. S. Mallidi, T. Larson, J. Tam, P.P. Joshi, A. Karpiouk, K. Sokolov and S. Emelianov, *Nano Lett.* **9** (2009) 2825–2831.
58. L.R. Hirsch, R.J. Stafford, J.A. Bankson, S.R. Sershen, B. Rivera, R.E. Price, J.D. Hazle, N.J. Halas and J.L. West, *PNAS* **100** (2003) 13549–13554.
59. P.K. Jain, K. Seok Lee, I.H. El-Sayed and M.A. El-Sayed, *Phys. Chem. B* **110** (2006) 7238–7248.
60. N. Harris, M. J. Ford and M.B. Cortie, *Phys. Chem. B* **110** (2006) 10701–10707.
61. K-S. Lee and M.A. El-Sayed, *Phys. Chem. B* **109** (2005) 20331–20338.
62. K. Lance Kelly, E. Coronado, L.L. Zhao and G.C. Schatz, *Phys. Chem. B* **107** (2003) 668–677.
63. A. Gaiduk, M. Yorulmaz, P.V. Ruijgrok and M. Orrit, *Science* **330** (2010) 353–355.
64. S. Berciaud, D. Lasne, G.A. Blab, L. Cognet and B. Lounis, *Phys. Rev. B* **73** (2006) 045424.
65. G. Baffou, R. Quidant and F.J.G. de Abajo, *ACS Nano* **4** (2010) 709–716.
66. G. Baffou, M.P. Kreuzer, F. Kulzer and R. Quidant, *Opt. Express* **17** (2009) 3291–2398.
67. G. Baffou, C. Girard and R. Quidant, *Phys. Rev. Lett.* **104** (2010) 136805.
68. D. Patrick O'Neal, L.R. Hirsch, N.J. Halas, J.D. Paynea and J.L. West, *Cancer Letters* **209** (2004) 171–176.
69. C. Loo, A. Lowery, N. Halas, J. West and R. Drezek, *Nano Lett.* **5** (2005) 709–711.
70. http://www.nanospectra.com/ (Accessed 10 May 2011).
71. S. Eunice Lee, G. Logan Liu, F. Kim and L.P. Lee, *Nano Lett.* **9** (2009) 562–570.
72. R. Huschka, O. Neumann, A. Barhoumi and N.J. Halas, *Nano Lett.* **10** (2010) 4117–4122.

Chapter 11
Gold Nanoparticles for Sensors and Drug Delivery

Christian Villiers

Institut Albert Bonniot Centre de recherche INSERM et Université
Joseph Fourier, U823 BP170, La Tronche, 38042 Grenoble cedex 9,
France. Email: christian.villiers@ujf-grenoble.fr

11.1 Introduction

The use of gold nanoparticles in biology has mainly been developed in
the last decade and benefits from the extraordinary development of nanotechnologies. Beyond the hype regarding the use of nano-materials at the
industrial level, the use of gold in biology has a peculiar status: it is not only
a well-known biocompatible material but gold nanoparticles (AuNPs) also
open the way to new diagnoses and new therapies that could not be envisioned by traditional therapeutic technologies. The objective of this chapter
is to highlight the importance of gold at the nano scale in biology and to
show that applications of such materials are emerging in many fields such as
real-time optical diagnoses, label-free detection, cellular tracking, tumour
treatments and drug delivery. We will first stress the importance of particle
functionalization to equip them with selected bio-molecular moieties and
specify their action in the biological medium. We will then discuss their
applications for diagnosis, targeting and, finally, medical treatment.

11.2 Gold Nanoparticle Surface Functionalization
for Biological Applications

Gold nanoparticles (AuNPs) are seldom used as bare nano-objects in biology. Indeed, surface functionalization, also termed surface coating, is essential to give specific properties to the AuNPs. Before discussing these properties in the next sections, it is crucial to pay attention to some potential side effects caused by surface functionalization.

Aggregation. The surface coating of AuNPs can affect the stability of the colloidal suspension and lead to aggregation and sometimes precipitation of gold. This effect must be absolutely avoided when particles are injected in the body for diagnosis or treatment. From a different perspective, AuNP aggregation may be the desired effect when they are used in some biomolecule detection schemes or protein quantification using *in vitro* techniques because these methods are based on particle/particle interactions. Such interaction modifies their optical response or affects other physical characteristics.

Elimination. After having fulfilled their mission, AuNPs need to be eliminated from the body and this elimination strongly depends on their surface coating. Requirements concerning their elimination rate from circulating fluids may vary according to the objective of the injection: it may be fast for diagnosis and slow for treatment (tumour imaging versus tumour treatment for example).

Stealth. It is very important that AuNPs remain invisible to the immune system. In these conditions, they are not treated as a foreign element, and no immune reaction will be triggered after the particle injection.

Targeting. Particles must specifically recognize their target, particularly in the case of treatment against tumours. This specific target recognition is ensured by proper surface functionalization.

Requirements may differ according to the use of AuNPs, and will be satisfied by modifying the surface, form and size of gold nanoparticles. At this stage, it is worth noting that AuNPs interact naturally with proteins when they are incubated in the presence of serum.[1] Although the formation of such a protein shell improves the stability of the colloidal gold suspension, it also has several drawbacks: proteins do not bind uniformly and hence some of them can interact with the AuNPs in a proportion not related to their amount in the serum; furthermore, the affinity of these proteins for the gold surface is weak and some exchange may occur when the AuNPs move in the body, meaning that the shell surrounding the AuNPs will not have a stable composition, which is unsuitable in the case of medical treatment. Moreover, whereas the affinity of free serum proteins for cell receptors is most often very weak, the concentration of a few identical proteins on the same particle can induce sufficient avidity to allow their interaction with the receptors present at the surface of many circulating cells. Such interaction may trigger adverse effects like cell activation or particle internalisation. The chemical composition, size, shape and surface characteristics of nanoparticles affect their binding to proteins, which may impact on the interaction of AuNPs with cells and tissues.[2] It has also been shown that such protein/particle interactions can induce protein unfolding, which may modify their capacity to bind with receptors and therefore their internalization and elimination by circulating cells.[3] In summary, to prevent their aggregation and to allow specific binding to target cells or receptors, it is very important to control the capacity of the nanoparticles to interact with their environment and the best way to achieve this is to cover gold nanoparticles with protective molecules which prevent further interaction with irrelevant proteins.

Surface modifications are usually realized in two steps: the first step is the binding of a molecule (the linker) to gold and will be discussed in Section 11.2.1, followed by the binding of the molecule of interest (spacer, targeting, coating etc.), as explained in Section 11.2.2.

11.2.1 *Surface modification of AuNPs*

As explained above, the purpose of surface modification is to form a protective layer that must help the AuNP fulfil its mission and be stable throughout the AuNP activity.

It has been shown that thiol (-SH) radicals bind to gold nanoparticles[4,5] with relatively high affinity (126–146 KJ/mol);[6] the nature of this link is not clear, but is often considered as nearly covalent because of the high electronegativity of gold (2.4 on the Pauling scale). The multiplication of

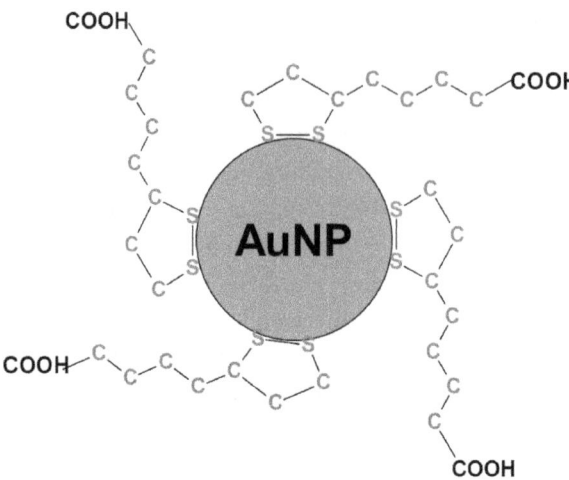

Fig. 11.1. Schematic representation of the interaction of dihydrolipoic acid with AuNPs. The good stability of the complex formed between AuNPs and the coating molecules is due to the presence of two thiol groups per molecule both of which interact with surface gold atoms.

anchors using high sulfur content molecules such as dihydrolipoic acid[7] increases the efficiency of grafting, and seriously limits the desorption of the coating (Fig. 11.1). This kind of bond is widely used to attach a large variety of molecules (biological components or linkers) to the AuNP surface. For example, gadolinium chelates were fixed to AuNP using this method via dithiolated derivatives of diethylenetriaminepenta acetic acid.[8] Indeed, most of the molecules used to create a protective monolayer around the AuNP are modified to bear a thiol moiety.

These modifications may be used to control particle reactivity and to induce either hydrophobicity or hydrophilicity to direct them to lipidic or aqueous areas respectively. By choosing the surfactant molecules, it is possible to adjust the surface properties of the particles. Many protocols have been developed for application in aqueous mediums and in these conditions SH bearing molecules possess carboxylic groups that give stability to the particles. Such molecules are particularly interesting because they cover the particles with stabilizing negative charges; furthermore, these groups can be used to attach other molecules of biological interest.

It is also possible to coat particles with a silica shell which prevents further interaction of other molecules or proteins with the gold surface. In this case, proteins interact with the silica shell. Poly-(ethylene-glycol) (PEG) is also often used to coat particles as this compound reduces non-specific adsorption of proteins and provides greater AuNPs stability by preventing interactions between particles by steric hindrance.[9]

11.2.2 *Functionalization of AuNPs with targeting molecules*

Very often, the use of gold nanoparticles in biology requires grafting molecules of interest onto their surface and most of the graftings are based on the use of the thiol end-group. Such strongly interacting groups may however substitute or complement the shell of stabilizing molecules. Peptides and oligonucleotides have been attached with such an approach. The functionalized nanoparticles can then be arranged according to the number of bound molecules with the help of techniques like metal ion affinity chromatography, which allows the selection of homo-functionalized nanoparticles.[10] When direct binding of the molecule of interest on gold is not possible, an alternative approach relies on linking this molecule to the shell formed by the stabilizing surfactant. The most common protocol consists of the formation of a covalent link between an amino group present on the biological molecule and a carboxylic group present on the shell, using a chemical reagent such as EDC (N-Ethyl-N′-(3-dimethylaminopropyl)-carbodiimide) as shown Fig. 11.2.[11]

In these conditions, virtually all molecules can be attached to gold nanoparticles. Even if this chemical reaction is relatively well understood, problems remain in adjusting the optimal conditions of the reaction and precisely characterizing the final product. In fact, two aspects have to be controlled when this technique is used: first, bridging between particles must be avoided because it would lead to the formation of aggregates; and second, concerning the homogeneity of the binding, as it is important to produce substituted particles as homogenous as possible. Moreover, it is sometimes crucial to be able to assess the relative amount of bound molecules on each gold particle,[12] though to date a method for controlling the number of

303

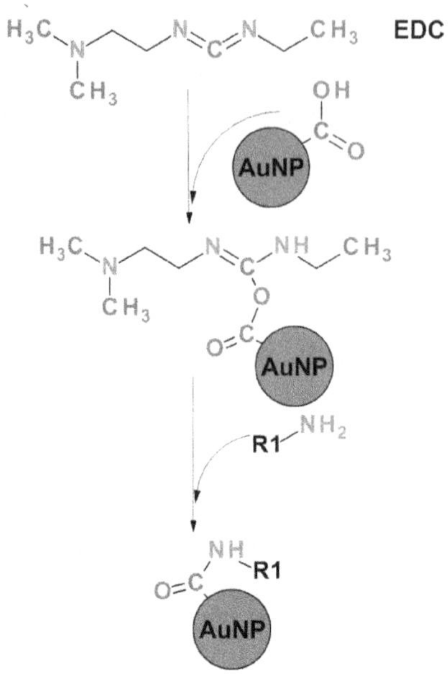

Fig. 11.2. Schematic representation of EDC reaction with a carboxyl group on AuNPs. This reaction is often used to attach proteins to AuNPs taking advantage of the presence of a carboxylic moiety in the stabilizing shell of the AuNP. EDC (N-Ethyl-N'-(3-dimethylaminopropyl)-carbodiimide) reacts with the carboxyl group on AuNPs and forms an O-acylisourea intermediate. This intermediate then reacts with the amine on another molecule (R1). This final reaction leads to the formation of an amine bond between the AuNPs and the molecule R1.

molecules bound to one AuNP has yet to be published. Even measuring this relative surface coverage of proteins is not straightforward.

11.2.3 *Biocompatibility*

Nanoparticles biocompatibility must be considered separately from their toxicity: this is addressed in Chapter 12. When particles are injected, it is important to avoid: (1) their recognition by immune cells as a foreign element; (2) their interaction with serum proteins which may lead to complement activation or blood coagulation; and (3) their detection by any cellular receptors and especially by phagocyte receptors which may induce their internalization by circulating cells like macrophages.

As far as biocompatibility is concerned, particles must be considered as a whole (core and shell). Even if AuNPs are often considered as non-cytotoxic[13] they may induce modification of the cell's functional activities.[14] As the shell is in direct contact with serum proteins and cells, this aspect is certainly the decisive factor for biocompatibility. By itself, the metal core may induce inflammation if there is a partial solubilization and release of ions or if an oxydo-reduction process occurs, but this is not the case for bulk metallic gold. Gold toxicity of AuNPs could be revealed if the shell is unstable, after its destruction by enzymes following endosomal internalisation for example, or in the case of accumulation in cells leading to steric hindrance.

11.3 Gold Nanoparticles for Diagnosis

The primary sense of diagnosis is the identification of a disease through evaluation of relevant elements: these elements can be biological molecules (proteins, hormones, mRNA) characterized or quantified with various measurement techniques or they can consist of cellular alterations subject to direct visualization. By extension, diagnosis is also used for the characterization of various elements such as proteins, or hormones (for the detection of pregnancy for example), or for *in vivo* intra-cellular or intra-corporal localization of molecules or cells respectively in non-pathological situations. For *in vivo* analysis, the techniques used for diagnosis must be non-destructive and non-toxic and AuNPs may be used for such purpose.[15]

11.3.1 *Optical techniques based on the use of gold nanoparticles*

Metallic nanoparticles dispersed in aqueous or biological media exhibit a strong optical contrast that make them easily detectable. This effect is related to the localized surface plasmon resonance (LSPR) of gold nanoparticles described in Chapter 3. When applying the coating strategies described above for AuNPs, it is possible to target specific regions in the body and localize a protein or a region of interest: the accumulation of particles in this region induces a strong optical contrast. Observation using an optical microscope in specific modes such as phase contrast or differential interference contrast makes it possible to visualize AuNPs with diameters above 30 nm by direct observation. Furthermore, as the scattered wavelength (colour) and scattering cross section (brilliance) varies with the shape and size of particles respectively, the simultaneous injection of AuNPs of different kinds can be used for labelling with different colours or brilliances. According to the Mie theory, small particles (less than 20 nm) exhibit poor scattering intensity such that only their optical absorption is measured.[16]

When particles are in suspension, their mutual distance is large (>1000 nm) and there is no dipolar interaction between them; their optical response is therefore identical to that of isolated particles. But when this mutual distance becomes less than their diameter, the wavelength of the absorption peak is modified and the particle accumulation can be monitored. This effect was used for the first time by Leuvering[17] for the quantification of biomolecules. This technique, named Sol Particles ImmunoAssay (SPIA), is based on the modification of absorption wavelength due to particle agglutination: polyclonal antibodies are grafted onto gold nanoparticles (50 nm) and the interaction of these antibodies with the different binding sites (epitopes) present on the corresponding antigen induces their aggregation and hence the bringing together of the particles. This method was applied for the detection of various molecules in urine or blood serum such as the hormone HCG (human chorionic gonadotrophin);[18] further developments based on the sensitivity of the plasmon resonance to mutual coupling of AuNPs can be found in Chapter 10.

Another consequence of light absorption is that it induces fluid heating in the gold nanoparticles' immediate environment, which can be measured by two methods: photothermal[19,20] and photoacoustic[21,22] imaging. In the first method, the ability to image these AuNPs results from the modification of the fluid density induced by heating; this leads to a change in the refractive index, visible by differential interference contrast microscopy. In the second method, the imaging results from the expansion of the liquid due to local heating around the AuNPs, which creates a microphone detectable wave.

Detection of AuNPs is also possible using near IR luminescence (800–1200 nm) for monolayer protected cluster (MPC) particles (2 nm).[23]

Gold nanoparticles also interact with electrons, because of the high molecular weight of Au; in consequence, these particles have a very considerable contrasting effect for electron-based technologies and are thus widely used for transmission electron microscopy (TEM).[24]

Furthermore, the ability of AuNPs to induce X-ray scattering is also sufficient to provide significant X-ray imaging.[25] Finally, gold (^{197}Au) captures neutrons very efficiently due to its very large cross section, and the resulting radioactive gold (^{198}Au) may be used as a beta emitter (99%, 0.96 MeV). The half life of this radioactive compound is, however, relatively short (t1/2 = 2.69 days).[26,27]

11.3.1.1 *SPR based techniques*

Surface Plasmon Resonance (SPR) is observed when light excites electrons at the surface of metals, inducing a non-propagative evanescent wave. The first instruments using SPR for the detection of interactions between biological molecules were developed in the 1990s (see Chapter 3 for a short overview on SPR). They generally use the surface of flat crystals covered with a thin layer of gold (50 nm) to detect biomolecular interactions which perturb the evanescent wave and induce a change in the refractive index at the interfacial layer. However, as the field associated with the evanescent light decays exponentially with the distance normal to the surface, SPR can only detect biomolecular interactions occurring near the metal surface (i.e. closer than 100 nm). For a wide range of molecules, the modification of the refractive index is linearly correlated to their molecular weight, such that the SPR signal depends almost exclusively on the mass of the ligands bound to the gold surface. Nevertheless, as the detection limit was shown to be approximately 1 pg (1 picogramme $= 1 \times 10^{-12}$g) per square millimetre, it is extremely challenging to detect small molecules at low concentration: for example, detection of molecules of 1 kDa is very difficult at concentrations below 0.1 to 0.2 μM. The sensitivity of the detection can be improved by an artificial increase of the molecular weight of the molecule; this is achieved by adding a label that interacts with the molecule of interest already bound to the sensor and eventually damps the plasmon waves. Therefore the plasmon signal is switched off if the biorecognition takes place (see Fig. 11.3). He *et al.* have used 12 nm gold nanoparticles, in such manner, to improve the detection of oligonucleotides,[28] reaching a threshold of 10 pM for a 24-mer oligonucleotide[i] (corresponding to a surface density less than

---

[i]Oligonucleotides are short nucleic acid polymers, typically with fifty or fewer bases. Their length is usually denoted by "mer" indicating the number of bases they comprise.

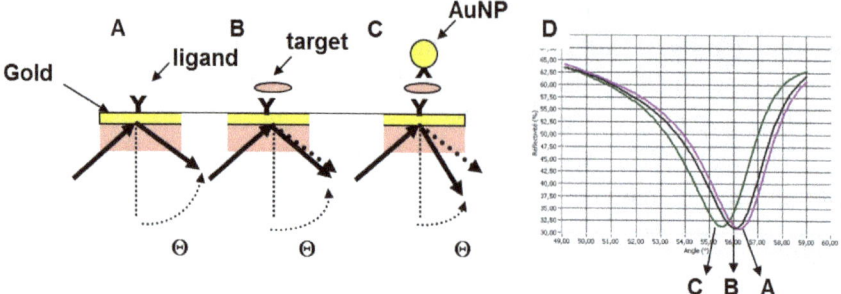

Fig. 11.3. Conventional surface plasmon resonance using a gold film. (a) A thin gold film at the interface between a crystal and a ligand of the molecule of interest. (b) When a small molecule (target) binds to the ligand, there is little change in the deflection angle (Θ) because the modification of the refractive index is too small. (c) The further binding of AuNPs to the target increases the modification of the deflection angle, allowing the detection of the molecule of interest. (d) Comparison of the deflection curves obtained in the three different conditions A, B and C.

8×10^8 molecules per square centimetre) which represents a 1000-fold improvement of the sensitivity compared to standard detection of oligonucleotides. Wang *et al.*[29] have also used gold nanoparticles to increase the sensitivity of the detection by SPR: they use an indirect competition assay to measure very low nucleotide concentration (10 nM) with a complex anti-adenosine aptamer conjugated to gold nanoparticles. However, real-time measurement is the great advantage of SPR technology, which is lost with this approach since detection is realised only after the formation of multi-layers made of the ligand fixed on the gold layer, the molecule of interest and the AuNPs. The same observation can be made in the case of the quartz-crystal balance[30] and micro-cantilevers,[31] for which the increased sensitivity obtained by addition of markers bearing gold nanoparticles is detrimental to real-time measurement.

Another approach is described by Matsui and colleagues[32] who used the molecular imprinted polymers (MIP) technology. They succeeded in increasing the sensitivity of SPR with gold nanoparticles while neither losing the ability of real time measurement nor needing any complex labelling. The molecular imprinted polymers are formed as follows:

- First, the imprinted polymer is formed by polymerization of an appropriate monomer in the presence of the target molecule.

- The target molecule, which is embedded in the matrix, is then extracted, leaving a complementary cavity.
- The imprinted polymer which also contains AuNPs is immobilized on a gold surface of a sensor chip and is ready to use.

Detection begins when the medium containing the target molecule to be quantified is added. The target molecule binds to the complementary cavity, and the local electromagnetic field between the nanoparticles and the gold film is expected to be enhanced by the binding of this molecule, making the sensor chip highly sensitive. This strategy was used to measure a very low concentration of a herbicide (atrazine, molecular weight: 215 Da), the authors being able to detect 5 pM of this very small molecule in acetonitrile.[32]

The association of gold nanoparticles with SPR is doubly interesting: in addition to the damping of the evanescent wave, there may be coupling between the gold layer and the particles. These results prompted Englebienne to go further and replace the gold layer with gold nanoparticles: by functionalizing them with antibodies, the binding of the corresponding antigen causes a change in the extinction spectrum which can be monitored at 600 nm with a conventional spectrophotometer.[33,34] In this case, the main difficulty is the heterogeneity of the particle size which reduces the sensitivity of the measurement. To circumvent this problem, gold nanorods were used as their shape minimizes the impact of the heterogeneity. Moreover, in addition to the conventional extinction peak at 530 nm, they absorb at greater wavelengths which may be in the near infrared range provided the aspect ratio of the nanorod (ratio of length over diameter) is significant. The advantage of measurement in the near infrared is that the absorption due to serum proteins or blood components is highly reduced and the sensitivity of the particles to the local refractive index changes is greater.[35]

11.3.1.2 *Gold nanoparticles and fluorescence*

Fluorescence is commonly used in biology for protein detection, cell or tissue characterization, etc. As shown for cyanine dyes[36] or quantum dots,[37] most fluorescent compounds are quenched in the vicinity of gold nanoparticles, and this effect can be used for sensing strategies. Thus, detection and quantification of molecules can be realised using competition assays based on quenching suppression: the molecule to be detected is linked to a fluorescent probe. If it is attached to the AuNP, the distance between the nanoparticle and the fluorophore is reduced and fluorescence is quenched. Before the detection steps, AuNPs are first functionalized with a ligand or an antibody which possesses a good affinity for the molecule. A molecule analogue to the target molecule and equipped with a different fluorophore is then used to specifically bind to the functionalized AuNP. Detection can now take place. As long as this analogue remains fixed near the NP, its associated fluorescence is quenched. When the target molecule is added to the solution, it competes with the analogue for the binding site, inducing the analogue's release and consequently the suppression of the fluorescence quenching. The higher the target molecule concentration, the more important the competition, and consequently, the stronger the fluorescent signal.[38]

Fluorescence quenching can also be used for protease activity assay: Ray and colleagues[39] used a fluorescent polynucleotide fragment (Cy3-labelled nucleic acid) covalently linked to gold nanoparticles; initially, the fluorescence is quenched, but, after addition of both a complementary DNA fragment and a nuclease, the double stranded DNA is cleaved, which releases the dye and suppresses the quenching. Nuclease activity can be monitored in real time through the augmentation of fluorescence; indeed, the rate of fluorescence increase is directly linked to the enzyme activity.

As the quenching of fluorescence by gold nanoparticles is strongly related to the distance between the probe and the particles,[39] this feature has been used to detect molecular interactions: If the binding of one molecule to another leads to a modification of the distance between a fluorescent dye and a gold particle, the interaction can be followed in real time by fluorescence recording. Dubertret and colleagues[40] described a hybrid material composed of (1) a single stranded DNA molecule, (2) a 1.4 nm gold nanoparticle and (3) a fluorescent dye that is highly quenched by gold: the DNA molecule forms a hairpin which brings together the nanoparticle and the dye inducing the quenching of the fluorescence. The addition of complementary DNA impairs hairpin formation, the consequence of which is an increase of the fluorescence by a factor of several thousand. The amount of complementary DNA is directly related to the quantity of measured fluorescence.

In addition to the quenching of fluorescence by metal nanoparticles, there are also recent findings on metal enhanced fluorescence[41] (see Chapter 10).

Recently, Martini *et al.*[42] have shown that AuNPs can be used to enhance the fluorescence yield of nanoparticles: indeed, fluorescent probes embedded in silica particles are often used for the study of intracellular or intracorporeal circulation. In order to increase the sensitivity, the amount of dye incorporated within one silica particle should be high. However, this method is limited because the increase of the concentration of fluorescent materials in the particles leads to a self-quenching of the luminescence. Nevertheless the incorporation of AuNPs as a central core in the silica shell together with the fluorescent dyes suppresses almost entirely the phenomenon of fluorescence quenching: such an architecture exhibits a quantum yield as high as 80% compared to isolated fluorescein for an interdye distance of 3 nm in such particles whereas it is less than 15% in the absence of gold.

11.3.1.3 *Examples of applications*

11.3.1.3.1 Detection of biological molecules by lateral diffusion

Many devices using gold nanoparticles have been developed for the detection of proteins, hormones and pesticides. The most popular system is based on a membrane sheet (made of nitrocellulose or of PVDF: polyvinylidene di-fluoride) where the solution to be analyzed undergoes lateral diffusion. The detection is associated with immune reaction with specific antibodies linked to nanoparticles (Fig. 11.4). When the solution containing target molecules is deposited on the membrane sheet, antibodies migrate with the mobile phase and bind their antigen in solution, which prevents them from binding to a test line where the same antigen is covalently fixed to the membrane. If there is no antigen in the solution, the binding sites of the antibody remain free to interact with the test line, and the resulting accumulation of nanoparticles induces the development of a coloured line: the increase of the amount of antigen in the medium induces a decrease in the amount of antibodies bonded to the test line and consequently a diminution of the coloration. These tests were used to characterize small molecules like hormones, pesticides or drugs.[43,44] In order to improve the sensitivity of the test, particles were directly linked to antigens instead of antibodies; in this case, there is competition between the

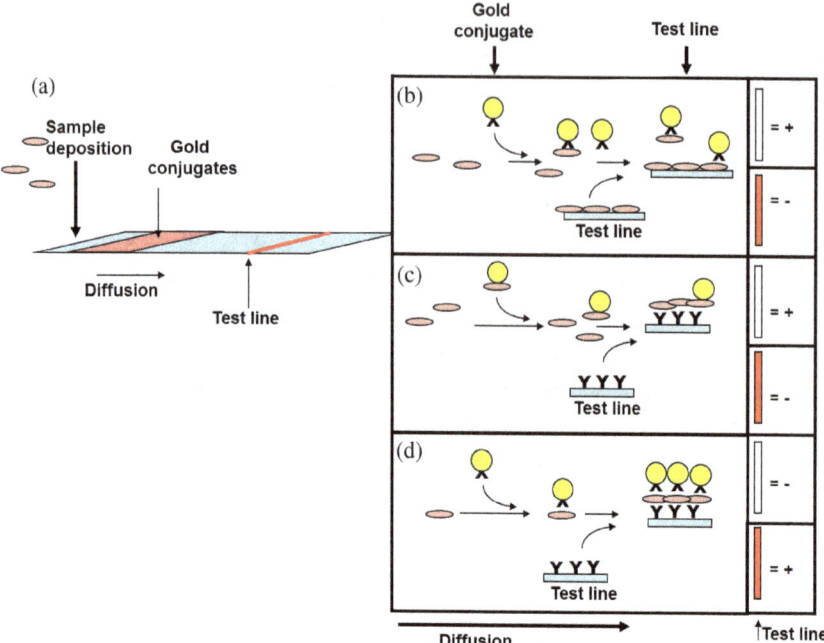

Fig. 11.4. Detection of biological molecules using lateral diffusion of AuNPs. (a) Principle: After sample deposition, lateral diffusion has two effects: first, the molecules of interest bind to the gold conjugates; then when reaching the test line, they bind to the ligand that was deposited on this line. The diffusion is performed on a nitrocellulose membrane. (b) Example of antibodies bound to AuNPs: in this case there is competition between the molecules in the sample and the same molecule on the test line for the binding of the antibodies. (c) Example of molecules of interest bound to AuNPs: there is competition between the molecules in the sample and the same molecules bearing AuNPs for the binding to antibodies fixed on the test line. (d) Example of antibodies bound to both the AuNPs and the test line: the molecules of interest in the sample react with the antibodies bound to AuNPs first and then with a second antibody fixed on the test line. These two antibodies are bound to the molecules of interest on two different and non-competitive binding sites. As indicated on the right side of the figure, the presence of the molecule to be analysed in the sample leads to the staining (C) or not (A–B) of the test line by gold nanoparticles.

antigens fixed to the particles and the same molecules free in the medium (the one to be quantified), for the binding to the antibodies on the test line. By adjusting the number of molecules fixed to the nanoparticles, the authors were able to improve the sensitivity compared to the traditional device.[45]

For the detection of large proteins like prostate antigen against which at least two different monoclonal antibodies are commercially available, a direct assay is used: one antibody is linked to the nanoparticles and the other one is fixed on the test line: in this case, the target molecules bind to both antibodies bearing nanoparticles and those fixed to the membrane leading to the formation of a coloured line, whereas, in the absence of antigen, there is no binding and no visible line.[46]

Development of molecular biology prompted several laboratories to build devices based on the same principle as those used for protein quantification but suitable for the rapid detection of polynucleotides. It is possible to attach a small oligonucleotide (A) to the membrane (test line) and another one (B) to gold nanoparticles. In theory, a third oligonucleotide containing complementary sequences for both oligonucleotides (A and B) should lead to the accumulation of gold on the test line;[47] unfortunately, the technique used to make copies of the nucleotidic sequence to be detected is based on polymerase chain reaction (PCR); this method used to rapidly produce many copies of a fragment of DNA for diagnostic or research purposes, generates a double stranded polynucleotide which blocks the reaction, making it necessary to perform an asymmetric PCR much more difficult to optimize (Fig. 11.5).

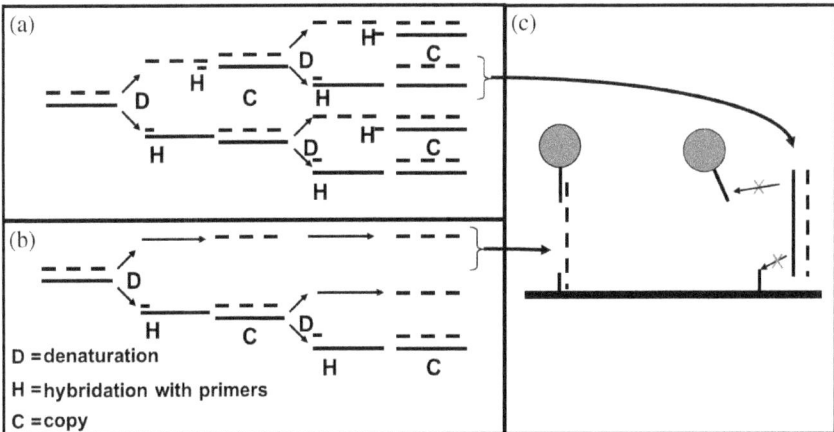

Fig. 11.5. Use of AuNPs for the dosage of oligonucleotides. The amount of oligonucleotides is usually increased by polymerase chain reaction (PCR). Normal PCR leads to the formation of double stranded polynucleotides (a) whereas asymmetric PCR leads to single stranded polynucleotides (b). Detection of polynucleotides by two complementary sequences is possible only with single stranded and not with double stranded polynucleotides (c).

For these reasons, lateral flow devices are not used for the measurement of oligonucleotides, as a number of technical problems are still unresolved. Another option is to incorporate a flag in the amplified oligonucleotide, and to use anti-flag antibodies fixed to both the test line and the nanoparticles.[48] The appeal of such a method is limited due to the small number of flags available, allowing only a few oligonucleotides to be detected in the same device.

The sensitivity of these methods may be increased by binding an enzyme to the gold nanoparticles: using HRP, He *et al.*[49] were able to lower the detection limit of nucleotides to a concentration of 0.01 pM, that is 1,000 times lower than in prior works, without modification of the instrumentation for reading.

These devices have many applications: they are compact, portable, easy to use by non-specialists without special material, and can be kept at room temperature as they are freeze-dried; furthermore, they are not expensive.

11.3.1.3.2 Gold nanoparticles and bio-barcodes

The goal of bio-barcodes is to identify, in one experiment, very low concentrations of different serums or cellular soluble proteins.

The bio-barcodes use a cascade of reactions for (1) specific detection, (2) transcription and (3) amplification of the signal. The first step is the recognition of the target protein: this is realized by specific antibodies bound to magnetic beads, which are retained by a permanent magnet, while unbound proteins are washed away. For the second step, gold nanoparticles bearing both specific antibodies and oligonucleotide sequences were added; the antibodies allow the specific binding of the gold nanoparticles to the proteins retained by the magnetic particles. After removal by washing of unbound material, bound gold nanoparticles are eluted and then further retained by an oligonucleotide fixed to the chip and which is complementary to the sequence fixed on the gold nanoparticles. This corresponds to the transcription stage of the detection: the binding, which is at the beginning specific for the protein, is transformed into a binding specific for an oligonucleotide. The last step is a silver amplification (Fig. 11.6) of the signal associated with the particles: Ag(I) reacts with the gold nanoparticle surface in the presence of reducing agents

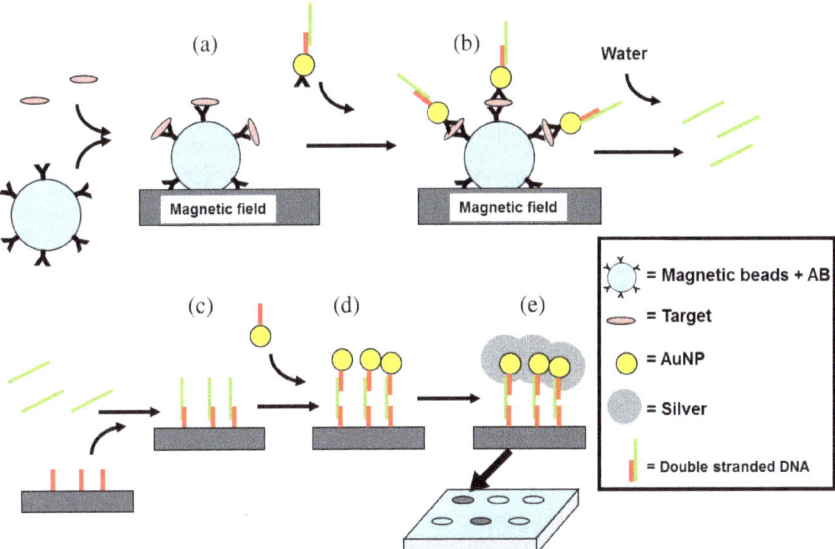

Fig. 11.6. Use of AuNPs for a bio-barcode. (a) The sample containing the molecule of interest is mixed with magnetic beads bearing antibodies specific for this molecule, and the formed complex is retained on a magnet. (b) AuNPs are added to the mixture; they are covered by two types of molecules: first, a double stranded polynucleotide, and second, an antibody able to fix the molecule of interest but using a binding site different and non-competitive with the one used by the first antibody. (c) Then, the polynucleotide which is released by addition of water migrates and may interact with another polynucleotide with a complementary sequence and fixed to a chip. (d) Gold nanoparticles are added; they are covered with a small polynucleotide also complementary to the first one. (e) Detection is then amplified by silver deposition. Multiple assays are possible using an array where various polynucleotides (one per molecule to be analyzed) are spotted on a chip.

such as hydroquinone, resulting in nanoparticle-promoted deposition of silver on the particles.[50] This method increases gold nanoparticle size and induces the augmentation of the signal by over five orders of magnitude. Simultaneous detection of different proteins is achieved thanks to a distribution on a two-dimensional array of oligonucleotides which are complementary to those attached to the gold nanoparticles, each oligonucleotide sequence corresponding to one specific antibody. Since the early approach in 2003,[51] different modifications of the protocol allow the proposal of disposable chips with a colorimetric reading[52] or in association to an evanescent wave fluorescent biosensor.[53] Nair *et al.*[54] have shown that in theory the limit of such detection is at the sub atto-molar level.

11.3.1.3.3 Gold nanoparticles and fingerprint identification

Detection of fingerprints must be performed on various substrates and should lead to the clearest result with a good contrast between the ridge pattern and the background. Researchers have shown that one such strategy is to immerse the substrate in colloidal gold at low pH; in these operating conditions, particles bound to the print. The signal may be further amplified by catalytic deposition of other metals.[55] Stauffer and colleagues have simplified the staining by using gold for both labelling and amplification,[56] whereas others have amplified the signal by multi-metal deposition[57] or luminescence.[58] These different techniques were recently reviewed by Becue.[59]

11.3.1.3.4 Gold nanoparticles and amperometric detection

Quantitative analysis of analytes by electric current is a very sensitive method. The detection is based on the modification of conductivity between two electrodes after the molecules of interest are bound. The use of gold nanoparticles as a conductance amplifier for the quantification of proteins was first published by Velev *et al.*[60] A full automatic platform was developed by Molecular Circuitry and the method was applied to detect nucleic acids.[61] The major problem with this technology is that the presence of nanoparticles between the electrodes (due to their binding to the target molecule) is not sufficient to amplify the electrical current; it requires several cycles of silver deposition for sufficient signal amplification (Fig. 11.7).

More recently Diessel[62] has shown that it is possible to reduce this coating by performing a real-time monitoring of the resistance; however, the gold nanoparticles were too far from each other on the chip and were surrounded by insulating biological molecules such that direct electrical measurement was not possible without silver amplification. Kim and his colleagues have introduced a conducting polymer (polyaniline) into the medium after immobilization of the gold nanoparticle conjugated antibodies on the target protein; in this case, the signal modification is 4.7 times faster when compared with plain gold and the maximum was 2.3 fold higher than that obtained using a photometric system under the same analytical conditions (Fig. 11.8).[63]

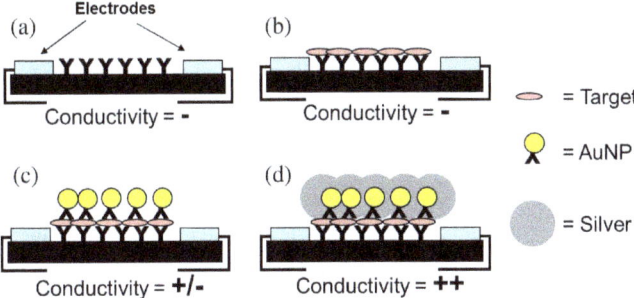

Fig. 11.7. Amperometric biosensors based on the use of AuNPs. (a) Antibodies specific of the molecule of interest are deposited between gold electrodes. In these conditions, the conductivity is very low. (b) The target molecule contained in the sample medium binds to the antibodies without modification of the conductivity. (c) AuNPs bearing a second anti-target antibody are then added, and bound to the chip: the presence of AuNPs between the electrodes is generally not sufficient to increase the conductivity. (d) Deposition of silver on the gold nanoparticles increases the conductivity, leading to a signal which is related to the presence of the target molecule.

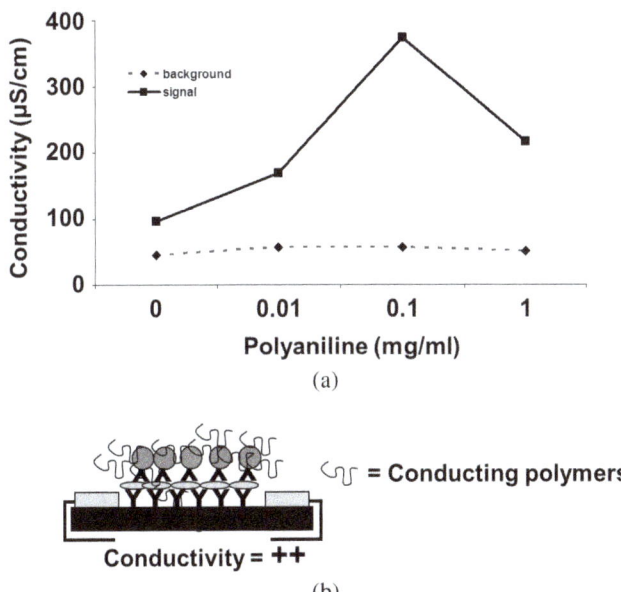

Fig. 11.8. Enhanced electron transfer in the presence of conducting polymers. Electron transfer is enhanced when a conducting polymer (polyaniline) is added to the medium; however an excess of polymer induces AuNP aggregation and thus causes an adverse effect (a). The polymer may enhance the conduction by forming a bridge between adjacent AuNPs leading to increased signal (b). Adapted from Kim *et al.*[63]

Gold can also be used as an electron carrier: the idea being to transport the electrons generated from Redox enzymatic reactions to electrodes via gold nanoparticles; measurement of the resulting current gives information on the presence and amount of the enzyme substrate. In these devices, the enzyme is directly conjugated to the surface of the particles which are fixed to the electrode of the chip. The enzyme binds preferentially to the particle and not to the electrode surface because the available surface is larger allowing a greater amount of enzyme to be accommodated; furthermore, it seems that the shape of the small particles facilitates a close contact between the enzyme and the conductive surface and thus the electronic exchanges.

11.3.2 *Gold nanoparticles as a cellular tracker*

The identification of specific cells in a complex mixture or the localization of intracellular compartments such as endosomes, lysosomes or mitochondria in cells is a permanent challenge in many biological fields. The use of electron microscopy is probably one of the best-known illustrations of intracellular protein localization based on AuNPs. AuNPs are providing a powerful contrast enhancement for transmission electron microscopy (TEM). Antibodies are covalently attached to AuNPs so that they can be easily traced. The interaction of these antibodies with cell proteins is investigated: the antibody can target a protein localized in a cellular or intracellular structure and due to the presence of the AuNP these events can be monitored by TEM. In the absence of such NP labelling, it is very difficult to characterize cellular compartments such as endosomes, lysosomes, Golgi apparatus, etc. By electron microscopy due either to very low contrast or to the absence of specific morphological characteristics. Cell or tissue slices are fixed on a support (cover slide or culture dish) and then permeabilized allowing particles to pass through the cellular membrane in order to interact specifically with the antigen against which the antibody is directed.

Gold nanoparticles provide an excellent contrast for observation using a transmission electron microscope, whereas it is much more difficult to observe the same nanoparticles with an optical microscopy due to the resolution limit which is theoretically a few hundredths of a nm. Recently, however, Klein *et al.* showed that the detection of particles of 60 nm is reliable, succeeding in counting such nanoparticles with a confocal microscope.

Furthermore, AuNPs observed in back-scattering mode induce a phase shift in the light signal, due to the cellular environment which thereby facilitates their observation.[64] Of course, cell labelling is also possible without permeabilization; in these conditions, nanoparticles do not enter the cells and can only interact with components localized on the outer face of the cellular membrane. The optical limitations mentioned above remain after permeabilization but this method has the advantage of monitoring proteins or structures in real-time since the cellular integrity is not altered, allowing dynamic analysis. For example, if an AuNP is attached to an antibody that targets a membrane protein, it will be fixed with high affinity to this membrane and will follow its movements in the cell. Therefore, if the nanoparticle is found inside the cell after incubation, it means that the protein has been internalized by the cell and in these conditions; a precise localization may be performed using TEM.[65] Dynamic analysis of the intracellular movements of the protein can be simply carried out by modifying the incubation time. In theory, these techniques combined with the small size of AuNPs, make it possible to monitor all the endocytic processes: phagocytosis, pinocytosis, receptor mediated endocytosis etc. But attention must be paid to the necessary absence of impact of this tracker on endocytosis: nanoparticles should neither perturb the process nor bind to proteins which could influence their internalization. Indeed, it is worth noting that despite their small size, nanoparticles could interfere with the internalization of receptors.[66] AuNPs may be a good alternative to fluorochromes for cellular imaging by confocal microscopy: fluorescent probes are often used because of their ease of use, but are very unstable due to photobleaching, whereas AuNPs are very stable.

Nanoparticle visualization is also possible by photo-acoustic imaging: when the distance between particles decreases, the wavelength of the localized plasmon resonance shifts to a higher value. When the beads are equally distributed at the cell surface, the distance between the particles is too great to induce interaction between the surface plasmons: in these conditions, when they are illuminated at a wavelength above the plasmon resonance frequency, no signal is recorded. Whereas, when antibody binding occurs on proteins which are aggregated at the cell surface, the gold nanoparticles are close to each other and their resulting interaction leads to light absorption and a photo-acoustic signal can be measured. Consequently, the gathering of receptors (capping) at the cell surface, which may be the result of the binding of their ligands, can be monitored by the variation of photo-acoustic signals.[21]

11.4 AuNPs Used for Treatment

As indicated in the first part of this chapter, many molecules can be bound to AuNPs, and this feature is widely exploited for medical treatment. In this case, AuNPs can be considered as a vector for a precise delivery of molecules at the required place. If immunization is the goal, antigens must be targeted to areas of high immunological activity; if tumour treatment is sought, active molecules must exclusively concentrate at the level of malignant cells in order first to facilitate their eradication and second to reduce adverse effects which may result from the presence of these very harmful molecules near non malignant cells. The delivery of such molecules can be realised using gold nanoparticles as carriers. For this purpose, they are fixed with a stable link to the particles that drive them into the area of interest and, at this stage, the molecule can be released by breaking the bond that links them to the particle either by proteolysis or by hydrolysis. These different possibilities are discussed in the following paragraphs.

11.4.1 *Gene gun*

The principle of the gene gun is to send gold nanoparticles with sufficient kinetic energy so that they penetrate inside the cell membrane, and carry into the cells the molecules fixed on their surface or on the shell. The main interest of this technique is to deliver molecules into the cells without receptor limitation and whatever the nature of the cells. Particle acceleration is obtained using various devices such as bullets, gas cartridges or electric discharges.[67] The gene gun is used for the introduction of coding DNA into the cellular cytoplasm. One of the most important applications of the gene gun is the introduction of plasmid in plants, the use of such material being particularly suitable for crossing highly resistant cellulosic walls found around plant cells.[68]

   The gene gun is also used to inject DNA into animal cells.[69] Indeed, skin is a highly immune-reactive tissue containing abundant antigen-presenting cells[70] and, consequently, it provides a favourable site for DNA immuniza-tion. Such immunization was obtained after injection of plasmide coding for the antigen alone or together with immune activators.[71]

11.4.2 *AuNPs for targeting cells*

The purpose of cell targeting is to precisely deliver a molecule to an organ or a group of cells and simultaneously avoid all aspecific interactions with non-targeted cells or tissues. Indeed, all cells can ingest nanoparticles but the amount of material found inside the cells may vary dramatically accord-ing to the cell type and the internalization pathway involved: pinocytosis, receptor mediated endocytosis, phagocytosis, etc. In the case of particle targeting, the interaction with the cells must be as specific as possible. The amount of material internalized by pinocytosis and phagocytosis is directly related to two factors: the concentration of the particles and their stealth, i.e. their capacity to be invisible in particular for immune cells. The biocompatibility of the particles has been previously documented in this chapter: when AuNPs are invisible to the immune system, the phagocytosis by macrophages or dendritic cells is reduced, and the particles remain in the circulating fluids allowing their routing to and interaction with the targeted cells. Specific interaction with targeted cells is much more important than aspecific endocytosis. Specific binding of AuNPs to membrane proteins expressed by the targeted cells ensures a great specificity but such inter-action does not systematically lead to the internalisation of the particles. For example, some membrane proteins induce an intracellular signaliza-tion after binding of their ligand but no internalization as is the case for toll-like receptors (the proteins in charge of the detection of foreign mate-rials; viral ARN non-human glycoproteins, etc). Whereas other proteins are internalized upon binding of their ligand, as is the case for transferring receptors for example. There is no possibility of inducing the internalization of a protein that normally does not have this capacity. However, according to the nature of tumour treatment performed using AuNPs, internalization is not always required: drug delivery may require internalization whereas hyperthermia is effective without internalization. The challenge is to target AuNPs to receptors present only on tumour cells, but the main problem is that only a few proteins are expressed exclusively by these cells and most of

Table 11.1. Cell surface proteins which may be used as docking receptors for particle targeting.*

Tumour or tissue	Targeted protein/receptor	References
Pancreatic adenocarcinoma	Cell surface plectin-1	Kelly et al.[72]
Tumour-associated lymphatic vessels	P32 or gC1q receptor	Fogal et al.[73]
Neuroblastoma	Aminopeptidase N (CD13)	Pastorino et al.[74]
Osteosarcoma	Interleukin-11	Lewis et al.[75]
Breast or head and neck tumours	CD44	Platt et al.[76]
Endothelium (liver cancer)	VEGF	Cheng et al.[77]
Endithelium (colon carcinoma)	MMP	Kondo et al.[78]
Endothelium (lung carcinoma)	VCAM	Gosk et al. [79]
Endotheliums (solid tumour) (pancreatic melanoma)	$\alpha v \beta 3$ and $\alpha V \beta 5$	Benezra et al.[80] Liu et al.[81]
Endothelium (solid tumour)	fibronectin	Nilsson et al.[82]
Endothelium (solid tumour)	Endosialin	Christian et al.[83]
Endothelium (solid tumour)	TEMs	Carson-Walter et al.[84]
Endothelium (solid tumour)	Annexin I	Oh et al.[85]
Endothelium (solid tumour)	Nucleolin	Christian et al.[86]

*The list of cell proteins which may be used as potential docking for AuNPs is not exhaustive and is adapted from the reviews of Ruoslahti et al.[87]

these are not accessible from the outside of the cells without permeabilization. Another possibility is to target proteins present on the endothelial cells forming the neovascularization of the tumour as many of them have already been characterized, some of the different possibilities are summarized in Table 11.1.

Some proteins, such as aminopetidase N[88] or endosialin[83] for example, were found to be over-expressed by tumours. In these cases, antibodies specific to the over-expressed proteins are attached to the nanoparticles, allowing them to target tumours.[89] Another possibility is to drive the AuNPs to structures related to tumours: for example, integrin $\alpha v \beta 3$ and $\alpha v \beta 5$ are highly expressed by endothelial tumours and simultaneously they are highly over-synthesised by the endothelial cells of the vessels irrigating malignant tumours.[90] These integrins exhibit a high affinity for peptides that contain the sequence of amino acid Arg-Gly-Asp (RGD), a property which is used for particle targeting: AuNPs grafted with a peptide containing this motif are rapidly found to associate with the cells or the vessels of the tumour.[91]

11.4.3 *AuNPs as carriers for tumour treatment*

Binding particles to membrane proteins at the surface of tumour cells does not systematically induce their internalization, but even in the case of endocytosis, the presence of particles in endosomes or lysosomes is not sufficient to induce cell death. Indeed, as documented previously, the toxicity of gold nanoparticles is too low to induce cell death *per se*. In order to kill the tumour, the particles have to be linked to active molecules. Two strategies are possible: either these molecules are able to correct the abnormality of the tumour by gene therapy, or they are cytotoxic and specifically eliminate tumour cells. In order to exercise their activity, particles must exit from the intracellular compartment to the cytosol or directly cross the extracellular membrane; for this, they can be coated by peptides like Tat that disrupt the membrane and allow their passage into the cytosol.[92]

For gene therapy, it has been shown that small interfering RNA (siRNA) could be bound to particles. Shim *et al.*[93] first attached siRNA to a polymer (polyethyleimine) via a pH sensitive link,[94] which is itself fixed to the gold nanoparticles. Once in the intracellular compartments, the link between siRNA and the polymer breaks, due to the environmental pH which is slightly acid. Then, the released siRNA can act and block the expression of the corresponding protein. RNA is also released in the vicinity of the tumour cells, indeed, it has been shown that the pH is slightly acid inside the tumour, and then the siRNA enters the cells and plays its role.

Particles can also be used for carrying anti-tumour drugs. These are attached either at the AuNP surface or to structures connected to the particles. Molecules are then transported into the cell together with the particles. Under these conditions, the accumulation is much greater and much faster than with molecules injected alone without the targeting element. Curnis *et al.*[95] have attached the tumour necrosis-factor (TNF) to the targeting peptide RGD on the AuNPs and injected the complex into animals, in combination with a chemotherapeutic treatment: this TNF strategy yields a very good anti-tumour effect. This cytokine is very toxic and usually difficult to use in systemic injection, but in this case, the targeting of the molecule contents itself with very low concentration (sub-nanograms), avoiding adverse effects after injection. Such treatment is now in phase 1 for the treatment of human cancer:[96] phase 1 corresponds to preliminary tests used to determine safety and absence of side effects of a drug injected usually into healthy volunteers. In this example, the use of gold nanoparticles provides no additional effect. However their optical properties open the way to simultaneous detection and localization of AuNPs.[97] Patra *et al.*[98] indicate that gold nanoparticles are particularly suitable for pancreatic cancer because of their unique physico-chemical properties, such as ultra small size, large surface area to mass ratio, high surface reactivity, presence of surface plasmon resonance (SPR) bands, biocompatibility and ease of surface functionalization. The use of AuNPs targeted to the epidermal growth factor receptor (EGFR) that is over-expressed on these tumours, leads to increased efficacy of traditional chemotherapeutics.

Kang *et al.*[99] use gold nanoparticles to induce DNA damage in tumour cells: two peptides are fixed on the particles, one containing the RGD sequence for tumour targeting, the second corresponding to a nuclear localization signal which induces the traffic of the particle to the nucleus: this particle localization leads first to the blockage of cell division and second to cell apoptosis. This work shows that nanoparticles localized at the cell nucleus can specifically affect cellular functions.

11.4.4 *AuNPs and hyperthermia for tumour treatment*

As indicated previously in this chapter, light at the wavelength corresponding to the plasmon resonance is absorbed by the AuNPs leading to the excitation of free electrons in the gold structure; the damping of these electron oscillations within the nanoparticles is accompanied by a thermal dissipation of energy and leads to a temperature increase within the nanoparticles and their environment (see Chapter 4). This phenomenon can be used for photothermal imaging as well as tumour treatment because of the impact of small temperature changes on cell viability.

11.4.4.1 *Hyperthermia treatment*

Cells are very sensitive to temperature variation and die above 42°C, the normal value for human cells being 37°C. This fact can be exploited to use localized hyperthermia to specifically kill tumour cells. To achieve this objective, there are two prerequisites: particle accumulation in the tumour cells and illumination of the nanoparticles with sufficient energy to induce heating (Fig. 11.9). As discussed previously, in theory, AuNPs can target cancer cells by encapsulating the particles with molecules which have a high affinity for the proteins expressed only by the tumour. This leads to particle accumulation in the tumour area due to their binding to the membrane with or without the ensuing internalization; in both cases, the consequence of the light absorption will be the same. As indicated previously, the absorption resulting from plasmon resonance occurs at a wavelength which varies according to the size,[100] shape[101] and structure[102] of the particles: the optimal parameters were found to be around 40 nm diameter for spherical gold nanoparticles, between 20 and 70 nm length for nanorods and 50 to 100 nm total diameter for core-shell structures where the gold outer shell

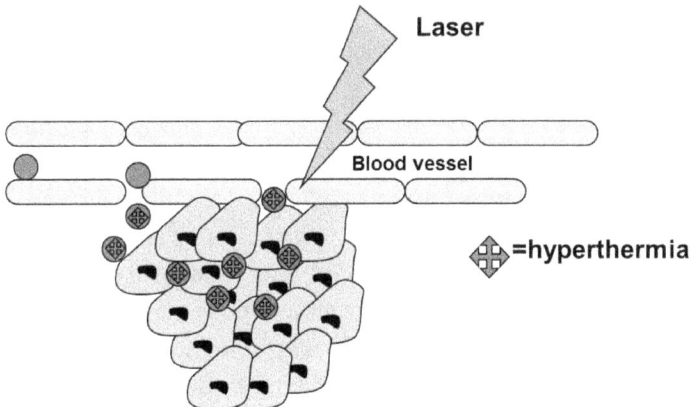

Fig. 11.9. Treatment of a tumour by hyperthermia using AuNPs. AuNPs bearing a tumour specific binding motif are injected; due to the increased permeability of the vessels surrounding the tumour, AuNPs interact with abnormal cells. Laser illumination of the particles induces the production of thermal energy sufficient to kill surrounding cells.

has a thickness of 7–10 nm. In these cases, the maximum absorption is measured at 800 nm. In conclusion, by modifying the physical characteristics of the nanoparticles, it is possible to modify the wavelength corresponding to the maximum light energy absorption (see Chapter 3). This property is very important because the main limitation of photothermal effects is the light absorption by animal tissues: a very small fraction of the light energy reaches the particles depending on the depth of the tumour. Tissue absorption varies according to the wavelength and is minimal in the infrared spectrum though, even in these conditions, the absorption remains fairly high. This means that such treatment can only be achieved for tumours close to the skin in the case of external illumination. For deep tumours, this technology is difficult to apply even using IR wavelengths; we may, however, imagine that in this case it would be possible to use intra-body laser illumination. The penetration of particles into a tumour is often facilitated by the so-called enhanced permeability and retention (EPR) effect observed in the vessels around tumours, but this permeability varies from tumour to tumour.[103]

11.4.4.2 Drug delivery by photo-induced heating

The temperature increase induced by illumination of gold nanoparticles can also be used to release drugs transported to the tumour area by AuNPs (Fig. 11.10).

Firstly, this technique can be used to release both proteins and polynucleotides attached to particles. Kogan et al. have shown that aggregated proteins at the particle surface may be solubilized by elevation of the temperature.[104] Several observations open new perspectives. For example, Stehr and colleagues have shown that AuNPs can be used to rapidly increase the temperature and liberate a polynucleotide;[105] to date, this approach has not yet been used for tumour treatment, but it has been shown, for example, that small hairpin RNA (shRNA) delivered near a tumour may control its proliferation.[106] On the other hand, it is well known that double stranded polynucleotide is dissociated by increasing the temperature; indeed, this technique is used for polynucleotide amplification by PCR. We can imagine that the combination of these two strategies (link disruption by heating and shRNA effect) may allow the delivery and release of polynucleotide by AuNPs for the treatment of tumours.

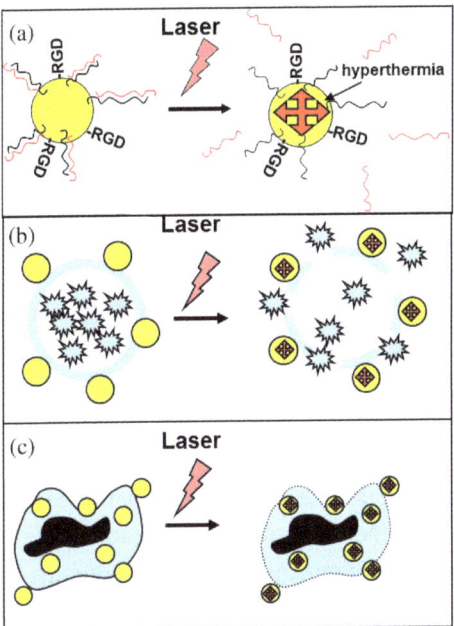

Fig. 11.10. Clinical use of hyperthermia induced by AuNPs. (a) Polynucleotide delivering in a specific place: aunps bearing both a targeting motif (for example a RGD sequence) and a double stranded polynucleotide are injected. The particles, which accumulate in the region of interest, are illuminated by a laser. The hyperthermia in this case is sufficient to liberate the polynucleotide but too low to kill the cells. (b) Drug delivery: drugs in containers bearing AuNPs with targeting motifs at their surface. Illumination of AuNPs induces hyperthermia leading to the rupture of the container and the liberation of the drug. (c) Cell killing: AuNPs bearing targeting motifs are injected. Their accumulation around the cells may be followed by endocytosis or not; in both cases, light illumination induces the production of heat energy; the number of laser pulses is determined to increase the temperature to a lethal level.

Second, the photo-induced heating can be used to liberate molecules from containers: the principle being to confine the molecule of interest in a container which is itself fixed to the AuNPs; the resulting hyperthermia induces the rupture of the wall of the container, and release of the molecules.[107] It can be performed outside or inside the cell after internalization:[108] a low number of laser pulses allows local heating of the particles and rupture of the container without any toxic effect on the cell, and cell apoptosis can be avoided: the thermal energy generated in these conditions is too low to have an impact on cell viability. The consequences of such a release range from specific inhibition of molecules or cell functions to cell labelling.[109]

11.5 AuNPs in the Future

The number of techniques based on the use of gold nanoparticles for diagnosis and tumour treatment is considerable and continues to increase. Of course, in some cases, gold could be replaced by other materials, however, contrary to many other metals, AuNPs are easy to manufacture, the particles are stable and most of the analysis shows that they are non-toxic. Because of its optical properties, gold remains the metal with the largest potential; there are many techniques based on the analysis of plasmon resonance modifications upon interaction of the particles with proteins, peptides or with other gold nanoparticles. For these experiments, modification of particle size, shape or structure induce changes in the optimum wavelength for plasmon resonance and allows the optimization of their use in relation to the operating conditions. Because of the large surface of the particles compared to their volume, there is a high number of binding sites for molecules of interest (cellular targeting, cell treatment) and for molecules inducing the particle stealth. The future of these technologies may be the use of particles as a platform for multipurpose: cellular targeting, specific delivery of molecules, hyperthermia, imagery, etc. Moreover, the structure of the particle may evolve to combine the advantages of various metals Si/Gd or Fe/Gd. The toxicity of such circulating complexes remains an important issue and needs complete and detailed analysis; it has been recently proposed to decrease the potential toxicity of nanoparticles by incorporation of proteins or peptides at their surface which may control the activation of the complement system,[110] for example, which may be a solution to reduce potential inflammation processes resulting from the injection of these materials.

All of these results indicate clearly that the use of AuNPs benefits from the optical and chemical characteristics which are specific to this metal, and we can assume that for both diagnosis and treatment, new applications will be developed to increase the sensitivity, the selectivity and the efficiency of these tools.

References

1. S.H. Lacerda, J.J. Park, C. Meus, D. Pristinski, M.L. Becker, A. Karim and J.F. Douglas, *ACS Nano* **4** (2010) 365.

2. Z.J. Deng, M. Liang, M. Monteiro, I. Toth and R.F. Minchin, *Nat. Nanotechnol.* **6** (2011) 39.

3. Z.J. Deng, M. Liang, M. Monteiro, I. Toth and R.F. Minchin, *Nat. Nanotechnol.* **6** (2010) 39.

4. R.G. Nuzzo and D.L. Allara, *J. Am. Chem. Soc.* **105** (1983) 4481.

5. A.C. Templeton, W.P. Wuelfing and R.W. Murray, *Acc. Chem. Res.* **33** (2000) 27.

6. R.G. Nuzzo, F.A. Fusco and D.L. Allara, *J. Am. Chem. Soc.* **109** (1987) 2358.

7. B. Garcia, M. Salome, L. Lemelle, J.L. Bridot, P. Gillet, P. Perriat, S. Roux and O. Tillement, *Chem. Commun. (Camb.)* (2005) 369.

8. C. Alric, J. Taleb, G. Le Duc, C. Mandon, C. Billotey, A. Le Meur-Herland, T. Brochard, F. Vocanson, M. Janier, P. Perriat, S. Roux and O. Tillement, *J. Am. Chem. Soc.* **130** (2008) 5908.

9. A.G. Kanaras, F.S. Kamounah, K. Schaumburg, C.J. Kiely and M. Brust, *Chem. Commun. (Camb.)* (2002) 2294.

10. R. Levy, Z. Wang, L. Duchesne, R.C. Doty, A.I. Cooper, M. Brust and D.G. Fernig, *Chembiochem.* **7** (2006) 592.

11. G.A. Craig, P.J. Allen and M.D. Mason, *Methods Mol. Biol.* **624** (2010) 177.

12. T. Pellegrino, R.A. Sperling, A.P. Alivisatos and W.J. Parak, *J. Biomed. Biotechnol.* **2007** (2007) 26796.

13. E.E. Connor, J. Mwamuka, A. Gole, C.J. Murphy and M.D. Wyatt, *Small* **1** (2005) 325.

14. C.L. Villiers, H. Fritas, R. Couderc, M.-B. Villiers and P. Marche, *J. Nanopart. Res.* **12** (2010) 55.

15. R. Wilson, *Chem. Soc. Rev.* **37** (2008) 2028.

16. M.A. van Dijk, A.L. Tchebotareva, M. Orrit, M. Lippitz, S, Berciaud, D. Lasne, L. Cognet and B. Lounis, *Phys. Chem. Chem. Phys.* **8** (2006) 3486.

17. J.H. Leuvering, P.J. Thal, M. van der Waart and A.H. Schuurs, *J. Immunoassay* **1** (1980) 77.

18. J.H. Leuvering, P.J. Thal, M. Van der Waart and A.H. Schuurs, *J. Immunol. Methods* **45** (1981) 183.

19. D. Boyer, P. Tamarat, A. Maali, B. Lounis and M. Orrit, *Science* **297** (2002) 1160.

20. L. Cognet, C. Tardin, D. Boyer, D. Choquet, P. Tamarat and B. Lounis, *Proc. Natl. Acad. Sci. USA* **100** (2003) 11350.

21. S. Mallidi, T. Larson, J. Aaron, K. Sokolov and S. Emelianov, *Opt. Express* **15** (2007) 6583.

22. S. Mallidi, T. Larson, J. Tam, P.P. Joshi, A. Karpiouk, K. Sokolov and S. Emelianov. *Nano Lett.* **9** (2009) 2825.

23. G. Wang, T. Huang, R.W. Murray, L. Menard and R.G. Nuzzo, *J. Am. Chem. Soc.* **127** (2005) 812.

24. J. Roth, *Histochem. Cell. Biol.* **106** (1996) 1.

25. J.F. Hainfeld, D.N. Slatkin, T.M. Focella and H.M. Smilowitz, *Br. J. Radiol.* **79** (2006) 248.

26. R.E. Gosselin, *J. Gen. Physiol.* **39** (1956) 625.

27. M.K. Khan, L.D. Minc, S.S. Nigavekar, M.S. Kariapper, B.M. Nair, M. Schipper, A.C. Cook, W.G. Lesniak and L.P. Balogh, *Nanomedicine* **4** (2008) 57.

28. I. He, D. Musick, S.R. Nicewarner, F.G. Salinas, S.J. Benkovic, M.J. Natan and C.D. Keating, *J. Am. Chem. Soc.* **122** (2000) 9071.

29. J. Wang, A. Munir, Z. Li and H.S. Zhou, *Biosens. Bioelectron.* **25** (2009) 124.
30. X.C. Zhou, S.J. O'Shea and S.F.Y. Li, *Chem. Commun.* (2000) 953.
31. M. Su, S. Li, V.P. Dravid, *Appl. Phys. Lett.* **82** (2003) 3562.
32. J. Matsui, M. Takayose, K. Akamatsu, H. Nawafune, K. Tamaki and N. Sugimoto *Analyst* **134** (2009) 80.
33. P. Englebienne *Analyst* **123** (1998) 1599.
34. P. Englebienne, A.V. Van Hoonacker and M. Verhas, *Analyst.* **126** (2001) 1645.
35. M.M. Miller and A.A. Lazarides, *J. Phys. Chem. B* **109** (2005) 21556.
36. R. Chhabra, J. Sharma, H. Wang, S. Zou, S. Lin, H. Yan, S. Lindsay and Y. Liu, *Nanotechnology* **20** (2009) 485201.
37. K.W. Kuo, T.H. Chen, W.T. Kuo, H.Y. Huang, H.Y. Lo and Y.Y. Huang, *J. Nanosci. Nanotechnol.* **10** (2010) 4173.
38. E. Oh, M.Y. Hong, D. Lee, S.H. Nam, H.C. Yoon and H.S. Kim, *J. Am. Chem. Soc.* **127** (2005) 3270.
39. P.C. Ray, A. Fortner and G.K. Darbha, *J. Phys. Chem. B* **110** (2006) 20745.
40. B. Dubertret, M. Calame and A.J. Libchaber, *Nat. Biotechnol.* **19** (2001) 365.
41. J.R. Lakowicz, *Plasmonics* **1** (2006) 5.
42. M. Martini, P. Perriat, M. Montania, R. Pansu, C. Julien, O. Tillement and S. Roux, *J. Phys. Chem. C* **113** (2009) 17669.
43. B.S. Delmulle, S.M. De Saeger, L. Sibanda, I. Barna-Vetro and C.H. Van Peteghem, *J. Agric. Food Chem.* **53** (2005) 3364.
44. G.P. Zhang, X.N. Wang, J.F. Yang, Y.Y. Yang, G.X. Xing, Q.M. Li, D. Zhao, S.J. Chai and J.Q. Guo, *J. Immunol. Methods* **312** (2006) 27.
45. J. Aveyard, P. Nolan and R. Wilson, *Anal. Chem.* **80** (2008) 6001.
46. C. Fernandez-Sanchez, C.J. McNeil, K. Rawson, O. Nilsson and H.Y. Leung, V. Gnanapragasam, *J. Immunol. Methods* **307** (2005) 1.
47. J. Aveyard, M. Mehrabi, A. Cossins, H. Braven and R. Wilson, *Chem. Commun. (Camb.)* (2007) 4251.
48. T. Suzuki, M. Tanaka, S. Otani, S. Matsuura, Y. Sakaguchi, T. Nishimura, A. Ishizaka and N. Hasegawa, *Diagn. Microbiol. Infect. Dis.* **56** (2006) 275.
49. Y. He, S. Zhang, X. Zhang, M. Baloda, A.S. Gurung, H. Xu, X. Zhang and G. Liu, *Biosens. Bioelectron.* **26** (2011) 2018.
50. T.A. Taton, C.A. Mirkin and R.L. Letsinger, *Science* **289** (2000) 1757.
51. J.M. Nam, C.S. Thaxton and C.A. Mirkin, *Science* **301** (2003) 1884.
52. E.D. Goluch, J.M. Nam, D.G. Georganopoulou, T.N. Chiesl, K.A. Shaikh, K.S. Ryu, A.E. Barron, C.A. Mirkin and C. Liu, *Lab. Chip* **6** (2006) 1293.
53. M. Trevisan, M. Schawaller, G. Quapil, E. Souteyrand, Y. Merieux and J.P. Cloarec, *Biosens. Bioelectron.* **26** (2010) 1631.
54. P.R. Nair and M.A. Alam, *Analyst* **135** (2010) 2798.
55. B. Schnetz and P. Margot, *Forensic Sci. Int.* **118** (2001) 21.
56. E. Stauffer, A. Becue, K.V. Singh, K.R. Thampi, C. Champod and P. Margot, *Forensic Sci. Int.* **168** (2007) e5.
57. M. Zhang and H.H. Girault, *Analyst* **134** (2009) 25.
58. A. Becue, A. Scoundrianos, C. Champod and P. Margot, *Forensic Sci. Int.* **179** (2008) 39.

59. A. Becue, S. Moret, C. Champod and P. Margot, *Biotech. Histochem.* (2010).
60. O.D. Velev and E.W. Kaler, *Langmuir* **15** (1999) 3693–3698.
61. S.J. Park, T.A. Taton and C.A. Mirkin, *Science* **295** (2002) 1503.
62. E. Diessel, K. Grothe, H.M. Siebert, B.D. Warner and J. Burmeister, *Biosens. Bioelectron.* **19** (2004) 1229.
63. J.H. Kim, J.H. Cho, G.S. Cha, C.W. Lee, H.B. Kim and S.H. Paek, *Biosens. Bioelectron.* **14** (2000) 907.
64. S. Klein, S. Petersen, U. Taylor, D. Rath and S. Barcikowski, *J. Biomed. Opt.* **15** (2010) 036015.
65. E. Onelli, C. Prescianotto-Baschong, M. Caccianiga and A. Moscatelli, *J. Exp. Bot.* **59** (2008) 3051.
66. S. Bhattacharyya, R. Bhattacharya, S. Curley, M.A. McNiven and P. Mukherjee, *Proc. Natl. Acad. Sci. USA* **107** (2010) 14541.
67. D.-R. Chen, C.H. Wendt and D.Y.H. Pui, *J. Nanopart. Res.* **2** (2000) 133.
68. V.M. Ramesh, S.E. Bingham and A.N. Webber, *Methods Mol. Biol.* **274** (2004) 301.
69. S. Kuriyama, A. Mitoro, H. Tsujinoue, T. Nakatani, H. Yoshiji, T. Tsujimoto, M. Yamazaki and H. Fukui, *Gene Ther.* **7** (2000) 1132.
70. P.W. Lee, S.H. Hsu, J.S. Tsai, F.R. Chen, P.J. Huang, C.J. Ke, Z.X. Liao, C.W. Hsiao, H.J. Lin and H.W. Sung, *Biomaterials* **31** (2010) 2425.
71. R. Weiss, M. Gabler, T. Jacobs, T.W. Gilberger, J. Thalhamer and S. Scheiblhofer, *Vaccine* **28** (2010) 4515.
72. K.A. Kelly, N. Bardeesy, R. Anbazhagan, S. Gurumurthy, J. Berger, H. Alencar, R.A. Depinho, U. Mahmood and R. Weissleder, *PLoS Med.* **5** (2008) e85.
73. V. Fogal, L. Zhang, S. Krajewski and E. Ruoslahti, *Cancer Res.* **68** (2008) 7210.
74. F. Pastorino, C. Brignole, D. Marimpietri, M. Cilli, C. Gambini, D. Ribatti, R. Longhi, T.M. Allen, A. Corti and M. Ponzoni, *Cancer Res.* **63** (2003) 7400.
75. V.O. Lewis, M.G. Ozawa, M.T. Deavers, G. Wang, T. Shintani, W. Arap and R. Pasqualini, *Cancer Res.* **69** (2009) 1995.
76. V.M. Platt and F.C. Szoka, Jr. *Mol. Pharm.* **5** (2008) 474.
77. C. Cheng, H. Wei, J.L. Zhu, C. Chang, H. Cheng, C. Li, S.X. Cheng, X.Z. Zhang and R.X. Zhuo, *Bioconjug. Chem.* **19** (2008) 1194.
78. M. Kondo, T. Asai, Y. Katanasaka, Y. Sadzuka, H. Tsukada, K. Ogino, T. Taki, K. Baba and N. Oku, *Int. J. Cancer* **108** (2004) 301.
79. S. Gosk, T. Moos, C. Gottstein and G. Bendas, *Biochim. Biophys. Acta.* **1778** (2008) 854.
80. M. Benezra, O. Penate-Medina, P.B. Zanzonico, D. Schaer, H. Ow, A. Burns, E. Destanchina, V. Longo, E. Herz, S. Iyer, J. Wolchok, S.M. Larson, U. Wiesner and M.S. Bradbury, *J. Clin. Invest.* **121** (2011) 2768–2780.
81. X. Liu and H.S. Qhattal, *Mol. Pharm.* (2011).
82. F. Nilsson, H. Kosmehl, L. Zardi and D. Neri, *Cancer Res.* **61** (2001) 711.
83. S. Christian, H. Ahorn, A. Koehler, F. Eisenhaber, H.P. Rodi, P. Garin-Chesa, J.E. Park, W.J. Rettig and M.C. Lenter, *J. Biol. Chem.* **276** (2001) 7408.
84. E.B. Carson-Walter, D.N. Watkins, A. Nanda, B. Vogelstein, K.W. Kinzler and B. St Croix, *Cancer Res.* **61** (2001) 6649.

85. P. Oh, Y. Li, J. Yu, E. Durr, K.M. Krasinska, L.A. Carver, J.E. Testa and J.E. Schnitzer, *Nature* **429** (2004) 629.
86. S. Christian, J. Pilch, M.E. Akerman, K. Porkka, P. Laakkonen and E. Ruoslahti, *J. Cell Biol.* **163** (2003) 871.
87. E. Ruoslahti, S.N. Bhatia and M.J. Sailor, *J. Cell Biol.* **188** (2010) 759.
88. R. Pasqualini, E. Koivunen, R. Kain, J. Lahdenranta, M. Sakamoto, A. Stryhn, R.A. Ashmun, L.H. Shapiro, W. Arap and E. Ruoslahti, *Cancer Res.* **60** (2000) 722.
89. P. Cherukuri and S.A. Curley, *Methods Mol. Biol.* **624** (2010) 359.
90. A. Erdreich-Epstein, H. Shimada, S. Groshen, M. Liu, L.S. Metelitsa, K.S. Kim, M.F. Stins, R.C. Seeger and D.L. Durden, *Cancer Res.* **60** (2000) 712.
91. Q.K. Ng, M.K. Sutton, P. Soonsawad, L. Xing, H. Cheng and T. Segura, *Mol. Ther.* **17** (2009) 828.
92. J.M. de la Fuente and C.C. Berry, *Bioconjug. Chem.* **16** (2005) 1176.
93. M.S. Shim, C.S. Kim, Y.C. Ahn, Z. Chen and Y.J. Kwon, *J. Am. Chem. Soc.* **132** (2010) 8316.
94. M.S. Shim and Y.J. Kwon, *Bioconjug. Chem.* **20** (2009) 488.
95. F. Curnis, A. Gasparri, A. Sacchi, R. Longhi and A. Corti, *Cancer Res.* **64** (2004) 565.
96. S.K. Libutti, G.F. Paciotti, A.A. Byrnes, H.R. Alexander Jr, W.E. Gannon, M. Walker, G.D. Seidel, N. Yuldasheva and L. Tamarkin, *Clin. Cancer Res.* **16** (2010) 6139.
97. S. Kumar, N. Harrison, R. Richards-Kortum and K. Sokolov, *Nano Lett.* **7** (2007) 1338.
98. C.R. Patra, R. Bhattacharya, D. Mukhopadhyay and P. Mukherjee, *Adv. Drug Deliv. Rev.* **62** (2010) 346.
99. B. Kang, M.A. Mackey and M.A. El-Sayed, *J. Am. Chem. Soc.* **132** (2010) 1517.
100. S. Link and M.A. El-Sayed, *J. Phys. Chem. B* (1999) 4212–4217.
101. B. Nikoobakht and M.A. El-Sayed, *Chem. Mater.* **15** (2003) 1957.
102. C. Loo, A. Lin, L. Hirsch, M.H. Lee, J. Barton, N. Halas, J. West and R. Drezek, *Technol. Cancer Res. Treat.* **3** (2004) 33.
103. H. Maeda, J. Wu, T. Sawa, Y. Matsumura and K. Hori, *J. Control Release* **65** (2000) 271.
104. M.J. Kogan, N.G. Bastus, R. Amigo, D. Grillo-Bosch, E. Araya, A. Turiel, A. Labarta, E. Giralt and V.F. Puntes, *Nano Lett.* **6** (2006) 110.
105. J. Stehr, C. Hrelescu, R.A. Sperling, G. Raschke, M. Wunderlich, A. Nichtl, D. Heindl, K. Kurzinger, W.J. Parak, T.A. Klar and J. Feldmann, *Nano Lett.* **8** (2008) 619.
106. S.M. Ryou, S. Kim, H.H. Jang, J.H. Kim, J.H. Yeom, M.S. Eom, J. Bae, M.S. Han and K. Lee, *Biochem. Biophys. Res. Commun.* **398** (2010) 542.
107. A.G. Skirtach, C. Dejugnat, D. Braun, A.S. Susha, A.L. Rogach, W.J. Parak, H. Mohwald and G.B. Sukhorukov, *Nano Lett.* **5** (2005) 1371.
108. A.G. Skirtach, J. Munoz, O. Kreft, K. Köhler, A.P. Alberola, H. Möhwald, W.J. Parak and G.B. Sukhorukov, *Angew. Chem. Int. Ed.* **45** (2006) 4612.
109. C.M. Pitsillides, E.K. Joe, X. Wei, R.R. Anderson and C.P. Lin, *Biophys. J.* **84** (2003) 4023.
110. R.B. Sim and R. Wallis, *Nat. Nanotechnol.* **6** (2011) 80–81.

Chapter 12

What About Toxicity and Ecotoxicity of Gold Nanoparticles?

Marie Carrière

Laboratoire Lésions des Acides Nucléiques, Commissariat à l'Energie Atomique (CEA), Université Joseph Fourier UMR_E3, 17 rue des Martyrs, 38054 Grenoble cedex 9 France.
Email: marie.carriere@cen.fr

12.1 Introduction

While production and use of nanoparticles in commercial products increase exponentially, the perception of risk also becomes more acute, as these new substances may generate new adverse effects, both on human health and on the environment. As mentioned in Chapters 10 and 11, gold nanoparticles (AuNPs) are promising tools for diagnostic and therapeutic purposes. Before launching any medical protocol using AuNPs, their innocuousness has to be proven. Gold colloids have been used for years for therapeutic purposes and this safe use suggests that AuNPs should also be safe. However the properties of materials at the nanoscale, i.e. in the 1–100 nm size range, are so different from the properties of the bulk material, that it is reasonable to revisit their toxicological and ecotoxicological impact. During the last decade several research groups have published valuable data proving that AuNPs exert moderate toxic effects on eukaryotic cells, on animal models and on several organisms representing different levels of ecosystems. These toxic effects greatly depend on AuNP size and surface coating, the coating itself often being more harmful than AuNPs *per se*. Note also that most of the data collected to date have been obtained after exposure of the organisms to very high concentrations of AuNPs, which do not reflect a real exposure of humans or a real release in the environment. The present chapter will survey these data and try to assess the risks generated by AuNP known to date.

12.2 Impact of Gold Nanoparticles on Human Health

12.2.1 *The toxicological approach, applied to nanoparticles*

Risk is commonly recognized as the product of hazard (H) and exposure (E). If there is no exposure, then even if the substance is hazardous there will be no risk. If exposure occurs, but the substance is safe, then there will also be no risk. Risk assessment and management are thus only possible if research data concerning these two elements are available. Hazard is an inherent property of the considered substance, while the extent of exposure is dependent of multiple variables and scenarios. There is an increasing literature related to NP hazard assessment, while the extent of data related to exposure is very low. Consequently, even if many advances have been made in the field of NP toxicology in the last two decades, it is today not possible to precisely answer the question of risk related to NP exposure.

Evaluating exposure to NPs is a challenging task. Several exposure scenarios can be ruled out: exposure of workers at their working place, accidental exposure of populations, and intentional exposure of patients for medical purpose. The intentional exposure of patients, for medical purpose, will probably concern AuNPs since they are seen as future therapeutic and diagnostic agents. In this scenario, exposure might be controlled, since the applied dose is known. However, depending on the route of application (intravenous injection, instillation, inhalation, skin deposition, etc.), AuNPs will have to cross different physiological barriers which have very different properties. The effective dose, reaching the target organ, will thus not strictly be the applied dose. For example, dermal penetration of NPs has been extensively studied, and it is now recognized that TiO_2 NPs do not cross an undamaged skin (whether this assumption is also true for AuNPs has not been reported to date). Conversely TiO_2 NPs have been shown to reach the brain when instilled in an animal's nose. Moreover, when directly injected intravenously, NPs diffuse through the whole body and reach various target organs. Each organ receives a particular quantity of NP, which depends both on the organ morphology and physiology, and on NP physico-chemical characteristics. Various degrees of toxic effects would then appear on all the reached organs. Depending on the required therapy, NPs then sometimes have to cross the cellular membrane to reach their final intracellular target. Several routes are available for NPs to reach the intracellular environment, from simple diffusion through cell membrane to cell uptake through specific transporters, if the NP is complexed to a specific ligand. Finally NPs might reach the intracellular compartment by endocytosis, which includes several modes: macropinocytosis; clathrin- or caveolae-mediated endocytosis; and clathrin/caveolae-independent endocytosis (for review, see Khalil *et al.*[1]). Exposure extent thus varies depending on the route of application and on the physico-chemical characteristics of applied NP.

Today, the question of **hazard** of chemical substances is regarded intensely, through various legislations such as the European regulation REACH which deals with Registration, Evaluation, Authorization and restriction of Chemical substances. This regulation, implemented on 1 June 2007, aims at identifying the intrinsic properties of chemical substances. Practically, manufacturers and importers must register their chemical substances, in a volume-triggered system: all substances produced and imported at more than one ton per year have to be registered. For substances produced or imported at more than ten ton per year, manufacturers and importers are also required to gather physicochemical, toxicological and ecotoxicological information, that they have to provide together with a chemical safety report. All this information is centralized on a database, run by the European Chemical Agency. Substances of "very high concern" are considered separately. They have to be registered even if their production or importation is lower than one ton per year. But sufficient toxicological data are needed to classify a substance as of "very high concern". Progressive substitution of the most dangerous substances is demanded, when substitutes are available. This regulation, in its present format, is partially applicable to NPs, since they are, *per se*, a chemical substance. In the present format of the regulation, AuNPs, if produced by more than one ton per year, will be registered and evaluated as a chemical substance and not as a nanoscaled particle, i.e. they will be registered if produced by more than one ton per year. However, toxicological data obtained to date are not sufficient to identify NPs as substances of "very high concern". More importantly, methods currently used for toxicity testing for regulatory purposes are not adapted and not validated for the assessment of NP toxicity, thus, avoiding their identification as substances of "very high concern". A working group was formed in July 2008, in order to study the possibility of adaptation of the present regulation, to improve it in terms of substances at the nanoscale.

Technically, studying the toxicity of materials at the nanoscale is a challenging task, and requires innovative strategies and methodologies. As compared to chemical pollutants, the specific properties of NPs, such as their high specific surface area, render them highly reactive. Specific surface area increases when NP size decreases. Consequently NP reactivity will depend on their size, and the smallest NPs will probably have a different toxicological profile than larger ones. Moreover their high surface reactivity would trigger interaction with various molecules; and among them with pollutants (Fig. 12.1). Indirect toxicity is thus likely to appear in addition to the inherent toxicity of the NP *per se*, if toxic pollutants such as synthesis residues are adsorbed on the surface of NPs.

Adsorbed molecules may modify NP fate/biodistribution in living organisms. NPs are unlikely to cross physiological barriers freely; rather their movement will rather be governed by their size and/or by the identity of functional molecules linked or adsorbed on their surface. In the specific case of AuNPs, their high affinity for thiol-bearing molecules would render them highly affine for some proteins or biologically important molecules.

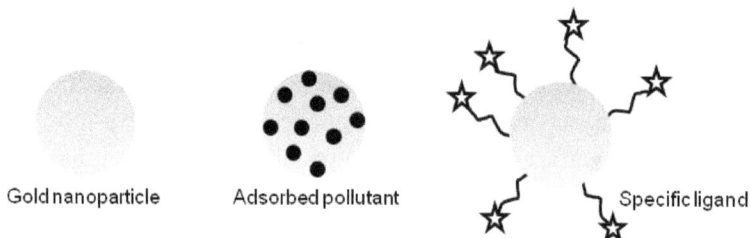

Gold nanoparticle   Adsorbed pollutant   Specific ligand

Fig. 12.1. Surface modification of AuNPs, either by adsorption of a pollutant, or by functionalization with a ligand, specific for a chosen cell membrane transporter.

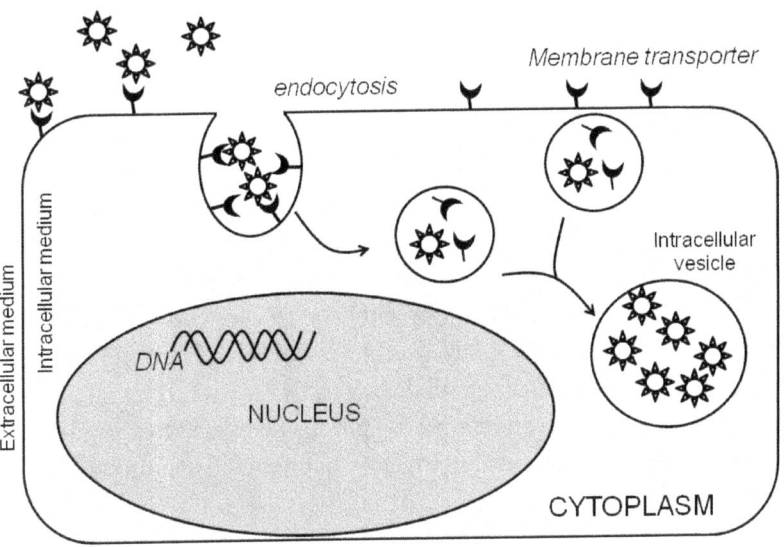

Fig. 12.2. Cell internalization of a AuNPs functionalized with a specific ligand, targeted to a cell membrane transporter.

This affinity is interesting for the construction of nanocomposites for therapeutic purposes; gaining affinity for some specific cell surface transporters would enable cellular internalization (Fig. 12.2).

High affinity of AuNP surface for thiol-bearing ligands may conversely cause problems if bare AuNPs reach physiological fluids or become internalized in cells, because they may then deplete fluids or cell cytoplasm from essential molecules, such as major molecular antioxidant of cells, namely glutathione. For all these reasons it is all the more important to deeply characterize AuNP physico-chemistry (size, porosity, specific surface area, surface charge, presence of synthesis residues on nanoparticle's surface, etc.) before launching any toxicology experiment.

Another point is that due to their high surface reactivity, NPs may adsorb some components of toxicology assay kits. This phenomenon has already been reported for carbon-based NPs such as carbon nanotubes,[2] and may also be true for AuNPs. Moreover, the plasmon resonance band of AuNP sometimes interferes with toxicity assays, which are often based on the measurement of absorbance of colorimetric dyes. For instance the classical MTT assay, used to quantify the activity of mitochondrial enzymes, and thus indirectly cell viability, ends up with the measurement of absorbance at 562 nm, which is in the range of some AuNP plasmon band. Interference between these assays and AuNPs would lead to misinterpretation of toxicological data, and again it is strongly recommended to precisely characterize nanoparticle physico-chemical parameters to properly address the question of their toxicity. It is also strongly recommended to verify toxicity data with at least two or more independent assays in order to have a clear proof of NP toxicity or harmlessness.

Lastly, the guidelines for toxicity testing published by national or international organizations (OECD, ISO, DIN etc.) have been validated and are considered valuable for chemical compounds, not for NPs. Several research consortiums are presently involved in working groups dedicated to the validation of tests for the assessment of NP toxicity, but no validated protocol exists at present.

Concerning the toxicity studies related to AuNPs, on *in vivo* (animal) or *in vitro* (cultured cells) models, inter-comparison of the published data is difficult, since different authors use different biological models, AuNPs with different shapes and diameters, coated with different ligands, or still covered with different synthesis residues. However, general trends can be ruled out, as described thereafter.

12.2.2 Biokinetics and target organs of AuNPs after systemic exposure

Several studies have attempted to identify the target organs and biokinetics of AuNPs. This is generally achieved by systemic injection in rodents. Two major parameters have been shown to play an important role in AuNP or Au nanorods biodistribution and biokinetics: their primary diameter and their surface coating.

Small AuNPs (diameter: 1.9 nm) have been tested by the Hainfeld team, whose first concern was to enhance radiotherapy treatments. After a single injection in mice of AuNPs, in suspension in phosphate buffer saline (PBS), NPs were rapidly cleared through the kidneys, without causing any harmful adverse effects to the animals.[3] Four years after, Semmler-Benke *et al.* showed that the biodistribution pattern of 1.4 nm AuNPs differed from that of 18 nm AuNPs, after intravenous injection in rats at 0.01 μg·g^{-1} of body weight.[4] The smaller NPs were preferentially eliminated via the urinary and hepatobiliary route, and thus recovered in urine and feces. 3.7% of 1.4 nm NPs still circulated in the bloodstream after 24 h. Conversely, within 24 h, almost all 18 nm NPs had been removed from the blood and accumulated in the liver and spleen. The biodistribution of larger AuNPs in rats (10–250 nm) also depend on their size. NPs 10 nm in diameter, in suspension in PBS (where they are agglomerated) distributed in the blood, liver, spleen, kidneys, testis, thymus, heart and lung and brain, 24 h after injection. Conversely, 50–250 nm NPs were majorly accumulated in the liver and spleen. In addition, the smallest NPs (10 nm) were shown to pass through the blood-brain barrier and locate in the brain.[5] This distribution pattern was also observed in mice after systemic injection of 15–200 nm NPs (at $1\,g\cdot kg^{-1}$ of body weight), and 15–50 nm NPs also passed the blood-brain barrier.[6] In the latter study, NPs were suspended in alginate, a biocompatible stabilizer, and thus not agglomerated. Liver, lung and spleen accumulation increased when AuNP diameter increased.

This biodistribution pattern, i.e. preferential storage in liver, spleen and lung, is classical for colloidal materials, due to their uptake by cells of the mononuclear phagocyte system (MPS), namely resident macrophages of the liver (Kupffer cells) and spleen. Strategies to avoid NP uptake by the MPS include their coating with a layer of amphiphilic polymer chains such as polyethylene glycol (PEG). This protective cover renders NPs more hydrophilic, and finally confers them a longer circulation time in the blood which is promising for therapeutic purposes.

PEG grafting to gold nanorods was achieved by Niidome *et al.*, by introducing PEG during the synthesis process.[7] The resulting nanorods had different zeta potentials: 41 mV for the original nanorods and −0.5 mV for the pegylated ones. The authors thus observed that 54% of the coated nanorods still circulated in the bloodstream 30 minutes after injection, whereas most of the original nanorods were immediately detected in the liver. This proportion then decreased with time, in favor of preferential accumulation in the liver, but 5% of the initial dose of pegylated nanorods was still circulating 24 h after injection. The same group published in 2009 another study, where gold nanorods grafted with different quantities of PEG were injected in tumor-bearing mice.[8] It was concluded from this study that a PEG:gold molar ratio of 1.5 was sufficient for AuNPs for MPS avoidance.

Nanoparticles finally accumulated in the liver, and when liver storage capacity was overwhelmed, AuNPs distributed in other tissues and among them in tumors. The authors state that these data would then be interesting in cancer-treatment strategies, although it would certainly have adverse effects on hepatic function.

To conclude, these data tend to demonstrate that systemically injected AuNPs are rather eliminated through the MPS, i.e. in feces, when their diameter is above 15–20 nm. When their diameter is below 15–20 nm, a proportion of the NPs are eliminated through the urinary tract. Moreover, blood retention of PEG-grafted nanorods is longer than that of plain gold nanorods.

12.2.3 *Translocation of gold nanoparticles through physiological barriers*

External exposure to NPs through inhalation, ingestion or contact with the skin, may lead to their internalization and subsequent redistribution in the body. This transfer through physiological barriers is also called translocation. Although translocation is required when AuNPs are intended to be used for therapeutic purposes, it can also cause harmful adverse effects both on the directly exposed organs and on secondary target organs where NPs accumulate after translocation.

In the particular case of AuNPs, they were shown to translocate through an *in vitro* reconstituted air-blood barrier, representing lung epithelium.[9] The model was composed of epithelial cells, macrophages and dendritic cells, and the authors showed that 25 nm gold NPs were accumulated in cells as free particles only, i.e. not agglomerated in subcellular vesicles. AuNP accumulation triggered an inflammatory response in cells, through the release of tumor-necrosis factor-α.[9] *In vivo*, the efficiency of translocation through the air/blood barrier of lungs depended on NP size.[4] The smallest NPs (1.4 nm) were efficiently translocated, whereas 18 nm NPs were almost entirely retained in the lungs after instillation in the trachea of rats. One day after instillation, 8.5% of the 1.4 nm NPs were found in secondary target organs, mainly the liver and kidneys.

Regarding the other exposure routes, translocation of 15 nm, 102 nm and 198 nm AuNPs through the skin and intestine was also reported.[6]

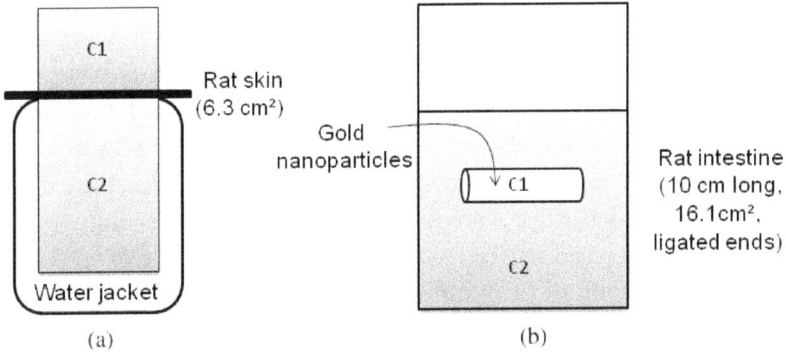

Fig. 12.3. Schematic representation of Franz diffusion cell and intestinal sac method, employed to study NP permeation through skin or intestine. Adapted from Ref. 6.

Two *ex vivo* models were used, namely Franz diffusion cell for permeation through the skin, and the "intestinal sac" method for assessing intestinal permeation (Fig. 12.3).

Permeation was demonstrated to be size dependent; it decreased as NP diameter increased. Permeation through the intestine was higher than permeation through the skin, where a time lag was observed before NPs diffused through the epithelium. In addition, NPs of smaller size penetrated deeper in the skin, whereas larger NPs were retained in the epidermis and dermis regions.[6]

Finally, the blood-retinal physiological barrier has received attention as it may be a route to achieve systemic treatment of eye pathologies, particularly those related to pathological angiogenesis.[10] It has been demonstrated that permeation of AuNPs through this barrier occurs with smaller NPs, namely 20 nm NPs, while 100 nm NPs did not cross the barrier. Transferred NPs were then taken up by neurons, endothelial cells and peri-endothelial glial cells, without causing any adverse effect.[10]

To conclude, the translocation of smallest AuNPs is likely to occur through most of the physiological barriers, including lungs, intestine, skin, blood-brain barrier and blood-retinal barrier. Moreover, some barriers are more permeable than the others, and logically the intestine seems to be the most permeable epithelium. Still, data collected from *in vivo* studies would be necessary to definitively conclude on AuNP translocation ability.

12.2.4 *Cellular toxicity, in vitro studies*

Cellular toxicity of AuNPs has been described to be low, which is why they are attractive candidates for therapeutic applications. Several studies attempted to identify the influence of AuNP physico-chemical properties, such as size, shape, agglomeration state, on the biological response of exposed cells, but failed to provide any clear conclusion. Generally, 15–100 nm AuNPs have been described to cause only a low decrease of cell viability,[11–13] independently of their surface coating. For instance biotin- and citrate-covered AuNPs were shown to display very low toxicity. Conversely, synthesis residues such as CTAB greatly increased cell death.[14] Consequently AuNPs were not inherently toxic to human cells, but surface pollutants possibly caused toxicity. This statement was also true for gold nanorods (also synthesized with CTAB), since washing nanorods, or overcoating them with polymers or polyelectrolytes drastically reduced their cytotoxicity.[7,11,15] Overcoating was presumed to prevent desorption of contaminant molecules from nanorods.[11] Conversely, Pernodet *et al.* showed that 13 nm citrate-coated AuNPs were taken up by human dermal fibroblasts from primary cultures, which in consequence showed impaired spreading, division and changes in cell morphology.[16] These modifications were attributed to a modification of actin structure, and observed only after long term exposure (four or six days). These discrepancies may be explained by the various sensitivities of different cell lines, but also by NP agglomeration state, which is not precisely described in the above-mentioned studies.

Concerning the effect of NP size on toxicity, poor correlation was found, except that a distinct behavior was noticed for 1.4 nm Au_{55} nanoclusters. These AuNPs have been shown to interact in a unique manner with DNA: due to their specific size, these clusters have been shown to fit into the major groove of DNA where they are irreversibly linked (Fig. 12.4).[17]

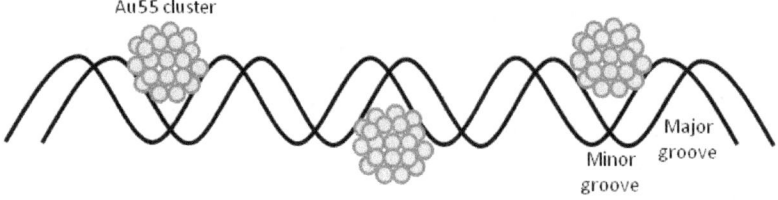

Fig. 12.4. Au_{55} cluster inserted in the major groove of DNA.

Whether this interaction with DNA and their cellular impact are related or not, these nanoclusters have been shown to be highly toxic, *in vitro*, to eleven cancerous or normal cell lines. The authors assumed that cytotoxicity was not due to contaminant molecules. Moreover 42.5% of the gold was detected in cell nuclei, and 21% of this proportion was shown to be bound to DNA.[18]

AuNPs are generally described to be taken up by cells, *in vitro*. Their uptake does not necessarily correlate with cytotoxicity since even the poorly toxic AuNPs tested by Connor *et al.*[14] were internalized and localized in intracytoplasmic vesicles. Nevertheless, AuNP internalization was shown to depend upon their size,[12] with 50 nm NPs being more efficiently internalized than their smaller or larger counterparts. The uptake mechanism is suspected to be through interaction of AuNP surface with serum proteins, which in turn are actively transported through cell membrane.[12]

12.3 Environmental Impact of AuNPs

12.3.1 *What can make NPs toxic for the environment?*

NPs are at present produced by the ton and used in many commercial products, which suggests that they will certainly be released into the environment. As discussed in the previous section dealing with human toxicology, the risk (R) is the potential that a substance causes harm. R is a function of two elements: the extent of exposure (E) and the type of harm, or hazard (H).[19] Information related to NP concentration in the environment is almost entirely absent, and consequently data concerning exposure (E) are cruelly lacking. This is due to the absence of a valuable and precise analytical method which would enable correct quantification of traces of NPs in the environment. Every part of the environment already contains high concentrations of nanosized matter; it is thus not possible to distinguish this background from traces of the so-called "engineered NPs". Today, the only way to access exposure data is modeling, and the reader must keep in mind that the parameters used for modeling are often very uncertain. For instance, estimations of worldwide NP production are not clear, and the behavior of NPs in the environment is poorly known, particularly regarding their potential dissolution. So far, two recent studies predict by modeling, TiO_2-, ZnO-,

AgNPs and carbon nanotubes concentrations in the different parts of the environment.[20,21] In all parts, NP concentration is predicted to exponentially increase in the coming years.[20] For example the concentration of AgNPs in U.S. sediments are predicted to rise from 0.7 to 2.2 μg·kg^{-1} between 2008 and 2012.[20] In the same conditions, the concentration of TiO$_2$NPs would rise from 0.2 to 0.6 mg·kg^{-1}. Together with TiO$_2$NPs, AgNPs are among the most produced NPs in the world, and thus certainly among the most released, hence the large amount of data on these NPs. Studies have not been conducted for AuNPs, which are produced in smaller quantities, but may follow the same dissemination trend. However, it is considered likely that AuNP agglomeration and/or mobilization in environmental conditions will occur (Fig. 12.5).

AuNPs agglomerate when ionic strength increases: it is immediate when ionic strength reaches 0.1 M.[22] Conversely, when naturally occurring organic substances such as humic substances are added, the colloidal stability of AuNP suspension increases, i.e. NP agglomeration is hindered.[22] In conclusion, NP fate in the environment, and consequently NP exposure, is greatly governed by agglomeration-sedimentation and/or disagglomeration-mobilization processes, themselves governed by changes of pH, ionic strength and the presence of organic substances.

Concerning data related to AuNP hazard (H), most of the studies published to date report their potential impact on living species, from unicellular

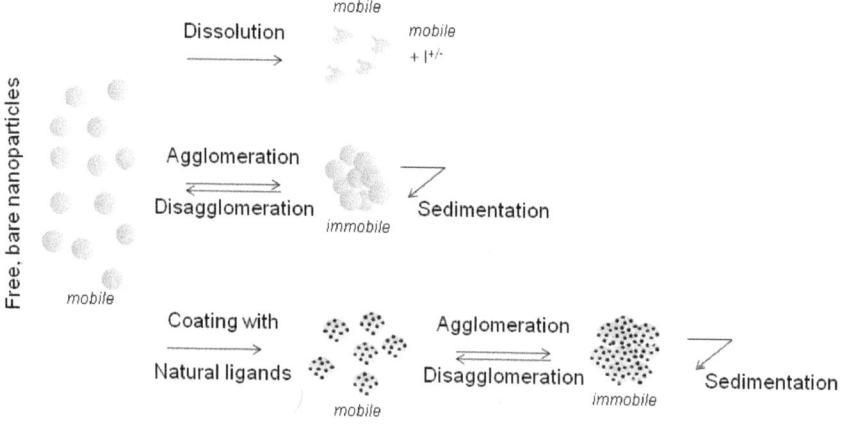

Fig. 12.5. Gold nanoparticles' fate in the environment.

organisms to animal and vegetal species. As already discussed in the human toxicology section, noxious NP effects may be direct, i.e. derived from the particle itself, but can also be indirect since NPs may vectorize pollutants by adsorbing them on their surface. Most of the studies related to AuNP toxicity to species of the environment have been, up until now, done on fresh water species, using standardized assays validated for regulatory toxicology, for instance OECD, ISO and DIN guidelines (for assays based on algae: ISO 8692, OECD 201, DIN 38412-33; for assays on daphnids: ISO 6341, OECD 202, DIN 38412-30). As underlined previously, these protocols are not yet validated for nanotoxicology applications, and some of the results described herein should be considered with caution.

12.3.2 *Impact of AuNPs on unicellular organisms: bacteria and algae*

Bacteria and algae constitute the lowest level of the ecosystems. They may therefore be a route of entrance of NPs to the food chain. Moreover, some NPs like AgNPs show efficient bactericidal and antifungal effects which have justified their use in first aid bandages, soap, self-cleaning textiles but also in consumers products such as washing machines or fridges. Due to these properties, they might also be lethal for some environmental bacteria, leading to ecotoxicological issues upon release in the environment. These properties have been attributed to various mechanisms, and among them the release of toxic silver ions, the impairment of proper regulation of transport through bacterial membranes due to NP adsorption, or interaction with membrane proteins or with DNA, which may induce bacterial cell death (for review, see Ref. 23). Considering these effects of AgNPs, a common view is that they may be extrapolated to AuNPs. However, most of the studies published to date instead show that AuNPs are quite inert for environmental species.

Firstly, some articles gave indirect evidence that AuNPs did not cause adverse effects when interacting with bacteria. Lin *et al.*[24] reported that mannose-coated AuNPs (6 nm diameter, ~200 mannose residues per nanoparticle) efficiently bound to *E. coli* type I pili, through specific interaction of mannose with FimH adhesin of bacterial pili. These NPs would then be used as efficient labeling probe. The authors did not mention any toxicity.

AuNPs have been used in several studies for photothermal killing of bacteria. Bare AuNPs are always shown to be non-toxic, but the surface coating that is added to NP to trigger their interaction with bacterial cell wall sometimes renders them slightly toxic. It should be mentioned here that the surface charge of AuNPs, dependent on their synthesis procedure, does not always allow their interaction with bacterial cell wall. Negatively charged AuNPs, for instance those synthesized using citrate as a reducing agent, do not adhere to the negatively charged bacterial cell wall (Fig. 12.6a). Conversely, when synthesizing AuNPs with $NaBH_4$ and cysteamine, AuNPs are positively charged and strongly adsorb onto bacterial cell wall (Fig. 12.6b).

Exposure of *Staphylococcus aureus* bacteria to AuNP-anti-protein A antibody conjugates (NP diameter was 10–40 nm) leads to efficient bounding of nanoconjugates on bacterial surface.[25] Then irradiation with focused laser pulses induces over-heating of AuNPs. This heating causes bubbling on bacterial surfaces, which physically damages them leading to their complete disintegration at the higher laser energies. Without any laser irradiation, the gold nanoconjugates cause a slight increase in bacterial death,

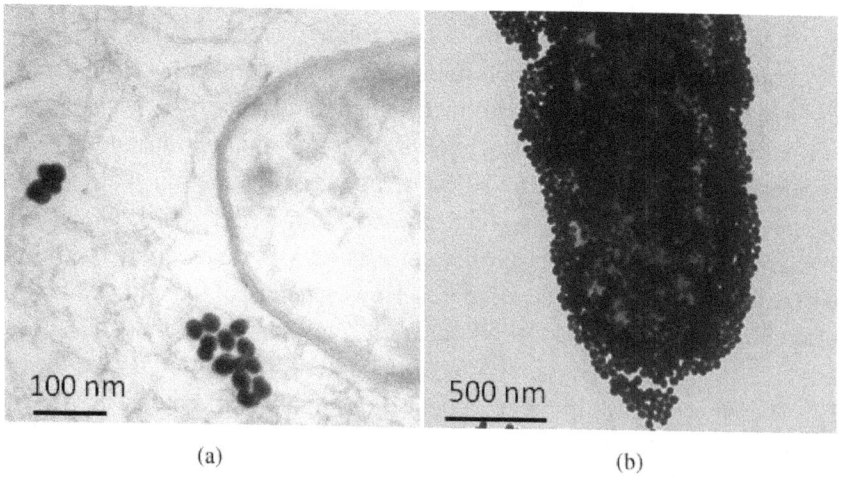

(a)   (b)

Fig. 12.6. Interaction between bacterial cell wall and AuNPs, depending on their surface charge. AuNPs have been synthesized either with citrate (A) or with NaBH4 and cysteamine (B), resulting in negatively charged (A) or positively charged (B) NP surfaces. This image was kindly provided by C. Sicard-Roselli, LPC, univ. Paris Sud, France.

whereas bare AuNPs are non toxic.[25] In another report, polygonal, 30 nm, vancomycin-coated AuNPs, were shown to induce the death of several bacterial strains, gram-positive or gram-negative, antibiotic-sensitive or resistant, after irradiation in the near-infrared region (808 nm). Without irradiation, the toxicity of vancomycin-coated AuNPs was lower but still significant. Conversely bare AuNPs were not toxic.[26] Finally, gold nanorods (68 nm × 18 nm, synthesized with CTAB) coated with an anti-*Pseudomonas aeruginosa* antibody, and irradiated in the near-infrared region (~785 nm), were shown to efficiently kill a multi-antibiotic resistant bacterial strain of *Pseudomonas aeruginosa*, sampled in the upper respiratory tract of sinusitis patients. Nanorods alone did not inhibit bacterial growth over a 24 h period, suggesting that they were not inherently toxic to the cells.[27] In these three studies, the toxicity of bare AuNPs is shown to be null, that of coated AuNPs is slightly higher, and a combination of AuNPs with irradiation (which would improbably occur in natural environments) causes cell death.

A direct proof of AuNPs harmlessness was published by Williams *et al.*[28] These authors exposed *E. coli* bacteria in normal growth conditions to several types of inorganic NPs, and among them PEG-coated, 30 nm-diameter AuNPs. They observed that these NPs caused no inhibition of bacterial cell multiplication, and that they were not agglomerated in the exposure medium (if agglomerated, bacteria-NP interaction would have been hindered).

Lastly, several microorganisms have been reported to produce biogenic AuNPs from gold ions ($AuCl_4^-$). For instance, in the metal-reducing bacterium *Shewanella algae*, gold deposits were formed in less than 30 minutes upon exposure to $AuCl_4^-$. After 90 minutes almost all the bacteria contained gold deposits of 5–15 nm, without evidence that these deposits induced any lethality.[29]

Among the unicellular organisms, algae are also one of the first steps in the food chain in aquatic ecosystems. In a study by Renault *et al.*, exposure for 24 h of the green algae *Scenedesmus subspicatus* to amine-coated 10 nm AuNPs induced a dose-dependent mortality, reaching 20–50% of the algae population.[30] AuNPs were strongly adsorbed on algae cell wall, but they were not observed in the cell cytoplasm, and may therefore be transferred to the food chain.

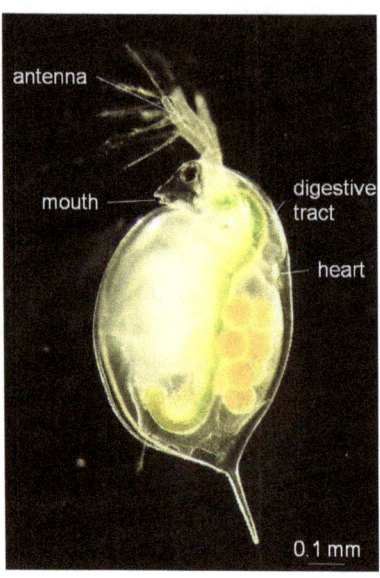

Fig. 12.7. A daphnid. This image was kindly provided by M. Floriani, LRE, IRSN, France.

12.3.3 *Impact of AuNPs on aquatic organisms: daphnids, bivalves, fishes*

Daphnids (*Daphnia magna*) (Fig. 12.7) are largely used in ecotoxicity tests due to their sensitivity to chemical toxicants. They are considered to be a valid test model species prior to mammalian testing.

When applied to daphnids at sublethal concentration, AuNPs (17–23 nm, 500 ppb) distributed through the gut, with highest gold accumulation after 12 h of exposure. AuNPs were observed in the lumen, but did not enter tissues.[31] A higher content of AuNPs was found in the mouth as compared to the tail region, and AuNPs were gradually excreted by the animal when placed in a NP-free environment.[31] The authors did not report any toxic effect of AuNPs in these conditions.

More recently, exposure of daphnids to spherical, 15 nm AuNPs was achieved by Li *et al.*,[32] who proved that 65–75 mg·l^{-1} of AuNPs caused the death of 50% of the population in 48 h. Daphnid motility was gradually reduced upon NP exposure, the animals finally sedimented and died. Daphnids exposed for eight days at sublethal dosage (10 mg·l^{-1}) did not show any decreased birth rates or embryo development, but their death rate was higher than that of control samples. The authors observed that AuNPs adhered to external appendages of daphnids, impairing their movements. These appendages have two functions for daphnids. Their movement creates a water-column around the animal, which brings food close to their mouth for their feeding, but also brings oxygen for their respiration. Inhibiting appendage movement would then cause starvation and oxygen deprivation. The authors also suggest that AuNPs likely inhibited food uptake by concentrating in the gut, as previously reported by Lovern *et al.*[31] In comparison, AgNPs were much more toxic, with a LC_{50} value (concentration leading to the death of 50% of daphnid population) $1,000\times$ lower.[32]

The impact of AuNPs has been tested on two bivalves, *Corbicula fluminea* and *Mytilus edulis*. Exposure to AuNPs of the blue mussel, *Mytilus edulis*, caused oxidative stress in several organs. Thirteen nm NPs led to the increase of catalase activity and protein ubiquitination in the digestive gland, and protein carbonylation in gills.[33] Five nm NPs induced a weak oxidative stress, but impaired thiol antioxidant defenses and oxidized protein thiols in the digestive gland, where they majorly accumulated.[34] This modest effect may impact the long-term ability to cope with traditional chemical pollutants. The report by Renault *et al.* shows the trophic transfer of AuNPs from green algae to *C. fluminea*. In their experiments *C. fluminea* bivalves were exposed to green algae, themselves having been previously exposed to AuNPs. As described in Section 12.3.2, AuNPs rapidly adsorbed onto the algae's cell wall. Two bivalves were then fed with these algae cells. Such exposure did not cause any structural disturbance of bivalves' branchial and digestive tissues, but AuNPs were shown to penetrate into the cells of these epithelia. They caused a three-fold increase of metallothioneins content in the gills and visceral mass at the highest exposure concentration. In addition, in the visceral mass the expression of metallothionein and several markers of response to oxidative stress (superoxide dismutase, glutathione S-transferase, cytochrome-c oxidase, catalase) were modulated by AuNP exposure.[30]

These data thus prove that AuNPs display pro-oxidant properties which may alter the function of filter-feeding organisms, but also that they may be transferred through the trophic chain, and may thus reach the food chain if disseminated in the environment.

Fig. 12.8. A male zebrafish. This image was kindly provided by M. Floriani, LRE, IRSN, France.

Lastly, the impact of AuNPs has also been studied on the zebrafish (*Danio rerio*) (Fig. 12.8). This fish is supported as one model to study environmental toxicity, but also as an alternative model to study development and physiology of higher organisms, including humans. The first article reporting the impact of AuNPs on this species showed almost no mortality, and no sublethal effects of the NPs on zebrafish embryos at 120 h post-fertilization, whatever their size (3, 10, 50, 100 nm), when exposed to concentrations reaching 250 μM. Conversely, numerous embryo malformations were observed after exposure to AgNPs, even if low mortality is reported.[35]

Another article reports that AuNPs (11 nm) caused a low mortality and an insignificant amount of deformed embryos after chronic exposure (120 h, 0.025–1.2 nM).[36] Again, AuNPs are much more biocompatible than AgNPs, and the authors show that AuNPs are accumulated in embryo, where they passively diffuse following a random route when embryonic cells divide, causing stochastic events altering embryo development.[36] Lastly, one study reports the impact of 5 nM AuNPs, microinjected in zebrafish embryos whose development was then followed. After seven days, normal looking larval fishes were observed, with ability to feed and responsive nervous systems. Most of their organs functioned normally (cardiovascular system, eye mobility, etc.), showing that physiological functions of embryos were not affected. Neither was affected the gene expression profile.[37]

12.3.4 *Impact of gold nanoparticles on plants*

Only one article reports the impact of AuNPs on plants, and shows that exposure to 10 nm AuNPs (62 g·l^{-1}) significantly increases plant germination and root elongation of cucumber and lettuce. The authors attributed this effect to the presence of citrate during the exposure (remaining from nanoparticle synthesis), which is also known as a food additive. Root elongation was induced, but root weight was not increased significantly. The hypothesis is that it might be an avoidance mechanism of the seed to a stress factor produced by the presence of NPs. Conversely, silver and Fe_3O_4 NPs caused the reduction of seed germination and root elongation.[38]

12.4 Conclusions

Most of the studies reported in this chapter have been achieved in acute exposure conditions, i.e. living organisms have been exposed to very high concentrations of AuNPs, during short incubation times. The obtained data prove that the lethality of AuNPs *per se* is relatively low. When lethality is observed, most of the time it may be related to toxic substances adsorbed onto NP surface. However, the sublethal effects of these NPs have been poorly reported. Poorly reported is also their long-term and chronic effects, which may occur if AuNPs persisted in living organisms after translocation/accumulation. These exposure conditions are more realistic and representative of environmental exposure, and future research in these areas would be required. Finally, even if a fair amount of work has been done in recent years, research into nanotoxicology and particularly nano-ecotoxicology, is still at an early stage, and more data is needed before gold nanoparticles can be used safely for diagnostic or therapeutic purposes.

References

1. I.A. Khalil, K. Kogure, H. Akita and H. Harashima, *Pharmacological Reviews* **58** (2006) 32.
2. J.M. Worle-Knirsch, K. Pulskamp and H.F. Krug, *Nano Lett.* **6** (2006) 1261.
3. J.F. Hainfeld, D.N. Slatkin and H.M. Smilowitz, *Phys. Med. Biol.* **49** (2004) N309.
4. M. Semmler-Behnke, W.G. Kreyling, J. Lipka, S. Fertsch, A. Wenk, S. Takenaka, G. Schmid and W. Brandau, *Small* **4** (2008) 2108.
5. W.H. De Jong, W.I. Hagens, P. Krystek, M.C. Burger, A.J. Sips and R.E. Geertsma, *Biomaterials* **29** (2008) 1912.
6. G. Sonavane, K. Tomoda, A. Sano, H. Ohshima, H. Terada and K. Makino, *Colloids Surf. B Biointerfaces* **65** (2008) 1.

7. T. Niidome, M. Yamagata, Y. Okamoto, Y. Akiyama, H. Takahashi, T. Kawano, Y. Katayama and Y. Niidome, *J. Control Release* **114** (2006) 343.
8. Y. Akiyama, T. Mori, Y. Katayama and T. Niidome, *J. Control Release* **139** (2009) 81.
9. B. Rothen-Rutishauser, C. Muhlfeld, F. Blank, C. Musso and P. Gehr, *Part Fibre Toxicol.* **4** (2007) 9.
10. J.H. Kim, K.W. Kim, M.H. Kim and Y.S. Yu, *Nanotechnology* **20** (2009) 19.
11. A.M. Alkilany, P.K. Nagaria, C.R. Hexel, T.J. Shaw, C.J. Murphy and M.D. Wyatt, *Small* **5** (2009) 701.
12. B.D. Chithrani, A.A. Ghazani and W.C. Chan, *Nano Lett.* **6** (2006) 662.
13. J.A. Khan, B. Pillai, T.K. Das, Y. Singh and S. Maiti, *Chembiochem.* **8** (2007) 1237.
14. E.E. Connor, J. Mwamuka, A. Gole, C.J. Murphy and M.D. Wyatt, *Small* **1** (2005) 325.
15. T.S. Hauck, A.A. Ghazani and W.C. Chan, *Small* **4** (2008) 153.
16. N. Pernodet, X. Fang, Y. Sun, A. Bakhtina, A. Ramakrishnan, J. Sokolov, A. Ulman and M. Rafailovich, *Small* **2** (2006) 766.
17. Y. Liu, W. Meyer-Zaika, S. Franzka, G. Schmid, M. Tsoli and H. Kuhn, *Angew. Chem. Int. Ed. Engl.* **42** (2003) 2853.
18. M. Tsoli, H. Kuhn, W. Brandau, H. Esche and G. Schmid, *Small* **1** (2005) 841.
19. G. Oberdorster, V. Stone and K. Donaldson, *Nanotoxicology* **1** (2007) 2.
20. F. Gottschalk, T. Sonderer, R.W. Scholz and B. Nowack, *Environ. Sci. Technol.* **43** (2009) 9216.
21. N.C. Mueller and B. Nowack, *Environ. Sci. Technol.* **42** (2008) 4447.
22. S. Diegoli, A.L. Manciulea, S. Begum, I.P. Jones, J.R. Lead and J.A. Preece, *Sci. Total Environ.* **402** (2008) 51.
23. M. Farre, K. Gajda-Schrantz, L. Kantiani and D. Barcelo, *Anal. Bioanal. Chem.* **393** (2009) 81.
24. C.C. Lin, Y.C. Yeh, C.Y. Yang, C.L. Chen, G.F. Chen, C.C. Chen and Y.C. Wu, *J. Am. Chem. Soc.* **124** (2002) 3508.
25. V.P. Zharov, K.E. Mercer, E.N. Galitovskaya and M.S. Smeltzer, *Biophys. J.* **90** (2006) 619.
26. W.C. Huang, P.J. Tsai and Y.C. Chen, *Nanomedicine* **2** (2007) 777.
27. R.S. Norman, J.W. Stone, A. Gole, C.J. Murphy and T.L. Sabo-Attwood, *Nano Lett.* **8** (2008) 302.
28. D.N. Williams, S.H. Ehrman and T.R. Pulliam Holoman, *J. Nanobiotechnology* **4** (2006) 3.
29. Y. Konishi, T. Tsukiyama, N. Saitoh, T. Nomura, S. Nagamine, Y. Takahashi and T. Uruga, *J. Biosci. Bioeng.* **103** (2007) 568.
30. S. Renault, M. Baudrimont, N. Mesmer-Dudons, P. Gonzalez, S. Mornet and A. Brisson, *Gold Bull.* **41** (2008) 116.
31. S.B. Lovern, H.A. Owen and R. Klaper, *Nanotoxicology* **2** (2008) 43.
32. T. Li, B. Albee, M. Alemayehu, R. Diaz, L. Ingham, S. Kamal, M. Rodriguez and S.W. Bishnoi, *Anal. Bioanal. Chem.* **398** (2010) 689.
33. S. Tedesco, H. Doyle, G. Redmond and D. Sheehan, *Mar. Environ. Res.* **66** (2008) 131.
34. S. Tedesco, H. Doyle, J. Blasco, G. Redmond and D. Sheehan, *Comp. Biochem. Physiol. C Toxicol Pharmacol.* **151** (2010) 167.

35. O. Bar-Ilan, R.M. Albrecht, V.E. Fako and D.Y. Furgeson, *Small* **5** (2009) 1897.
36. L.M. Browning, K.J. Lee, T. Huang, P.D. Nallathamby, J.E. Lowman and X.H. Xu, *Nanoscale* **1** (2009) 138.
37. Y. Wang, J.L. Seebald, D.P. Szeto and J. Irudayaraj, *ACS Nano* **4** (2010) 4039.
38. R. Barrena, E. Casals, J. Colon, X. Font, A. Sanchez and V. Puntes, *Chemosphere* **75** (2009) 850.

Chapter 13
Technological Applications of Gold Nanoparticles

Michael Cortie

Institute for Nanoscale Technology, University of Technology Sydney,
PO Box 123, Broadway, NSW 2007, Australia.
Email: michael.cortie@uts.edu.au

13.1 Introduction

This chapter surveys the existing and potential commercial applications of
AuNPs. This is an area in which there is currently rapidly developing activity.
This new-found interest is ironic, given that gold was historically the first of
the metals to be exploited[1] and that AuNPs themselves were first used over
1,500 years ago.[2] These earlier applications are discussed in greater detail
in Chapter 1 of this book. Here I consider only the modern applications. The
chapter starts with a brief review of why AuNPs are useful building blocks
in various nanotechnologies, before considering in turn their electronic,
optical, catalytic, decorative, biotechnological and medical applications.

The ever-expanding applications of nano-particulate gold are due to gold
in this form possessing a unique cluster of desirable material properties: it is
metallic and relatively inert; the optical properties are conducive to a plasmon resonance in the visible region of the electromagnetic spectrum (see
Chapter 3); there is a useful tendency to bond selectively to sulfur and some
other chalcogens; it is an excellent electrical conductor; and, very importantly, AuNPs can be readily manufactured in a wide range of shapes and
sizes (see Chapter 5). These attributes are exploited either individually or in
combinations in the various technological applications to be discussed here.

The commercial market for AuNPs is satisfied by a number of vendors, who can supply either the basic colloidal gold nanosphere or nanospheres that have been functionalised in various ways. Very small nanoparticles, more akin to clusters because they contain less than 100 gold atoms, are also available for specialised applications. Gold nanorods also have become available recently from a few suppliers. In general, the commercial customers for these products are in the decorative or biomedical industries for, as we shall see, these are areas in which the technological uses of AuNPs are best established. However, there is some up-take in the electronics industry for use in conductive inks. Of course, part of the 'market' for AuNPs is not strictly commercial in nature and I refer here to the sale of nanoparticles to researchers at universities and elsewhere for use in pursuits of a rather scientific nature. Nevertheless, as will be shown in this chapter, a great diversity of other *potential* technological applications for AuNPs are being researched, and it seems likely that some of these will also become genuine commercial products in due course. Therefore they are described here too.

13.2 Electronic and Opto-Electronic Applications

Thin films, wires and electronic and soldering pastes, which are not the subject of this book, are important industrial uses of gold in the electronics industry, normally consuming about 300 tons of the element per year.[3] Here, however, we consider only the technological applications of discrete AuNPs. The market for these latter products currently consumes a rather smaller mass of Au, of the order of several tons, almost all of which is currently used in decorative inks or glazes, but the added value is extremely high. A nominal breakdown of the industrial uses of gold at the time of writing is provided in Fig. 13.1.

13.2.1 *Applications of the optical and electronic properties of gold*

The physical and optical properties of gold and AuNPs were discussed in Chapters 3 and 4. The electronic and optoelectronic applications of AuNPs rely mainly on only two of these properties: the dielectric function (i.e. the optical properties or refractive index as a function of frequency) and

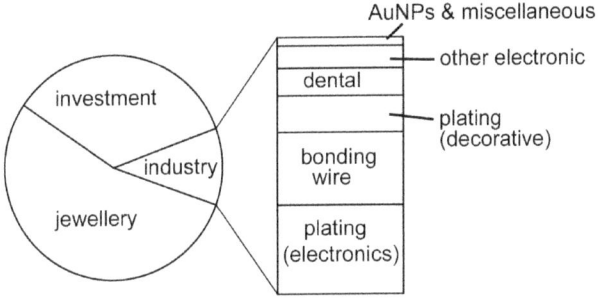

Fig. 13.1. Approximate proportion of gold consumed in various applications in 2010. Data drawn from diverse internet sources (including www.wgc.org and the *GFMS Gold Survey 2011*). About 3,800 tons of gold were 'consumed' in 2010. 'Other electronic' includes sputtering targets and pastes. The importance of electroplating as an end use of gold within the 'industry' sector is obvious. Note that dental use has been declining steadily due to substitution with other materials. The author estimates that about 15 tons of the total gold consumed in 2010 were in the high tech areas ('AuNPs & miscellaneous').

the d.c. electrical conductivity. Of course, other metals are electrically conductive too, but, with the exception of platinum, nanoscale structures of these other elements would be rapidly destroyed by oxidation under ambient atmospheric conditions. Similarly, although it can be argued that the dielectric function of some other elements, such as Al, Ag, K, Na or Li, are superior to Au for the development of localised plasmon resonances at most wavelengths,[4] the oxidation resistance of the alternative elements is either relatively poor or non-existent, which complicates their use in technological applications.

13.2.2 *Sinter inks*

The oxidation resistance of AuNPs makes them very suitable for the manufacture of conductive inks. The principle here is that an electric circuit can be 'printed' or 'painted' using readily available digital printing technologies, using an ink that is loaded with an appreciable volume fraction of AuNPs. The AuNPs are prevented from agglomerating prematurely because they are individually coated with some suitable surface ligand, such as a thiol, mercaptide, or alkyl amine.[5] The potential flexibility in manufacturing that can be achieved is obvious. However, in general, a thermal treatment ('metallising' or 'sintering') is required to destroy the protective ligand so that the AuNPs can weld together, an essential prerequisite to yield the desired electrical conductivity or reflective optical properties. In the past this has limited the application of this technology to substrates capable of withstanding at least 200°C. Ways to induce sintering without appreciably heating the substrate have been investigated and include laser, microwave or UV irradiation, and chemical treatment.[6,7] Of course, conductive inks based on copper and silver are also used in the field, but gold has the advantage of providing the ultimate in oxidation resistance and durability.

Recently, improved formulation of the gold-based inks has led to the 'sintering' temperature being routinely brought to between 120 and 180°C,[5,8] (Fig. 13.2), while sintering temperatures of below 100°C have been demonstrated under laboratory conditions.[7] The rapid low temperature sintering of AuNPs is driven by their enormous surface energy, which is released as heat when the particles fuse to form a solid mass. A strong exothermic signal has been observed from this process.[7] New formulations of inks containing AuNPs now make it possible to even coat polymer artefacts with conductive gold coatings by this technique. (Note, however, that electroless deposition of Au films or nanoparticles from solution[9] is a competing technology.)

13.2.3 *Spectrally selective coatings*

Continuous coatings of gold of 20–50 nm thickness are relatively transparent to light in the visible part of the electromagnetic spectrum but are highly reflective in the infrared region. Therefore, the application of such coatings

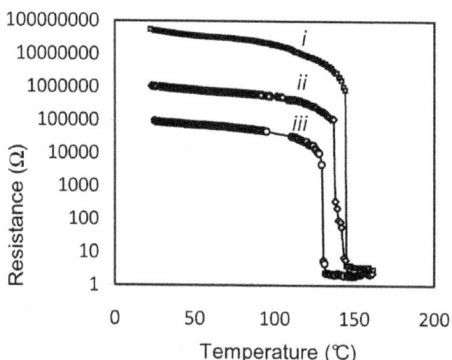

Fig. 13.2. Sintering of films made with AuNP inks that have been stabilised with various ligands. *i* 1-hexanethiol, *ii*. 1-butanethiol, *iii*. 2-propanethiol. (Data courtesy of M. Coutts, University of Technology Sydney. More information on these inks may be found in Coutts *et al.*[7])

onto glass or another transparent substrate produces a spectrally-selective surface which blocks infrared radiation. This has been used in the past to produce energy-efficient windows, but gold has now been surpassed in this application by other materials, such as multilayer stacks comprising silver and transparent conducting oxides (TCOs). However, gold coatings are still used in some specialised optical applications.

More recently, the possibility of using coatings of discrete AuNPs in a spectrally selective role has been considered.[3,9,10] These block light by absorption (rather than by reflection) and the position of the absorption maximum corresponds to the position of the plasmon resonance of the nanostructure. A reasonable proportion of solar infrared can be attenuated by this means, as seen in Fig. 13.3. In the case of glass coated with gold nanorods, the pronounced extinction in the upper visible and near-infrared between 750 and 1000 nm blocks a significant proportion of incoming solar heat. The advantage for architectural use would be that AuNPs can be readily deposited onto glass by wet chemical means, rather than by the expensive vacuum coating technologies currently used for Ag/TCO stacks. The potential reduction in processing cost appears to be large enough to compensate for the cost of the gold used.[9] Gold nanorods should give even better performance than nanospheres because their resonances can be tuned into the near-infrared region of the spectrum[11] but, so far, trials on prototype glass panes have yielded only modest solar screening figures-of-merit.[12]

13.2.4 *Non-linear optical applications*

The fluctuation of the electric field of light with time, $E(t)$, induces a fluctuating dipole moment in the material through which the light is passing. At low intensities the relationship between electric field and dielectric polarisation, P, is linear. However, accurate description of the dielectric polarisation at high intensities of illumination requires the introduction of higher order terms, so that

$$P(t) \propto \chi^{(1)}E(t) + \chi^{(2)} \cdot (E(t))^2 + \chi^{(3)} \cdot (E(t))^3 + \cdots.$$

where the material parameters $\chi^{(1)}$, $\chi^{(2)}$ and $\chi^{(3)}$ are the first-, second- and third-order electric susceptibilities respectively. $\chi^{(2)}$ has non-zero values only for some materials and is not usually of interest in the context of AuNPs. On the other hand, values of $\chi^{(3)}$ in the range of 1×10^{-6} to 1×10^{-13}

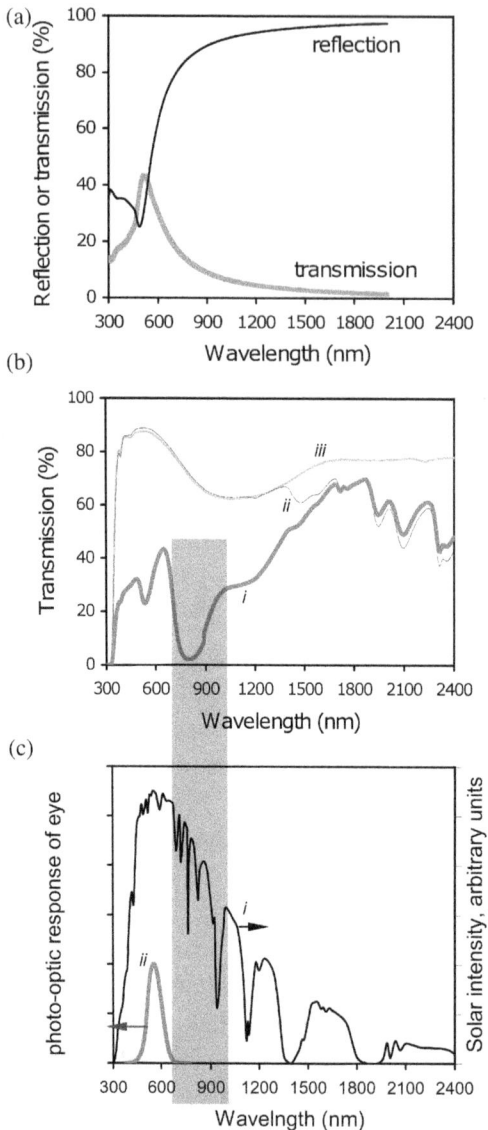

Fig. 13.3. (a) Simulated transmission and reflection spectra of glass coated with a continuous nanoscale film of 28 nm thick gold. (b) *i*. Measured transmission of 3 mm window glass coated with a dispersion of gold nanorods in poly vinyl alcohol (PVA). *ii*. plain glass coated with PVA film only, *iii*. plain glass only. (c) Energy distribution in standard solar spectrum (*i* black) and the photo-optic response of the human eye (*ii* indicated with a grey stripe). (Figure 3(b) modified from *J. of Nanoparticle Research*, 12, 2010, 2821 and reproduced here with kind permission from Springer Science+Business Media.)

e.s.u are reported for various materials containing AuNPs,[13–18] although it should be noted that the technological figure-of-merit for exploitation of these phenomena also requires the lowest possible linear absorption coefficient too[14] (which is problematic for metallic nanoparticles). Non-linear optical properties have existing or potential applications in devices like holography, frequency converters, optical limiters, signal processing, and phase conjugators.[14] There is a degree of non-linearity in the optical properties of AuNPs,[13] and in particular the $\chi^{(3)}$ parameter of composites containing AuNPs has attracted attention.[13–18] So far, however, there seem to have been few or no actual practical applications for AuNPs in this domain.

13.2.5 *Data storage*

There are two means by which AuNPs can be used to store digital data. The first is capacitive: an isolated AuNP is used to store or release electric charge in the manner of the now familiar 'flash memory'.[19,20] Nanoparticles produced by the citrate method are suitable because they are in the optimum 10–20 nm diameter size range for this application. In contrast, very small particles (<5 nm) are subject to quantum size effects which would complicate the operation of the device.

The other storage scheme makes use of the anisotropic optical properties of gold nanorods, and their dependence on both the aspect ratio of the nanoparticle and the polarisation of the light. This mode could find application in the CD or DVD type of recordable storage medium, although only in WORM mode ('Write Once Read Many'). Considerable density of data storage can be achieved by using more than one layer of nanorods of different aspect ratios embedded in a polymer matrix, with so-called 'five-dimensional' optical recording having been demonstrated.[21,22]

Neither technology has been commercialised yet but they remain under active development.

13.2.6 *Single electron conductivity and quantum devices*

Capacitance is a measure of the ease with which electric charge can be stored on or in an object. Because AuNPs are so small, adding an extra electron causes a relatively significant increase in Coulomb repulsion between the free electrons on the particle.[23] There is therefore a significant energy cost associated with such a charge transfer process. The capacitance of an individual AuNP varies from between about 1×10^{-18} F for a 2 nm particle to about 2×10^{-16} F for an 11 nm particle.[24] Because charge itself is discretised into units of about 1.6×10^{-19} C, and the voltage on a capacitor is given by Q/C, where Q is the charge and C the capacitance, the transfer of a single quantum of charge will cause a voltage step of 0.16 V over a 1×10^{-18} F capacitor.[24] This means that the voltage on a sufficiently small metallic nanoparticle changes in a step-wise fashion as it is charged or discharged. Once one unit of charge is placed on the nanoparticle, no further addition of charge can take place until the applied potential difference is further increased to exceed the voltage step. Only then can a second unit of charge be forced onto the particle. This phenomenon is termed the Coulomb blockade. The step-wise change in voltage that accompanies a series of charge transfer events is termed a Coulomb staircase.

A single-electron transistor can be obtained by exploiting this phenomenon, using a third electrode which applies an external electric field to modulate the step size of the Coulomb blockade. In general these devices need to operate at very low temperatures because thermal noise at room temperature readily obscures the phenomenon. On the other hand, devices of this type should be scalable to extremely small size. However, they have yet to be implemented in any commercially significant devices.

13.3 Catalytic Applications

Chapter 7 has described the scientific basis of catalysis by AuNPs. Here, some recent technological examples of these principles will be presented. Although many reactions can be catalysed by AuNPs, it is the oxidation of carbon monoxide that has received the largest share of scientific attention. In principle, the lower operating temperature of an AuNP-based catalyst should make the application of this technology in automobile exhaust catalysts very attractive. Surprisingly, however, there have been proportionally few actual transfers of this knowledge to industry, so far. One issue is that the potentially high temperature of exhaust gases in a petrol-fuelled automobile running at full power will sinter and destroy a catalyst system based on AuNPs.

Diesel-fuelled vehicles, which have a lower exhaust gas temperature, might be a more forgiving application and a commercial application for gold-catalysts in this area has been claimed (see www.nanostellar.com). Gold catalysts have also reportedly been implemented for production for vinyl acetate monomer by BP Chemicals,[25] and for gas masks[26] and, at pilot plant scale, for the one-step direct production of methyl glycolate from ethylene glycol and methanol,[27] for the oxidation of mercury in the gas exhaust of coal fired power stations,[28] and in fuel cells.[29]

13.4 Decorative Applications

13.4.1 *Historic uses in ceramics and glass*

The bright colour of gold itself, and of the dispersions of its nanoparticles, have been used in decorative applications for thousands of years. Besides the well-known red colour of colloidal suspensions of AuNPs, which have been used to colour glass as far back as Roman times,[2] there is also a range of purple-hued pigments based on an intimate mixture of Au and tin oxide nanoparticles. These are known as the *Purple of Cassius* and have been used since 1679 to colour high quality ceramic ware.[30] More information on the historical uses of gold may be found in Chapter 1.

13.4.2 *Colouring textiles*

It is certainly possible to stain textiles with AuNPs, gold-silver nanoparticles or the Purple of Cassius-type nanoparticle, to produce a range of red to purple colours. The principles were already established in 1794.[31] In recent times there has been some experimentation with dying silk or wool by these means[32-34] (see Fig. 13.4) and there is currently some interest from the commercial sector.[3] Such a product would provide a means to add value to woollen fashion items intended for the high end of the market.

13.4.3 *Use in paint and polymers*

The vivid colours of colloidal gold are occasionally exploited in specialised paints or to colour polymers. A range of reddish to purple colours are obtained, with the exact colour depending on the angle-of-incidence of the

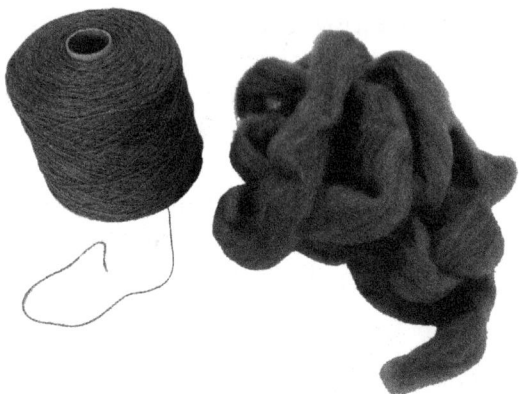

Fig. 13.4.   High quality wool that has been dyed purple using AuNPs that have been grown in situ on the fibre. Photo courtesy of Prof. J. Johnston, Victoria University, New Zealand. Further detail may be found in Ref. 34.

viewer and light source.[35] Gold loading is high, however, so these materials are relatively expensive.

13.5 Use in Sensors and Biomedical Diagnostics

Considerable information on this topic may be found in Chapters 10 and 11. Here only a rather applied summary of the topic is provided.

13.5.1 *Refractometric sensors*

Refractometric sensors using gold are based on either of two related optical phenomena. In the older and more developed scheme, the characteristics of a surface plasmon polariton (SPP) propagating on the surface of a nanoscale gold film are monitored using suitable instrumentation. The details of the SPP can be tuned by controlling the average refractive index in the region of medium immediately adjacent to the surface of the gold. Generally, this is exploited to make a sensor that is selective to specific biological molecules or protein fragments. This is done by first coating the surface of the gold film with a molecule that will selectively bind to the analyte. When the analyte is present, it coats the film, thereby raising the average refractive index of the immediately adjacent medium and changing the critical angle and absorption maximum of the SPP. More information on this technique is available in Chapter 10. Several instruments that exploit this system are commercially available but, as the measurement does not usually involve nanoparticles, this scheme will not be discussed further here.

The other type of refractometric sensor exploits the localised plasmon surface plasmon resonance of AuNPs. (Plasmon resonances were described in Chapter 3.) The important point here is that the position of the resonance peak or peaks on the optical extinction spectrum of the sensor is influenced by the magnitude of the refractive index of the surrounding medium, in particular the resonance is shifted to longer wavelengths with an increase in the refractive index. This phenomenon can be exploited in two main ways. In the simplest version, the refractive index of a liquid medium can be sensed directly, or, similarly to the case of the SPP sensor described above, the surfaces of the AuNPs are functionalised so that they selectively bind to specific biological molecules or protein fragments. When the analyte is present, it attaches to the nanoparticle, thereby raising the average refractive index of the immediately adjacent medium and redshifting the particle's plasmon resonance. Generally the sensor system is designed to measure the magnitude of the redshift,[36,37] but a monochromatic interrogation of the optical absorbance also appears to be feasible.[38] Although the scheme does work with gold nanospheres, greater sensitivity is obtained when using other shapes, such as nanorods or nano-bipyramids.[38–40] The capability to detect a single protein molecule by these means has recently been claimed.[41]

Of course, refractometric sensors can be constructed out of arrays of AuNPs too, with the advantage being that constructive inter-particle interactions can in some cases give sharper extinction peaks (and hence ultimately greater sensitivity) and the arrays can be used to collect and focus light onto a central 'hot spot'. However, precise positioning of the AuNPs is required for array devices and they are currently made by 'top down' lithographic techniques and, as such, do not involve the chemically-produced AuNPs that are the focus of this book.

13.5.2 *Colorimetric assays and related diagnostic techniques*

A rather different platform for sensing an analyte using AuNPs is based simply on a colour change. This has the huge advantage that suitable test kits can be used in the field by a person with only basic training. The scheme is based either on a red-to-blue colour change that occurs when two different populations of AuNPs aggregate together, or on the development of a red colour on a white background when one kind of Au nanoparticle binds to a white surface. Detailed information on the operation of these sensor schemes is available in the literature in Refs. 31, 42–46; here only a general overview will be presented.

In the first scheme, as the particles of Au approach closely to one another, the electric fields of their plasmon resonances interact, a factor which lowers (redshifts) their resonant frequency. This generates a new resonance peak in the optical extinction spectrum which mainly absorbs the red wavelengths. The result is that only the blue wavelengths are transmitted and the colloidal suspension turns blue to the eye (Fig. 13.5). A sensitive and selective sensor platform can be designed using this basic principle by arranging for the

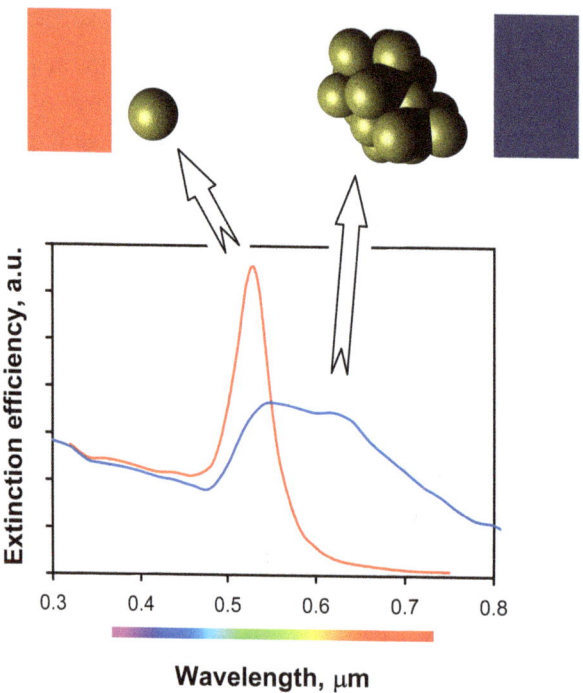

Fig. 13.5. Isolated spherical AuNPs have a strong plasmon resonance with 520 nm light, causing extinction of green colours but transmission of red. When the AuNPs agglomerate, the plasmon resonance shifts to longer wavelengths and broadens, causing red and orange colours to be absorbed. Now the AuNPs seem blue in colour to the eye. Reprinted from Ref. 67, with permission from Elsevier.

colloidal suspension to contain two populations of nanoparticles, which we will designate here as A and B. The 'A' and 'B' nanoparticles each bind to some different part of an analyte molecule, C. When C is added to the suspension, the A and B nanoparticles are brought into close proximity because both bind to the same C molecule, and a strongly redshifted plasmon resonance is developed. The concept was elegantly demonstrated for polynucleotides in the 1990s,[42] and has been developed for a wide variety of analytes since then. Recently, it has been shown that a one-pot detection of two target sequences can be achieved,[47] which further extends the range of possibilities.

On the other hand, in the so-called lateral flow sensor, a suspension of AuNPs to which a secondary antibody has been attached flow over a white surface which has been functionalised to previously selectively collect the primary analyte. The AuNPs bind to the immobilised analyte (if it is present), with the appearance of a red line after ten minutes or so indicating a positive analysis. More information on this technique is available in Chapter 11.

A very large variety of colorimetric tests using Au nanoparticles has been successfully demonstrated, ranging from pregnancy tests, to microbiological pathogens such as *Salmonella*, *Streptococcus*, viral agents such as herpes, specific genetic sequences, levels of specific human cell lines in the blood, or the proteins that signal the onset of prostrate cancer.[3,48–50]

13.5.3 *Assays based on quartz microbalance*

Because Au nanoparticles can be functionalised so that they will bind selectively to specific biological molecules, it follows that they can also be used to enhance the sensitivity of detection schemes based on the piezoelectric quartz microbalance. In one version the AuNPs are attached to the surface of the microbalance itself,[51] in the other they are in the solution.[52] In either event a significant improvement in sensitivity is reported.

13.5.4 *Contrast enhancement in electron and optical microscopy*

There is a specialised market for AuNPs as a contrast enhancer for electron microscopy. Here the high electron density and chemical properties of Au are exploited. If the AuNPs are suitably functionalised to bind to a specific site or molecule, then they will facilitate the detection and imaging of such a site in a transmission electron microscope section[53] due to their high atomic number. Further information is available in a comprehensive text on the subject.[54] Resolution and functionality can be improved by use of very small nanoparticles, more of the nature of clusters. The NanoGold™ particle is an example of these and can be purchased pre-functionalised to bind to a wide range of targets.[55]

Use of AuNPs may also be advantageous in optical microscopy[56] and optical coherence tomography.[57] Even though the individual nanoparticles are much smaller than the wavelength of light used to illuminate the sample, they can, if big enough, scatter light and thereby modify the optical image in a beneficial manner. Enhanced detection of cancerous cells, e.g. Ref. 58, would be a typical objective of work in this domain.

13.5.5 *Bifunctional metallo-dielectric hybrids for microscopy*

It is possible to prepare hybrid nanoparticles, consisting for example of a gold part and a second material which exhibits some additional functionality. For example, a magnetic material (e.g. Fe_3O_4) or a fluorescent material (e.g. a semiconducting quantum dot) can be attached to an AuNP to produce a hybrid that has multiple functionalities. In this way the electron density of Au can be combined with the light-emission of a semiconductor, for example, to facilitate optical microscopy. Or, the presence of the magnetic material can be used to manipulate the position of the hybrid nanoparticle. Recent reviews on this topic provide more information; see Refs. 59, 60.

13.5.6 *Surface enhanced Raman spectroscopy*

Although Ag is historically the better established substrate for SERS, there has also been interest from the very earliest days of SERS research in nanostructured Au substrates.[61] One reason for this is that silver nanostructures are susceptible to oxidation and this changes or compromises their performance. On the other hand, the plasmon resonances of the gold nanostructures are at lower energies (longer wavelengths) than those of Ag, and so somewhat different results are obtained. However, these older types of substrates, whether of Ag or Au, are produced by etching and are of an essentially random morphology. Interest in using geometrically well-defined configurations, such as nanoparticle dimers, for SERS enhancement is more recent and appears to date from 1999.[62] Good results are reported on specially prepared configurations of AuNPs, see for example Refs. 39, 63–65. Gold nanoshells, semi-shells, or crescents seem most suitable.

13.5.7 *Two photon technologies*

Various types of non-linear optical phenomena are possible on gold nanostructures but of course losses are high due to the opacity of metals. Four-wave mixing, for example to produce 600 nm photons from a mixture of 800 and 1,200 nm photons, is possible, as is broadband two photon luminescence ('TPL'). In general, however, complex geometries are needed, and the output from ordinary gold nanospheres is poor. Gold nanorods, on the other hand, have moderately good non-linear properties and have been used in biomedical contexts as a type of two photon marker.[66] More information on this topic is available in Chapter 10.

13.6 Potential or Actual Therapeutic Applications

The reader is referred to some recent publications on the use of AuNPs in medical contexts for more detailed information (see Refs. 67–69), and to Chapter 11 in the present book. Despite occasional reports to the contrary,[70–73] it is generally believed that pure gold colloid itself has a negligible biological effect. This would be the expected situation considering the chemical unreactivity of AuNPs in physiological environments. Therefore, the use of AuNPs in medical contexts is more usually predicated upon the incorporation of some additional active functionality. Most commonly, this is the attachment of a pharmaceutically active molecule[74] or genetic material,[75] energisation of the colloid by plasmonic heating,[76] or, rarely, exploitation of the radioactive decay of ^{198}Au.[77] The presence of AuNPs can also improve the efficacy of X-ray radiotherapy by a sensitising mechanism that appears not to be fully understood as yet.[78]

13.6.1 *Drug delivery*

Delivery of a pharmaceutical compound can be, in principle, facilitated by conjugating it to the surface of a AuNP. The argument in favour depends on whether a more selective targeting of the compound to the desired region of the body can be obtained when using a nanoparticle vector. Certainly, there are some specific mechanisms by which such selective targeting can be achieved. For example, the vascular structure around rapidly growing tumours is relatively porous and so nanoparticles of less than 300 nm diameter diffuse out of the vascular system and collect in the tumour.[79-81] This mechanism of 'passive targeting' is known as extravasation. Delivery of a AuNP-drug conjugate by exploiting endocytosis is a less well-established process, but can certainly occur in principle because cells such as macrophages have evolved to engulf foreign particles in a certain size range. Furthermore, in principle, the drug component of the conjugate could be released or activated once in the desired location by the application of illumination and hence plasmonic heating, thereby invoking two functionalities simultaneously.

The argument against drug delivery by means of a nanoparticle vector is that only a minute dose can be delivered, of the order of a few tens of drug molecules per AuNP, because the drug necessarily forms only a surface coating on the nanoparticle core. Drug delivery by this means is therefore by necessity restricted to very potent substances, such as TNF (tumour necrosis factor)[74] or photodynamic sensitisers such as a phthalocyanine.[81] At least one of these prototype drug delivery schemes is now in Phase 1 clinical trials.[3]

13.6.2 *Gene therapy*

The penetration of an AuNP through a cell membrane, whether by endocytosis or some other means, provides a way to move foreign genetic material into a cell. Of course, the DNA or other payload must escape being destroyed in an endosome within the cell or degradation by a nuclease, and must somehow progress from there through the internal nuclear membrane and into the nucleus proper, before the transfective effect of its payload can be realised. Considerable progress has been made, and transfection using AuNPs as a vector has been frequently demonstrated.[75,82,83] Since an AuNP can undergo plasmonic heating, the additional possibility of controlling the release of genetic material by application of light has also been considered.[84−88] This provides an interesting and unique functionality, and permits better targeting since the genetic payload will be effectively unloaded in the illuminated cells only.

A related idea uses DNA-coated AuNPs of 2–3 μm diameter[89] to deliver a dose of genetic material that can (if successfully transfected) cause the recipient's body to express a target protein to which an immune response is desired.[90] This neatly overcomes the problem of the very small chemical payloads that are possible when using AuNPs because the patient's own body is manipulated into generating the desired protein. In this case the particles are projected with high kinetic energy against the skin of the patient and carry their payload into the patient. This indirect form of vaccine and its delivery using gold particles is currently being commercialised by Pfizer.[3]

13.6.3 *Radiotherapy*

The ^{198}Au isotope, which has a half life of 2.7 days, can be used in cancer therapy,[77,91] and has in the past been occasionally administered in the form of colloidal gold nanospheres. Targeting the radioactive material to the desired location is the key challenge, but can in principle be accomplished by extravasation, as discussed in Section 13.5.1. It is also known that the presence of AuNPs increases the efficacy of therapies using ionising electromagnetic radiation[78,92,93] but the reasons for this are, however, not yet well understood.

13.6.4 *Hyperthermal techniques*

The energy of any light that is absorbed during a plasmon resonance in an AuNP (see Chapter 4) makes the particle a potent point source of heat when appropriately illuminated. Local temperature increases of a few tens of kelvin[94,95] are obtained with comparatively low-power laser diodes, quite adequate to kill mammal cells or single-cell pathogens. More intense illumination causes the AuNPs to melt or even vaporise, and

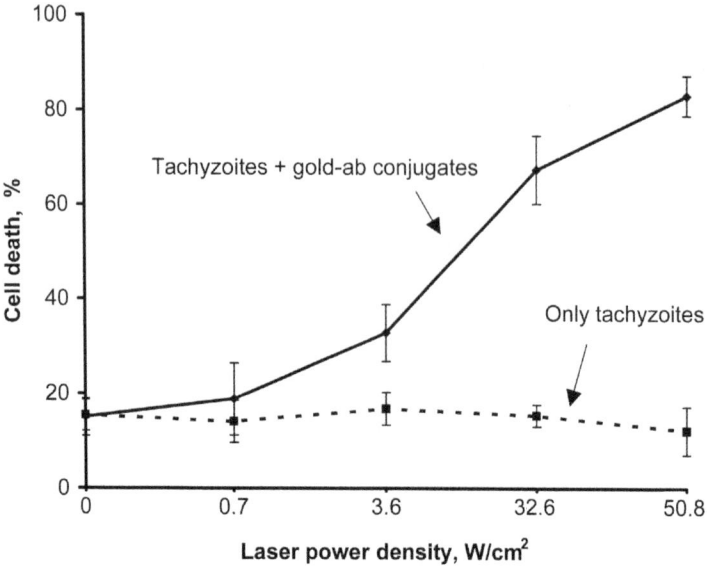

Fig. 13.6. Death rates of *tachyzoites* of *Toxoplasmosis gondii* that had been labelled with gold nanorod-antibody conjugates compared to those of non-labelled *T.gondii tachyzoites* (a *tachyzoite* is one of the mobile stages in the pathogen's complex life cycle). Reprinted with permission from Pissuwan *et al.*[97] Copyright 2007 American Chemical Society.

the resulting pressure impulse can perforate a nearby cell membrane.[96] Gold nanospheres, nanoshells and nanorods are most frequently mentioned in respect of these proposed treatment schemes. Medical trials using nanoshells have progressed the furthest, and are now in Phase 1 clinical trials[3] for head and neck cancer. However, bacteria[67] and protozoa[97] may, in principle, also be targeted (Fig. 13.6). It has been argued that nanorods would provide a similar or superior platform on which to develop technologies based on plasmonic heating than nanoshells.[98,99] In either case, however, it is desired that the plasmonic heating be applied by a laser located outside the body. Therefore, since the body is reasonably transparent in the near-infrared region of the spectrum (the so-called 'tissue window'[100]), it is most helpful if the AuNP has a strong plasmon resonance in this part of the spectrum. This is readily achieved by adjusting the aspect ratio of gold nanoshells or nanorods. Ordinary gold nanospheres, on the other hand, absorb most strongly in the mid-visible, and would, at first sight, seem to be unsuitable for such an application. Nevertheless, it has been noted that usable optical extinction in the tissue window can be obtained from Au nanospheres provided that they are allowed to aggregate.[101,102] This collapses the single particle plasmon and replaces it with a broad, redshifted resonance which has an increased extinction cross-section at wavelengths within the tissue window.

13.7 Environmental Remediation

The chemical affinity of AuNPs for elements such as sulfur and mercury have also suggested their use as an absorbent for specific pollutants. For example, it has been reported that Hg ions can be readily removed from water by exposure to Au colloid.[103] Of course, the colloid would have to be collected and purged of its mercury in a fully reversible process before such a scheme could be technologically feasible and, even then, the commercial feasibility of removing mercury contamination this way is doubtful. It has also been suggested that AuNPs could be useful in photocatalytic applications with the potential to remove some environmentally deleterious contaminant such as a volatile organic compound.[104]

13.8 Conclusions and Outlook

AuNPs have unique and useful chemical and physical properties. For this reason, they are currently used in commercial quantities in biological diagnostic assays, as specialised stains for electron microscopy, as the colouring agent in niche glassware and ceramics, in high-end decorative inks, and in some electronics as sinter inks. There are also, however, a very large range of potential applications under development, which this chapter has attempted to cover too. With so many examples under development it is likely that several will be successful, but of course, many others will fail to reach the level of commercial application. Only a very small quantity of gold is used in any individual device or kit in most of the applications or potential applications described here. Furthermore, in many cases these applications are in the high-technology domain and have very high intrinsic added value. Therefore, the value of the gold that they contain is relatively insignificant as a proportion of the total cost.

References

1. R.F. Tylecote, *History of Metallurgy*, Institute of Materials, London, 1992.
2. I. Freestone, N. Meeks, M. Sax and C. Higgitt, *Gold Bull.* **40** (2007) 270.
3. T. Keel, R. Holliday and T. Harper, *Gold for Good. Gold and Nanotechnology in the Age of Innovation*, World Gold Council, London, 2010.
4. M.G. Blaber, M.D. Arnold, N. Harris, M.J. Ford and M.B. Cortie, *Phys. B (Amsterdam, Neth.)* **394** (2007) 184.
5. P.T. Bishop, L.J. Ashfield, A. Berzins, A. Boardman, V. Buche, J. Cookson, R.J. Gordon, C. Salcianu and P.A. Sutton, *Gold Bull.* **43** (2010) 181.
6. S. Sun, P. Mendes, K. Critchley, S. Diegoli, M. Hanwell, S.D. Evans, G.J. Leggett, J.A. Preece and T.H. Richardson, *Nano Lett.* **6** (2006) 345.
7. M. Coutts, M.B. Cortie, M.J. Ford and A.M. McDonagh, *J. Phys. Chem. C* **113** (2009) 1325.
8. T. Bakhishev and V. Subramanian, *J. Electronic Mater.* **38** (2009) 2720.
9. X. Xu, M. Stevens and M.B. Cortie, *Chem. Mater.* **16** (2004) 2259.
10. H. Chowdhury, X. Xu, P. Huynh and M.B. Cortie, *J. Solar Energ. T. ASME* **127** (2005) 70.
11. X. Xu, T. Gibbons and M.B. Cortie, *Gold Bull.* **39** (2006) 156.
12. N. Stokes, A. McDonagh and M.B. Cortie, *J. Nanopart. Res.* **12** (2010) 2821.
13. M.-J. Kim, H.-J. Na, K.C. Lee, E.A. Yoob and M. Lee, *J. Mater. Chem.* **13** (2003) 1789.
14. D. Compton, L. Cornish and E. van der Lingen, *Gold Bull.* **36** (2003) 10.
15. M. Li, Z.S. Zhang, X. Zhang, K.Y. Li and X.F. Yu, *Optics Express* **16** (2008) 14288.
16. E. Cattaruzza, G. Battaglin, F. Gonella, R. Polloni, B.F. Scremin, G. Mattei, P. Mazzoldi and C. Sada, *Appl. Surf. Sci.* **254** (2007) 1017.
17. Y. Hamanaka, K. Fukuta, A. Nakamura, L.M. Liz-Marzán and P. Mulvaney, *Appl. Phys. Lett.* **84** (2004) 4938.
18. C. Pecharromán, A. Esteban-Cubillo, H. Fernández, L. Esteban-Tejeda, R. Pina-Zapardiel, J.S. Moya, J. Solis and C.N. Afonso, *Plasmonics* **4** (2009) 261.
19. J.-S. Lee, J. Cho, C. Lee, I. Kim, J. Park, Y.-M. Kim, H. Shin, J. Lee and F. Caruso, *Nature Nanotech.* **2** (2007) 790.
20. L.D. Bozano, B.W. Kean, M. Beinhoff, K.R. Carter, P.M. Rice and J.C. Scott, *Adv. Funct. Mater.* **15** (2005) 1933.
21. J.W.M. Chon, C. Bullen, P. Zijlstra and M. Gu, *Adv. Funct. Mater.* **17** (2007) 875.
22. P. Zijlstra, J.W.M. Chon and M. Gu, *Nature* **459** (2009) 410.
23. T. Laaksonen, V. Ruiz, P. Liljeroth and B.M. Quinn, *Chem. Soc. Rev.* **37** (2008) 1836.
24. M.B. Cortie, M.H. Zareie, S.R. Ekanayake and M.J. Ford, *IEEE Trans. Nanotech.* **4** (2005) 406.
25. anon., in *CatGold News*, World Gold Council, London, 2003, p. 1.
26. anon., in *CatGold News*, World Gold Council, London, 2008, p. 1.
27. anon., in *CatGold News*, World Gold Council, London, 2006, p. 1.
28. anon., in *CatGold News*, World Gold Council, London, 2005, p. 1.
29. anon., in *CatGold News*, World Gold Council, London, 2004, p. 1.

30. J. Carbert, *Gold Bull.* **13** (1980) 144.
31. M.-C. Daniel and D. Astruc, *Chem. Rev.* **104** (2004) 293.
32. Y. Nakao and K. Kaeriyama, *J. Appl. Polymer Sci.* **36** (1988) 269.
33. M. Richardson and J. Johnston, *J. Colloid Interface Sci.* **310** (2007) 425.
34. J.H. Johnston and K.A. Lucas, *Gold Bull.* **44** (2011) 85.
35. A. Iwakoshi, T. Nanke and T. Kobayashi, *Gold Bull.* **38** (2005) 107.
36. A.J. Haes and R.P.V. Duyne, *Expert Rev. Mol. Diagn.* **4** (2004) 527.
37. M. Himmelhaus and H. Takei, *Sens. Actuators B* **63** (2000) 24.
38. C.S. Kealley, M.D. Arnold, A.J. Porkovich and M.B. Cortie, *Sens. Actuators, B* **148** (2010) 34.
39. K.-S. Lee and M.A. El-Sayed, *J. Phys. Chem. B* **110** (2006) 19220.
40. H. Chen, X. Kou, Z. Yang, W. Ni and J. Wang, *Langmuir* **24** (2008) 5233.
41. K.M. Mayer, F. Hao, S. Lee, P. Nordlander and J.H. Hafner, *Nanotechnol.* **21** (2010) 255503.
42. R. Elghanian, J.J. Storhoff, R.C. Mucic, R.L. Letsinger and C.A. Mirkin, *Science* **277** (1997) 1078.
43. E. Hutter and J.H. Fendler, *Adv. Mater.* **16** (2004) 1685.
44. S.-Y. Lin, S.-H. Wu and C.-H. Chen, *Angew. Chem. Int. Ed.* **45** (2006) 4948.
45. J. Zhang, L. Wang, D. Pan, S. Song, F.Y.C. Boey, H. Zhang and C. Fan, *Small* **4** (2008) 1196.
46. N.T.K. Thanh and Z. Rosenzweig, *Anal. Chem* **74** (2002) 1624.
47. G. Doria, M. Larguinho, J.T. Dias, E. Pereira, R. Franco and P.V. Baptista, *Nanotechnol.* **21** (2010) 255101.
48. E.I. Laderman, E. Whitworth, E. Dumaual, M. Jones, A. Hudak, W. Hogrefe, J. Carney and J. Groen, *Clin. Vaccine Immunol.* **15** (2008) 159.
49. C.R. Martin and D.T. Mitchell, *Analytical Chemistry News & Features* **70** (1998) 322A.
50. A.M. Horgan, J.D. Moore, J.E. Noble and G.J. Worsley, *Trends Biotechnol.* **28** (2010) 485.
51. L. Wang, Q. Wei, W. ChunSheng, Z. Hu, J. Ji and P. Wang, *Chin. Sci. Bull.* **53** (2008) 1175.
52. H. Wang, C. Lei, J. Li, Z. Wu, G. Shen and R. Yu, *Biosens. Bioelec.* **19** (2004) 701.
53. W.P. Faulk and G. Taylor, *Immunochem.* **8** (1971) 1081.
54. M.A. Hayat, *Colloidal Gold: Principles, Methods, and Applications*, Academic Press, San Diego, CA, 1989.
55. R. Powell and J. Hainfeld, *Micron* **42** (2011) 163.
56. L. Tong, Q. Wei, A. Wei and J.-X. Cheng, *Photochem. Photobiol.* **85** (2009) 21.
57. M. Hu, J. Chen, Z.-Y. Li, L. Au, G.V. Hartland, X. Li, M. Marquez and Y. Xia, *Chem. Soc. Rev.* **35** (2006) 1084.
58. X. Huang, I.H. El-Sayed, W. Qian and M.A. El-Sayed, *J. Am. Chem. Soc.* **128** (2006) 2115.
59. N. Sanvicens and M.P. Marco, *Trends Biotechnol.* **26** (2008) 425.
60. M.B. Cortie and A.M. McDonagh, *Chem. Rev.* (2012) in press.
61. C.G. Blatchford, J.R. Campbell and J.A. Creighton, *Surf. Sci.* **120** (1982) 435.
62. H. Xu, E.J. Bjerneld, M. Käll and L. Börjesson, *Phys. Rev. Lett.* **83** (1999) 4357.

63. C.L. Haynes, A.D. McFarland and R.P.V. Duyne, *Anal. Chem.* (2005) 338A.
64. J.-F. Li, Z.-L. Yang, B. Ren, G.-K. Liu, P.-P. Fang, Y.-X. Jiang, D.-Y. Wu and Z.-Q. Tian, *Langmuir* **22** (2006) 10372.
65. F. Le, D.W. Brandl, Y.A. Urzhumov, H. Wang, J. Kundu, N.J. Halas, J. Aizpurua and P. Nordlander, *ACS Nano* **2** (2008) 707.
66. H. Wang, T.B. Huff, D.A. Zweifel, W. He, P.S. Low, A. Wei and J.-X. Cheng, *Proc. Natl. Acad. Sci.* **102** (2005) 15752
67. D. Pissuwan, C.H. Cortie, S.M. Valenzuela and M.B. Cortie, *Trends Biotechnol.* **28** (2010) 207.
68. D. Pissuwan, T. Niidome and M.B. Cortie, *J. Controlled Release* (2010) in press.
69. D. Pissuwan, S.M. Valenzuela and M.B. Cortie, *Biotechnol. Genet. Eng. Rev.* **25** (2008) 93.
70. G.E. Abraham and P.B. Himmel, *Nutri. Env. Med.* **7** (1997) 295.
71. N. Pernodet, X. Fang, Y. Sun, A. Bakhtina, A. Ramakrishnan, J. Sokolov, A. Ulman and M. Rafailovich, *Small* **2** (2006) 766.
72. C.L. Brown, M.W. Whitehouse, E.R.T. Tiekink and G.R. Bushell, *Inflammopharmacology* **16** (2008) 133.
73. C.L. Brown, G. Bushell, M.W. Whitehouse, D. Agrawal, S. Tupe, K. Paknikar and E.R. Tiekink, *Gold Bull.* **40** (2007) 245.
74. G.F. Paciotti, L. Myer, D. Weinreich, D. Goia, N. Pavel, R.E. McLaughlin and L. Tamarkin, *Drug Delivery* **11** (2004) 169.
75. T. Niidome, K. Nakashima, H. Takahashi and Y. Niidome, *Chem. Commun.* (2004) 1978.
76. D. Pissuwan, S. Valenzuela and M.B. Cortie, *Trends Biotechnol.* **24** (2006) 62.
77. A. Sherman and M. Ter-Pogossian, *Cancer* **6** (1953) 1238.
78. J.F. Hainfeld, D.N. Slatkin and H.M. Smilowitz, *Phys. Med. Biol.* **49** (2004) N309.
79. D.P. O'Neal, L.R. Hirsch, N.J. Halas, J.D. Payne and J.L. West, *Cancer Lett.* **209** (2004) 171.
80. S.M. Moghimi, A.C. Hunter and J.C. Murray, *FASEB J.* **19** (2005) 311.
81. Y. Cheng, C.S. Anna, J.D. Meyers, I. Panagopoulos, B. Fei and C. Burda, *J. Am. Chem. Soc.* **130** (2008) 10643.
82. A.C. Bonoiu, S.D. Mahajan, H. Ding, I. Roy, K.T. Yong, R. Kumar, R. Hu, E.J. Bergey, S.A. Schwartz and P.N. Prasad, *Proc. Natl. Acad. Sci. U. S. A.* **106** (2009) 5546.
83. N.L. Rosi, D.A. Giljohann, C.S. Thaxton, A.K.R. Lytton-Jean, M.S. Han and C.A. Mirkin, *Science* **312** (2006) 1027.
84. Y. Niidome, T. Niidome, S. Yamada, Y. Horiguchi, H. Takahashi and K. Nakashima, *Mol. Cryst. Liq. Cryst.* **445** (2006) 201/[491].
85. C.C. Chen, Y.P. Lin, C.W. Wang, H.C. Tzeng, C.H. Wu, Y.C. Chen, C.P. Chen, L.C. Chen and Y.C. Wu, *J. Am. Chem. Soc.* **128** (2006) 3709.
86. H. Takahashi, Y. Niidome and S. Yamada, *Chem. Commun.* **17** (2005) 2247.
87. A. Wijaya, S.B. Schaffer, I.G. Pallares and K. Hamad-Schifferli, *ACS Nano* **3** (2008) 80.
88. S.E. Lee, G.L. Liu, F. Kim and L.P. Lee, *Nano Lett.* **9** (2009) 562.
89. C.D. Medley, B.K. Muralidhara, S. Chico, S. Durban, P. Mehelic and C. Demarest, *Anal. Bioanal. Chem.* **398** (2010) 527.
90. D. Tang, M. DeVit and S.A. Johnston, *Nature* **356** (1992) 152.

91. P. Rubin and S.H. Levitt, *J. Nuclear Med.* **5** (1964) 581.
92. P. Diagaradjane, A. Shetty, J.C. Wang, A.M. Elliott, J. Schwartz, S. Shentu, H.C. Park, A. Deorukhkar, R.J. Stafford, H. Cho, J. W. Tunnell, J. D. Hazle and S. Krishnan, *Nano Lett.* **8** (2008) 1492.
93. A. Simon-Deckers, E. Brun, B. Gouget, M. Carrière and C. Sicard-Roselli, *Gold Bull.* **41** (2008) 187.
94. L.R. Hirsch, R.J. Stafford, J.A. Bankson, S.R. Sershen, B. Rivera, R.E. Price, J.D. Hazle, N.J. Halas and J.L. West, *Proc. Natl. Acad. Sci. U. S. A.* **100** (2003) 13549.
95. D. Pissuwan, S. M. Valenzuela, M.C. Killingsworth, X. Xu and M.B. Cortie, *J. Nanopart. Res.* **9** (2007) 1109.
96. C.M. Pitsillides, E.K. Joe, X. Wei, R.R. Anderson and C.P. Lin, *Biophys. J.* **84** (2003) 4023.
97. D. Pissuwan, S. Valenzuela, C.M. Miller and M.B. Cortie, *Nano Lett.* **7** (2007) 3808.
98. N. Harris, M.J. Ford, P. Mulvaney and M.B. Cortie, *Gold Bull.* **41** (2008) 5.
99. P.K. Jain, K.S. Lee, I.H. El-Sayed and M.A. El-Sayed, *J. Phys. Chem. B.* **110** (2006) 7238.
100. R. Weissleder, *Nature Biotechnol.* **19** (2001) 316.
101. V.P. Zharov, E.N. Galitovskaya, C. Johnson and T. Kelly, *Lasers in Surgery and Medicine* **37** (2005) 219.
102. D. Pissuwan, S.M. Valenzuela, C.M. Miller, M.C. Killingsworth and M.B. Cortie, *Small* **5** (2009) 1030.
103. K.P. Lisha, Anshup and T. Pradeep, *Gold Bull.* **42** (2009) 144.
104. X. Chen, H.-Y. Zhu, J.-C. Zhao, Z.-F. Zheng and X.-P. Gao, *Angew. Chem. Int. Ed.* **47** (2008) 5353.

Biographies of the Authors

1.1 Catherine Louis

Catherine Louis is Research Director at the Laboratoire de Réactivité de Surface of the Université Pierre et Marie Curie. She directs the "Catalysis: from Materials to Reactivity" research team. She has been with the academic group since 1982, when she was appointed by the French National Centre of Research (CNRS). She received a diploma in Chemical Engineering (1979), from the Ecole Supérieure de Physique et Chimie Industrielle of Paris, and her PhD in Chemistry in 1985 (prepared under the direction of Prof. Michel Che). From 1986 to 1988 she was a post-doctoral fellow at the University of Berkeley with Prof. Alex Bell. She is a specialist in catalyst preparation and has worked on gold catalysts since 2000. She has authored around 120 publications. She co-authored the book *Catalysis by Gold* (Imperial College Press, 2006) with Geoffrey C. Bond and David T. Thompson. She is also the author of three book chapters on synthesis of supported metal catalysts, and recent advances in CO oxidation of gold nanoparticles. She is the director of Or-Nano (www.or-nano.org), a CNRS network gathering around three hundred French researchers (physics, chemists and biologists) working with gold nanoparticles.

1.2 Geoffrey Bond

Geoffrey Bond held academic positions at the Universities of Leeds and Hull before being appointed Head of the Johnson Matthey Research Group on Catalysis (1962–1970). He then became Professor of Applied Chemistry at Brunel University, Uxbridge, where he held various posts (Head of the Chemistry Department, Dean of the Faculty of Science, Vice-Principal) until his retirement in 1992. His research has mainly concerned supported metal catalysts for hydrogenation and hydrogenolysis, and supported oxides for selective oxidation. He has published more than 250 scientific papers and review articles. Since retiring he has worked on gold catalysts, and has co-authored several review articles as well as the book *Catalysis by Gold* published by Imperial College Press. Earlier books include *Catalysis by Metals* (1962), *Heterogeneous Catalysis, Principles and Applications* (2nd edn, 1987), and *Metal-Catalysed Reactions of Hydrocarbons* (2005).

1.3 Olivier Pluchery

Olivier Pluchery graduated from the Ecole Normale Supérieure de Cachan (Paris, France) in 1997 with a specialization in laser physics. He obtained his PhD in chemical physics from University Paris-Sud in 2000 and was interested in the investigation of the electrochemical reactions on a gold interface, monitored with sum frequency generation, a nonlinear optical spectroscopy. In 2001 he joined Yves Chabal's team at Bell-Labs (USA) to work on semiconductor interfaces. In 2002, he obtained a position as associate professor at University Pierre et Marie Curie (Paris) where he is currently developing several research programs dealing with the control of the adsorption of organic molecules on silicon and with gold nanoparticles.

1.4 Bruno Palpant

Bruno Palpant is a Professor at Ecole Centrale Paris. He leads research activities in the Quantum and Molecular Photonics Laboratory (LPQM, belonging to both CNRS and Ecole Normale Supérieure de Cachan). He is in charge of a group devoted to the study of the thermal and optical ultrafast transient responses of metal nanoparticles and their applications. He was awarded his PhD in 1998 from University of Lyon (France), which focused on quantum size effects in the optical properties of matrix-embedded noble metal clusters, before joining Keio University (Japan) for one year. He was an assistant professor at the Institut des NanoSciences de Paris (CNRS-UPMS) for ten years, with an interest in the linear and nonlinear optical responses of noble metal nanoparticles, as well as their link with thermal transport at small space and time scales. More recently, his activities have focused on the applications of nanoscale energy conversion processes in biology, chemistry and photonics.

1.5 Dahea Seo

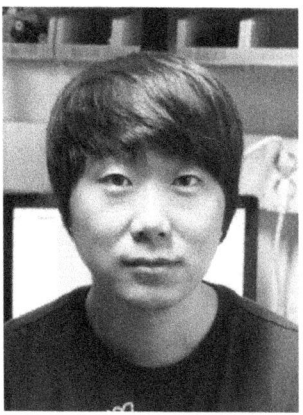

Daeha Seo received his PhD in Chemistry at the Korea Advanced Institute of Science and Technology (KAIST) in 2010, under the supervision of Professor Hyunjoon Song. He is now working as a postdoctoral fellow for the joint project between Professor Paul A. Alivisatos at University of California, Berkeley and Professor Young-Wook Jun at University of California, San Francisco. His research has centered on the synthesis of gold based-hybrid nanostructures and their plasmonic applications for catalysis and biology.

1.6 Hyunjoon Song

Hyunjoon Song is an Associate Professor of Chemistry at Korea Advanced Institute of Science and Technology (KAIST). He received his M.S. and PhD degrees from KAIST in 1996 and 2000. He worked as a postdoctoral fellow at KAIST for two years, and spent another postdoctoral period in Chemistry at University of California, Berkeley from 2002–2004. His research interests are morphology control of metal and metal oxide nanocrystals, and their applications for surface plasmon monitoring, photoactive energy catalysts, and electroactive materials.

1.7 Shamil Shaikhutdinov

Shamil Shaikhutdinov received his PhD (1986) in physics at the Moscow Institute of Physics and Technology studying the microwave properties of bio-organic materials and their models under the supervision of Prof. E.M. Trukhan. Then he joined the group of Prof. K.I. Zamaraev at the Boreskov Institute of Catalysis at Novosibirsk to carry out surface science studies of catalytic systems, in particular with a scanning tunneling microscope he designed there. In addition, he has been working as a postdoctoral fellow in several research centres in Germany and France. Since 2004 he has been a group leader at the Department of Chemical Physics (headed by Prof. H.-J. Freund) of the Fritz-Haber Institute at Berlin. His current research interest is focused on an understanding of structure-reactivity relationships in heterogeneous catalysis.

1.8 Hannu Häkkinen

Prof. Hannu Häkkinen gained his PhD in physics in 1991 at the University of Jyväskylä, Finland. After his PhD he worked for several years as postdoctoral researcher, senior research scientist and Academy of Finland Research Fellow at Georgia Institute of Technology, Atlanta and in University of Jyväskylä. Since 2007, he has been a professor in computational nanoscience at the University of Jyväskylä in a joint position at Physics and Chemistry Departments and at the Nanoscience Center. He leads a group of about ten researchers focusing on computational studies of electronic, optical, magnetic and catalytic properties of various metal nanoclusters and nanostructures. His teaching curriculum includes solid state physics, physical chemistry and computer simulation methods. He has co-authored about 130 peer-reviewed articles.

1.9 Romain Quidant

Professor Romain Quidant got his PhD in Physics in 2002 from the University of Burgundy (France). In 2006, he was appointed junior Professor and group leader at ICFO and became a tenured Professor both at ICFO and ICREA in 2009. The same year, he received the Fresnel prize from the European Physics Society for his outstanding contribution to the field of plasmon optics. In 2011 he was awarded the Prize of the City of Barcelona and of the Prince of Girona. Since January 2010, he serves as the coordinator of the European FP7-STREP project 'SPEDOC'. His research focuses on the study of the optical properties of metal nano-structures, and their use in the elaboration of future miniaturized optical functionalities and devices. In particular, his group investigates news strategies to control light and heat at the nanometer scale for biomedical applications, including early detection and photothermal therapy of cancer. Other research interests are optical trapping, nonlinear optics, metamaterials and quantum optics.

1.10 Christian Villiers

Christian Villiers obtained his biochemistry degree from a Technological High Education Establishment (INSA — Lyon, France) and received his PhD in biology and immunology from the University of Grenoble in 1984. He worked as a postdoctoral fellow at the MRC centre of Cambridge (UK) in 1985. He is a now a research member of INSERM (National Institute for Health and Medical Research). His current interest concerns the modification of the immune response resulting from inflammation response with a special focus on the evaluation of the impact of nano-particles on the cellular behaviour and on the immune system. He is the author of more than 75 original articles.

1.11 Marie Carrière

Marie Carrière is a junior research scientist at the Atomic Energy Commission (CEA) in Grenoble, France. She recently joined the Nucleic

Acid Lesions Laboratory in the CEA Nanoscience and Cryogeny Institute (INAC). She received her PhD from the National Institute of Agronomics at Paris in February 2002 studying the efficiency, cell distribution and metabolization of lipidic vectors developed for gene therapy applications. She then did postdoctoral work at CEA Saclay centered on the study of toxicological impact of heavy metals on cultured animal and human cell lines. Her current research interests are toxicology, ecotoxicology and bioavailabillity of metal and metal oxide nanoparticles as well as carbon nanotubes.

1.12 Michael Cortie

Michael Cortie is Professor of Nanotechnology and Director of the Institute for Nanoscale Technology at the University of Technology Sydney, in Australia. He has a BSc(Engineering) degree in physical metallurgy from the University of the Witwatersrand, South Africa (1978), a Masters in Engineering degree on from the University of Pretoria, South Africa (1983), and a PhD degree on metal fatigue at high temperatures from the University of the Witwatersrand (1987). He joined Mintek, the South African national minerals and metals research organization in 1987. He was a Senior Engineer there before becoming head of its Physical Metallurgy Division between 1997 and 2002. While at Mintek he consulted widely to the international precious metals industry in the areas of nanotechnology, catalysis and physical metallurgy. He relocated to the University of Technology Sydney in July 2002. Michael's main research interests are the nanoparticles and intermetallic compounds of the precious metals, with a particular focus on how these relate to optical properties.

Index